Science Horizons Year Book

1990

MACMILLAN EDUCATIONAL COMPANY
A Division of Macmillan, Inc.
New York

COLLIER MACMILLAN CANADA
Toronto

MAXWELL MACMILLAN INTERNATIONAL PUBLISHING GROUP
New York   Oxford   Singapore   Sydney

Published by Macmillan Educational Company 1990

This book is also published under the title Science Annual 1991

Copyright © 1990 by Franklin Watts

Library of Congress Catalog Card Number 64–7603

ISBN 0-02-942485-2

All rights reserved. No part of this book may be reproduced or transmitted in any form or by any means, electronic or mechanical, including photocopying, recording, or by any information storage and retrieval system, without permission in writing from the Publisher.

Printed in the United States of America

ACKNOWLEDGMENTS

Sources of articles appear below, including those reprinted with the kind permission of publications and organizations.

HARVESTING THE NEAR-EARTHERS, Page 16: Reprinted with permission of the authors; article first appeared in the November 1989 issue of *Ad Astra*.

THE SCIENCE OF SPACE SUITS, Page 21: Reprinted with permission of the author; article first appeared in the October/November 1989 issue of *Air & Space/Smithsonian*.

MOONSTRUCK, Page 30: Tom Waters/© 1989 Discover Publications.

FINDING THE ZONE, Page 40: © 1989 by Larry Shainberg. Reprinted by permission of John Farquharson Ltd. Article originally appeared in the April 9, 1989, issue of *The New York Times Magazine*.

WOMAN IN A CAVE, Page 48: Copyright 1989 by Alcestis Oberg and reprinted with the permission of Omni Publications International, Ltd.

DANGEROUS DELUSIONS, Page 51: Copyright © 1989 by The New York Times Company. Reprinted by permission.

PANIC DISORDER, Page 56: Reprinted with permission from *Insight*. © 1989 Insight. All rights reserved.

SHOCK INCARCERATION, Page 60: Reprinted with permission from *Insight*. © 1989 Insight. All rights reserved.

BIOCHEMICAL CODES: THE LANGUAGE OF LIFE?, Page 70: Reprinted with permission of the author; article first appeared in the October 1989 issue of SMITHSONIAN.

FARMERS LEARN NEW TRICKS FROM MOTHER NATURE, Page 79: Reprinted from November 6, 1989, issue of *Business Week* by special permission, copyright © 1989 by McGraw-Hill, Inc.

THE WONDER OF BIOMECHANICS, Page 84: Reprinted by permission of the author; article first appeared in the June 1989 issue of SMITHSONIAN.

BIOSENSORS, Page 92: Reprinted with permission from *Insight*. © 1990 Insight. All rights reserved.

INSIDE ARTIFICIAL REALITY, Page 102: Reprinted with permission of the author; article first appeared in the November 1989 issue of *PC/Computing*.

SMART TV, Page 117: Reprinted with permission, FORTUNE, © 1989 The Time Inc. Magazine Company. All rights reserved.

COMPUTING AT THE SPEED OF LIGHT, Page 124: Reprinted by permission of *The Wall Street Journal*, © Dow Jones & Company, Inc. 1990. All Rights Reserved Worldwide.

STOPPING BEACH EROSION, Page 141: Copyright © 1989 by The New York Times Company. Reprinted by permission.

TORNADO!, Page 147: Reprinted from *Popular Science* with permission © 1988 Times Mirror Magazines, Inc.

ICE CYCLES, Page 154: Robert Kunzing/© 1989 Discover Publications.

AUTOMOTIVE FUELS OF THE FUTURE, Page 166: Reprinted with permission, FORTUNE, © 1989 The Time Inc. Magazine Company. All rights reserved.

THE FUTURE OF FUSION, Page 173: Reprinted with permission, FORTUNE, © 1989 The Time Inc. Magazine Company. All rights reserved.

COAL MINING GOES HIGH TECH, Page 180: Copyright © 1990 by The New York Times Company. Reprinted by permission.

FUNERAL FOR A REACTOR, Page 184: Reprinted with permission of the author; article first appeared in SMITHSONIAN, October 1989.

TOURISTS IN ANTARCTICA, Page 196: Reprinted by permission of the authors; article first appeared in the September/October 1989 issue of *Sea Frontiers*.

EAVESDROPPING IN THE WILDS, Page 203: Reprinted by permission of the author; article first appeared in the November 1989 issue of *Audubon*.

THE DWINDLING DRY FOREST, Page 208: Copyright 1989 by the National Wildlife Federation. Reprinted from the January-February 1989 issue of *International Wildlife*.

HUGO'S ENVIRONMENTAL AFTERMATH, Page 215: Reprinted from *Audubon*, the magazine of the National Audubon Society.

NEW ADVANCES IN THE CANCER WAR, Page 235: Copyright © 1989 by The New York Times Company. Reprinted by permission.

WAITING FOR THE BIONIC MAN, Page 242: Reprinted by permission of *Forbes* magazine, September 18, 1989. © Forbes, Inc., 1989.

CRYING THE WEIGHT-LOSS BLUES, Page 245: Reprinted with permission from CHANGING TIMES Magazine, © Kiplinger Washington Editors, Inc., April 1989.

THE BLACK DEATH, Page 254: Reprinted by permission of the author; article first appeared in the February 1990 issue of SMITHSONIAN.

THE SEARCH FOR INTELLIGENT LIFE, Page 263: Copyright © 1990 by The New York Times Company. Reprinted by permission.

WORDS OF LOST WORLDS, Page 276: Reprinted with permission from *Insight*. © 1989 Insight. All rights reserved.

COSMIC TIME TRAVEL, Page 284: David H. Freedman/© 1989 Discover Publications.

FAREWELL TO THE WOODEN BAT?, Page 292: By Peter Gammons, reprinted courtesy of *Sports Illustrated* from the July 24, 1989, issue. Copyright © 1989, The Time Inc. Magazine Company. (Original title: "End of an Era.") ALL RIGHTS RESERVED.

BEHIND THE LABORATORY DOOR, Page 299: Reprinted through the courtesy of Halsey Publishing Company, publishers of Delta Air Lines' *Sky* magazine.

THE NEW SCIENCE OF SOUND, Page 304: Reprinted by permission of the author; article originally appeared in the September 1989 issue of *World Monitor*.

A NEW LIFT FOR HELICOPTERS, Page 328: Reprinted with permission from *Insight*. © 1989 Insight. All rights reserved.

SPACE SPIES, Page 332: Reprinted from *Popular Science* with permission © 1990 Times Mirror Magazines, Inc.

VIEWING VIEWERS, Page 341: Reprinted from *Popular Science* with permission © 1990 Times Mirror Magazines, Inc.

SAVING THE ELEPHANTS, Page 348: Copyright © 1990 by The New York Times Company. Reprinted by permission.

THE DAY OF THE DOLPHINS, Page 356: Copyright 1989 by Justine Kaplan and reprinted with the permission of Omni Publications International, Ltd.

AMERICA'S OTHER EAGLE, Page 364: © 1989 by the National Wildlife Federation. Reprinted from the October/November 1989 issue of *National Wildlife*.

THE DISARMING ARMADILLO, Page 369: © 1989 by the National Wildlife Federation. Reprinted from the October/November 1989 issue of *National Wildlife*.

RETURN OF A REPTILE, Page 373: Reprinted with permission of the author; article first appeared in the March 6, 1989, issue of *Sports Illustrated*.

Pages 2–3: © Paul DiMare
Pages 36–37: © G. V. Faint/The Image Bank
Pages 66–67: © Stephen Frink/The WaterHouse
Pages 98–99: © Michel Tcherevkoff/The Image Bank
Pages 128–129: © Jerry Jacka Photography
Pages 162–163: © Hank Morgan
Pages 192–193: © Wendy Shattil
Pages 222–223: © Vladimir Lange/The Image Bank
Pages 250–251: © M. Funk/The Image Bank
Pages 280–281: © Dominique Sarraute/The Image Bank
Pages 314–315: © Tom Tracy/The Stock Market
Pages 344–345: © Frans Lanting/Minden Pictures

SCIENCE HORIZONS STAFF

EDITORIAL

Editorial Director
Lawrence T. Lorimer

Executive Editor
Joseph M. Castagno

Managing Editor
Doris E. Lechner

Copy Editors
David M. Buskus
Meghan O'Reilly LeBlanc

Proofreader
Stephan Romanoff

Chief Indexer
Pauline M. Sholtys

Editorial Assistant
Carol B. Cox

Manuscript Typist
Susan A. Mohn

Art Assistant
Elizabeth Farrington

Contributing Editor
Philip A. Storey

Editorial Librarian
Charles Chang

Art Director
Eric E. Akerman

Production Editor
Diane L. George

Production Assistant
Sheila Rourk

Manager, Picture Library
Jane H. Carruth

Chief, Photo Research
Ann Eriksen

Photo Assistant
Linda R. Kubinski

Financial Manager
Jean Gianazza

Staff Assistants
Jeanne A. Schipper
Freida K. Jones

Manager, Electronics
Cyndie L. Cooper

MANUFACTURING

Director of Manufacturing
Joseph J. Corlett

Senior Production Manager
Christine L. Matta

Assistant Production Manager
Barbara L. Persan

Production Assistant
Pamela J. Murphy

CONTRIBUTORS

MICHAEL ABRAHAM, Free-lance writer; history teacher
　　　　　　　　　　　　　　IN MEMORIAM

LARRY ARMSTRONG, Senior correspondent, *Business Week*
　　　　　　　　　　　　　　BUG-EATING BUGS

WILLIAM BARTON, Free-lance writer; coauthor, *Fellow Traveler*
　　　Coauthor, HARVESTING THE NEAR-EARTHERS

HELLE BERING-JENSEN, Writer, *Insight*
　　　　　　　　　　　　WORDS OF LOST WORLDS

BRUCE BOWER, Staff writer, *Science News*
　　　REVIEW OF THE YEAR: BEHAVIORAL SCIENCES

WILLIAM J. BROAD, News science reporter, *The New York Times*
　　　　　　　　THE SEARCH FOR INTELLIGENT LIFE

WILLIAM E. BURROWS, Contributor, *Popular Science*
　　　　　　　　　　　　　　　　SPACE SPIES

GENE BYLINSKY, Board of editors, *Fortune*
　　　　　　　　　　　　THE FUTURE OF FUSION

MICHAEL CAPOBIANCO, Free-lance writer; coauthor, *Fellow Traveler*
　　　Coauthor, HARVESTING THE NEAR-EARTHERS

YVETTE CARDOZO, Writer specializing in adventure and travel
　　　　　　　Coauthor, TOURISTS IN ANTARCTICA

ANTHONY J. CASTAGNO, Energy consultant; manager, nuclear information, Northeast Utilities, Hartford, CT
　　　　　　　　　REVIEW OF THE YEAR: ENERGY

GLENN ALAN CHENEY, Adjunct professor, Fairfield University, Fairfield, CT; senior writer, Benchmark Communications, Inc., West Redding, CT; author, *Environmental Refugees*
THE 1989 NOBEL PRIZE FOR MEDICINE OR PHYSIOLOGY
　　　　　　　AMERICA'S BICENTENNIAL COUNTDOWN
THE 1989 NOBEL PRIZES FOR PHYSICS AND CHEMISTRY

THEODORE A. REES CHENEY, Associate professor, Fairfield University, Fairfield, CT; author, *Land of the Hibernating Rivers* and *Living in Polar Regions*
　　　　　　　　　　　　　　EARTHQUAKES

DONALD CUNNINGHAM, Washington-based free-lance writer; former associate editor, *American Scientist*
　　　　　　　　REVIEW OF THE YEAR: TECHNOLOGY

GODE DAVIS, Meteorologist; associate member, American Meteorological Society
　　　　　　　　　　　　　　　TORNADO!

CORY DEAN, News science and health deputy editor, *The New York Times*
　　　　　　　　　STOPPING BEACH EROSION

GINO DEL GUERCIO, Free-lance writer and television producer specializing in science and technology; contributor, *World Monitor* magazine
　　　　　　　　THE NEW SCIENCE OF SOUND

STEVE DITLEA, Free-lance journalist, based in Armonk, NY, specializing in personal computer technology; columnist, *Omni* magazine
　　　　　　　　　INSIDE ARTIFICIAL REALITY

BRUCE FELLMAN, Naturalist, natural history columnist, and science writer
　　　　　　　　THE WONDER OF BIOMECHANICS

JOHN FREE, Senior editor, *Popular Science*
　　　　　　　　　SPACE WAR AGAINST DRUGS

DAVID H. FREEDMAN, Contributor, *Discover*
　　　　　　　　　　　COSMIC TIME TRAVEL

PETER GAMMONS, Senior writer, *Sports Illustrated*; 1989 National Sportswriter of the Year
　　　　　　　FAREWELL TO THE WOODEN BAT?

DANIEL GOLEMAN, News science reporter, *The New York Times*
　　　　　　　　　　DANGEROUS DELUSIONS

JAMES GORMAN, Coauthor of *Digging Dinosaurs*, the 1989 winner of the New York Academy of Sciences' Children's Science Book Award
　　　　　　　　　　　RETURN OF A REPTILE

FRANK GRAHAM, JR., Field editor, *Audubon*
　　　　　　　HUGO'S ENVIRONMENTAL AFTERMATH

JOHN GROSSMANN, Free-lance writer; contributing editor, *Health* magazine
　　　　　　　　THE SCIENCE OF SPACE SUITS

JEFFREY H. HACKER, Free-lance writer; editorial director, Grey Castle Press
　　　　　　　　　　PORTHOLES TO THE PAST
　　　　　　　　NEW LIGHT ON THE OLD MASTERS

STEPHEN S. HALL, New York-based writer; author, *Invisible Frontiers*
　　　BIOCHEMICAL CODES: THE LANGUAGE OF LIFE?

KATHERINE HARAMUNDANIS, Free-lance writer; former research associate, Smithsonian Astrophysical Laboratory, Cambridge, MA; coauthor, *An Introduction to Astronomy*
　　　　　　　REVIEW OF THE YEAR: ASTRONOMY

WILLIAM J. HAWKINS, Senior editor, *Popular Science*
　　　　　　　　　　　　　VIEWING VIEWERS

NANCY HENDERSON, Associate editor, *Changing Times*
　　　　　　　CRYING THE WEIGHT-LOSS BLUES

GLADWIN HILL, National environmental correspondent—retired, *The New York Times*
　　　　　　　REVIEW OF THE YEAR: THE ENVIRONMENT

BILL HIRSCH, Writer specializing in adventure travel
　　　　　　　Coauthor, TOURISTS IN ANTARCTICA

DAVID HOLZMAN, Writer, *Insight*
　　　　　　　　　　　　　　　BIOSENSORS

LAURENCE HOOPER, Staff reporter, *The Wall Street Journal*
　　　Coauthor, COMPUTING AT THE SPEED OF LIGHT

DANIEL KAGAN, Writer, *Insight*
　　　　　　　　　　　SHOCK INCARCERATION

JUSTINE KAPLAN, Research editor, *Omni*
　　　　　　　　　THE DAY OF THE DOLPHINS

SUSAN KATZ KEATING, Writer, *Insight*
　　　　　　　　A NEW LIFT FOR HELICOPTERS

ROBERT KUNZIG, Senior editor, *Discover*
    ICE CYCLES

ANDREW KUPFER, Associate editor, *Fortune*
    AUTOMOTIVE FUELS OF THE FUTURE

MARC KUSINITZ, Free-lance science and medical writer
    REVIEW OF THE YEAR: PHYSICAL SCIENCES

THOMAS A. LEWIS, Roving editor, *International Wildlife*
    THE DWINDLING DRY FOREST

DENNIS L. MAMMANA, Resident astronomer, Reuben H. Fleet Space Theater and Science Center, San Diego, CA
    REVIEW OF THE YEAR: SPACE SCIENCE
    VOYAGER II: NEPTUNE AND BEYOND

WILLIAM H. MATTHEWS III, Regents' professor of geology emeritus, Lamar University, Beaumont, TX
    REVIEW OF THE YEAR: EARTH SCIENCES

MARTIN M. McLAUGHLIN, Free-lance consultant; former vice president for education, Overseas Development Council
    Coauthor, REVIEW OF THE YEAR: PAST, PRESENT, AND FUTURE

CHARLES L. MEE, JR., Author and Obie-winning playwright; former editor-in-chief, *Horizon* magazine
    THE BLACK DEATH

JOHN NIELSEN, Knight science journalism fellow, MIT
    REVIEW OF THE YEAR: WILDLIFE

ALCESTIS OBERG, Writer; journalist-in-space finalist
    WOMAN IN A CAVE

JANE PERLEZ, Chief, Nairobi (Kenya) bureau, *The New York Times*
    SAVING THE ELEPHANT

ELISABETH ROSENTHAL, Writer and emergency-room physician
    NEW ADVANCES IN THE CANCER WAR

BRENTON R. SCHLENDER, Associate editor, *Fortune*
    SMART TV

JACOB M. SCHLESINGER, Staff reporter, *The Wall Street Journal*
    Coauthor, COMPUTING AT THE SPEED OF LIGHT

LAWRENCE SHAINBERG, Writer; contributor, *The New York Times Magazine*; author, *Memories of Amnesia*
    FINDING THE ZONE

SETH SHULMAN, Science, technology, and environment writer; Boston correspondent, *Nature*
    FUNERAL FOR A REACTOR

JOANNE SILBERNER, Associate editor, *U.S. News & World Report*
    REVIEW OF THE YEAR: BIOLOGY

RUTH SIMON, Staff writer, *Forbes*
    WAITING FOR THE BIONIC MAN

EMILY SMITH, Science editor, *Business Week*
    FARMERS LEARN NEW TRICKS FROM MOTHER NATURE

PETER STEINHART, Contributing editor, *Audubon*; author, *Tracks in the Sky*
    EAVESDROPPING IN THE WILDS

PHILIP A. STOREY, Staff editor, *Academic American Encyclopedia*
    REVIEW OF THE YEAR: COMPUTERS AND MATHEMATICS

JENNY TESAR, Free-lance science and medical writer; author, *Parents as Teachers, Introduction to Animals, Preparing for the SAT and Other Aptitude Tests*
    REVIEW OF THE YEAR: HEALTH AND DISEASE
    THE FDA: NEW CONTROVERSIES OVER DRUG APPROVAL

GARY TURBAK, Montana-based journalist; contributor, *National Wildlife* magazine
    AMERICA'S OTHER EAGLE

DINA VAN PELT, Reporter, *Insight*
    PANIC DISORDER

MATTHEW L. WALD, News financial reporter, *The New York Times*
    COAL MINING GOES HIGH TECH

BERNIE WARD, Professional writer; author and coauthor of books on sports and organized crime
    BEHIND THE LABORATORY DOOR

TOM WATERS, Associate editor, *Discover*
    MOONSTRUCK

JIM WATSON, Senior editor, *National Wildlife* magazine
    THE DISARMING ARMADILLO

PETER S. WELLS, Professor of anthropology and director, Center for Ancient Studies, University of Minnesota, Minneapolis, Minnesota
    Coauthor, REVIEW OF THE YEAR: PAST, PRESENT, AND FUTURE

CONTENTS

ASTRONOMY AND SPACE SCIENCE	Review of the Year	4
	Voyager II: Neptune and Beyond	8
	Harvesting the Near-Earthers	16
	The Science of Space Suits	21
	Moonstruck	30
BEHAVIORAL SCIENCES	Review of the Year	38
	Finding the Zone	40
	Woman in a Cave	48
	Dangerous Delusions	51
	Panic Disorder	56
	Shock Incarceration	60
BIOLOGY	Review of the Year	68
	Biochemical Codes: The Language of Life?	70
	Farmers Learn New Tricks from Mother Nature	79
	The Wonder of Biomechanics	84
	Biosensors	92
	The 1989 Nobel Prize for Physiology or Medicine	95
COMPUTERS AND MATHEMATICS	Review of the Year	100
	Inside Artificial Reality	102
	America's Bicentennial Countdown	110
	Smart TV	117
	Computing at the Speed of Light	124

EARTH SCIENCES

Review of the Year	130
Earthquakes	132
Stopping Beach Erosion	141
Tornado!	147
Ice Cycles	154

ENERGY

Review of the Year	164
Automotive Fuels of the Future	166
The Future of Fusion	173
Coal Mining Goes High Tech	180
Funeral for a Reactor	184

THE ENVIRONMENT

Review of the Year	194
Tourists in Antarctica	196
Eavesdropping in the Wilds	203
The Dwindling Dry Forest	208
Hugo's Environmental Aftermath	215

HEALTH AND DISEASE

Review of the Year	224
The FDA: New Controversies Over Drug Approval	228
New Advances in the Cancer War	235
Waiting for the Bionic Man	242
Crying the Weight-Loss Blues	245

PAST, PRESENT, AND FUTURE

Review of the Year	252
The Black Death	254
The Search for Intelligent Life	263
Portholes to the Past	268
Words of Lost Worlds	276

PHYSICAL SCIENCES

Review of the Year	282
Cosmic Time Travel	284
Farewell to the Wooden Bat	292
Behind the Laboratory Door	299
The New Science of Sound	304
The 1989 Nobel Prizes for Physics and Chemistry	311

TECHNOLOGY

Review of the Year	316
New Light on the Old Masters	320
A New Lift for Helicopters	328
Space Spies	332
Viewing Viewers	341

WILDLIFE

Review of the Year	346
Saving the Elephants	348
The Day of the Dolphins	356
America's Other Eagle	364
The Disarming Armadillo	369
Return of a Reptile	373

IN MEMORIAM	380
INDEX	383

ASTRONOMY AND SPACE SCIENCE

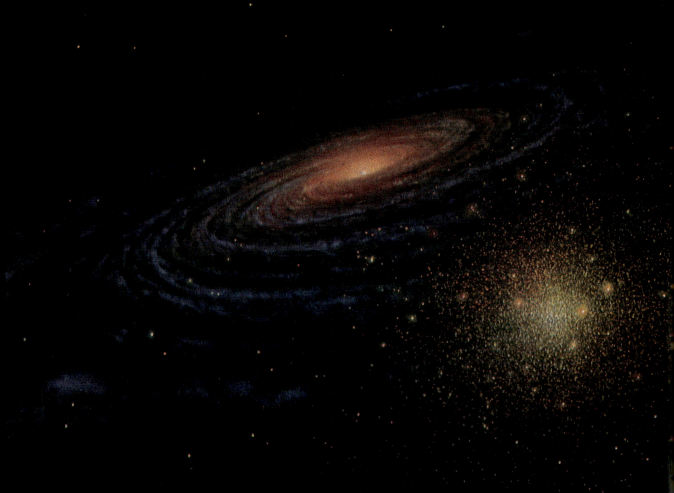

Review of the Year

ASTRONOMY AND SPACE SCIENCE

Advances in instrumentation, a new understanding of heavenly bodies, and remarkable feats achieved by the shuttle and both American and foreign probes have led us to the beginning of a second golden age of space science.

The recently launched Hubble Space Telescope must have a flawed mirror repaired before it can focus properly.

ASTRONOMY
by Katherine Haramundanis

The Universe

The remnants of Supernova 1987A may contain a newly formed pulsar. ■ Supernova 1986J is the brightest supernova known, and may be observable for years to come. ■ Sixty-five million light-years from Earth, a rotating cloud of hydrogen gas shows indications of being an embryonic galaxy. If true, this finding would contradict the notion that all galaxies formed soon (astronomically speaking) after the Big Bang. ■ Astronomers have observed the largest coherent structure yet seen in the universe. Dubbed the "Great Wall," it appears to be a thin, sheetlike structure that follows the edges of large voids. ■ A relationship may exist between the "Great Wall" and the "Great Attractor," an area some 50 to 100 megaparsecs in diameter into which some faraway galaxies seem to be falling. ■ Normal galaxies and quasars may be physically associated and perhaps clustered. ■ The most distant object ever detected, a quasar in the constellation Ursa Major, probably originated about 1 billion years after the Big Bang. Its early formation may cause some rethinking of the timetable of galaxy formation.

■ New theories of galaxy formation will have to account for a wide variety of astronomical discoveries, including the existence of Cold Dark Matter, the large-scale flow of matter in the universe, and the vast distribution of galaxies in the universe.

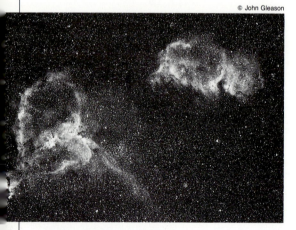

Giant molecular clouds are vast complexes of gas and dust that contract from the forces of gravity, break apart into smaller clouds, and ultimately collapse to form new stars. Astronomers believe this process is the primary means by which stars come into existence, although they understand few specifics of the mechanism.

The Milky Way

In the constellation Centaurus, an unusual object dubbed the "Southern Crab" may be evolving into a planetary nebula. The object's shape, however, cannot be explained by current theory. ■ No brown dwarfs have yet been positively identified, although astronomers have found several candidates. These bodies are sometimes called failed stars because they resemble stars in formation, but lack the mass to emit light. ■ A variable star in the constellation Cygnus suddenly brightened by 2,000 times. Astronomers will continue to observe the phenomenon. ■ On May 22, 1989, a gamma-ray source in the center of the Milky Way abruptly disappeared. Some astronomers think it may have been swept into the black hole thought to lie at the center of the galaxy; others expect it to reappear, as it has in the past. ■ A new model of the Milky Way suggests that it is a barred spiral rather than a circular galaxy.

The Solar System

High levels of solar activity will continue into 1991. The most dramatic manifestation of the cycle thus far occurred on March 6, 1989, when a massive solar flare emitted a series of high-energy discharges. The final outburst, a shock wave in the solar wind, created dramatic aurora displays throughout much of North America. For several days afterward, the sunspot group that spawned the flare remained unusually active. ■ During such periods of increased sunspot activity, the Earth can become significantly warmer. Although astronomers cannot yet reliably predict variations in the Sun's energy output, interest in the greenhouse effect and global warming has helped galvanize research efforts. ■ Scientists will have a prime opportunity to study the Sun during the total solar eclipse scheduled for July 11, 1991, and visible from Hawaii, Baja California, and south to Brazil. ■ Mercury, the planet closest to the Sun, has a thin atmosphere of sodium and potassium, perhaps diffused through fractures in its crust. ■ The Magellan spacecraft will soon provide new information on the volcanic activity that shaped the surface of Venus. Previous mapping of the planet's surface suggests Venus' volcanic plains are younger than its cratered uplands. ■ In South Africa an outcropping nearly as big as Rhode Island was apparently formed by a series of massive meteoric impacts about 3.4 billion years ago; this is the oldest such deposit identified. ■ In Iowa an impact crater some 22 miles (35 kilometers) in diameter has been identified as the spot where a flaming meteorite smashed into Earth about 65 million years ago. ■ Last year saw the closest approach to the Earth by an asteroid in 50 years. The asteroid, designated 1989 FC, belongs to the so-called Apollo asteroids that regularly orbit the Sun. It came as close as 500,000 miles (800,000 kilometers). ■ Observers discovered an unusual double-lobed asteroid that may have formed when two asteroids joined gently. ■ Laser reflectors left on the Moon 20 years ago have determined that the Moon is receding from the Earth by about 1.5 inches (3.8 centimeters) per year. ■ The reflectors, together with Earthbound instruments, have measured the Earth-Moon distance to within 1.2 inches (3 centimeters).

■ Soviet observations have determined that the solar wind blew away most of the atmosphere of Mars, and that the red planet's moons, Phobos and Deimos, were originally asteroids. ■ Phobos' icy interior could be a potential source of hydrogen, oxygen, and water for future explorers. ■ Io, one of Jupiter's moons, contains hydrogen sulfide both in its atmosphere and its surface. ■ Saturn's moon Titan has a surface of icy rocks and solid carbon-dioxide continents floating in a sea of liquid hydrocarbons.

■ Voyager II has added immeasurably to our knowledge of Neptune and its satellites (see p. 8). ■ Pluto has an atmosphere of carbon dioxide and methane. ■ No protocomets have yet been discovered in a survey of the outer reaches of the solar system.

Had last year's close call with an asteroid been an actual collision, it could have caused a catastrophe as destructive as the collision that some scientists think wiped out the dinosaurs.

Both illustrations: NASA/JPL

When the Magellan probe (left) reaches Venus (below) in August 1990, it will use radar waves to produce extraordinarily detailed maps of the planet's surface.

SPACE SCIENCE
by Dennis L. Mammana

U.S. Space Program

In 1989 the National Aeronautics and Space Administration (NASA) flew 25 astronauts aboard five successful space-shuttle missions. And, on May 10, 1989, President Bush chose the name *Endeavour* for a new space-shuttle orbiter to replace the *Challenger*, which was destroyed in a tragic accident in 1986. The name was selected from a nationwide competition involving 71,650 students and faculty members from kindergarten through 12th grade. The name was proposed in honor of the first ship commanded by Lieutenant James Cook, an eighteenth-century British explorer, navigator, and astronomer.

The new orbiter is now being built by Rockwell International of Downey, California. It is expected to be completed in 1991, and should fly by March 1992.

Voyager II's fantastic journey to Neptune and its moon Triton (see p. 8) overshadowed another important milestone in the space program: the launching of the first U.S. planetary mission in 11 years. On May 4, 1989, the Magellan spacecraft began its 15-month odyssey to the planet Venus when it was deployed from the cargo bay of the space shuttle *Atlantis*. It was the first-ever planetary mission launched from the shuttle.

The U.S. celebrated its second straight planetary-launch success last October, when the Galileo spacecraft was launched from the space shuttle *Atlantis* on a six-year journey to Jupiter. Its circuitous 2.5-billion-mile (4-billion-kilometer) route will take the craft past Venus once and the Earth twice, the second time within only 200 miles (320 kilometers) of our atmosphere.

At Venus, Galileo will look for lightning flashes in the Venusian clouds. When it passes the Earth, it will map the atmospheric distribution of chlorofluorocarbons and other gases thought to contribute to global warming, search for thunderstorm-generated clouds, study the ozone hole over Antarctica, look at the mineralogical differences on the near and far sides of our Moon, and photograph a previously unmapped strip of lunar terrain south of the huge Orientale basin. And on its way outward toward Jupiter, Galileo will photograph an asteroid as it whips by at 28,000 miles (45,000 kilometers) per hour.

Galileo will arrive at the giant planet Jupiter in 1995. In July of that year, a squat, ice-cream-cone-shaped probe will separate from the craft and dive hundreds of miles into the Jovian atmosphere, gathering data about temperatures and chemical composition before being destroyed by the planet's heat and pressure.

Five months later the remaining vehicle will become the first to orbit one of the giant outer planets. It will make an infrared map of Jupiter, watch the changing weather and cloud formations, and study the planet's complex magnetic field. It will swing close enough to the planet's four largest moons to radio back to Earth photos as much as 1,000 times more detailed than those taken by Voyager a decade earlier.

The $1.5-billion mission represents an international effort between scientists in the U.S. and those in West Germany. The Galileo craft was named for the 17th-century Italian scientist who was the first to study Jupiter and its moon with a telescope.

The mission is not without controversy, however. On September 28, 1989, several groups filed an unprecedented lawsuit in federal court to stop its launch. Some protesters even threatened to infiltrate the space center and physically disrupt the lift-off. At issue was the 48 pounds (22 kilograms) of plutonium 238 dioxide carried aboard the spacecraft. It formed the basis for two radioisotope thermoelectric generators, also known as RTGs, which convert heat from natural radioactive decay into electricity to power the spacecraft's experiments and radio transmitter.

Both photos: AP/Wide World Photos

Technicians monitor every aspect of each space-shuttle mission (left). Nearly every shuttle crew now includes at least one woman (below).

The safety of the astronauts remains NASA's top priority. Rescue vehicles are being developed to evacuate the proposed space station in an emergency.

Opponents argued that a launch-pad explosion or an unexpected reentry of Galileo into the Earth's atmosphere would spread the highly poisonous plutonium, and, therefore, the health risk was too great to justify the mission.

Scientists maintained that such a power source was needed because sunlight at Jupiter, some 500 million miles (800 million kilometers) away, is too faint to allow efficient operation of solar cells. They showed that similar generators had been launched on 22 previous flights—including the Voyager and Apollo missions—with never a problem. They calculated the odds of an accident crushing the lead casing holding the plutonium as only 1 in 25,000. And they argued that the casing was strong enough to survive even a fiery reentry into our atmosphere, should it occur. A federal district judge ultimately ruled that NASA had complied with environmental requirements, and refused to block the launch.

Last November the sky lit up over the Southern California desert when a two-stage Delta rocket left Space Complex 2 at the Vandenberg Air Force Base. The 180-foot (55-meter)-tall rocket carried a $150-million satellite named the Cosmic Background Explorer (COBE) into a 215-mile (347-kilometer)-high polar orbit.

When COBE began observing the heavens in December, it became the first spacecraft to probe the earliest stages of the universe. During its one-year mission, the $330 million COBE will use two cryogenically cooled infrared imaging systems and a microwave receiver to study the cosmic background radiation left over from the Big Bang, the explosion from which the universe is believed to have originated some 15 billion years ago. It is expected to produce data 100 or 1,000 times better than that produced by ground-based telescopes.

For nearly a decade, the Solar Max Mission (Solar Max) satellite orbited the Earth and watched the powerful Sun, recording and studying 12,500 explosive flares that blasted high-energy particles into space. Then, on December 2, 1989, the nearly 2.5-ton spacecraft plunged into the atmosphere over the Indian Ocean and burned up.

Ironically, its demise was caused by the very phenomena it was charged with studying. The rise of solar activity during the year heated our atmosphere enough for it to expand outward, causing the drag to increase on the orbiting satellite.

Launched in 1980, Solar Max made headlines four years later, when space-shuttle astronauts visited the crippled spacecraft, hoisted it into the shuttle's cargo bay, replaced some failed components, and then released it into a higher, safer orbit. Solar Max continued its detailed observations of the Sun until the satellite's death in early December.

Private Launches

On March 29, Space Services, Inc., of Houston, Texas, became the nation's first private company to launch a rocket large enough to require a federal license. On that day the rocket roared into space from the White Sands Missile Range in New Mexico, on its way to a 15-minute suborbital flight.

Five months later an 11-story Delta rocket, built and launched by McDonnell Douglas Space Systems Company of Huntington Beach, California, carried into orbit a TV broadcasting satellite for a British company—the first satellite ever launched by a privately owned rocket.

On October 5, the American Rocket Company (AMROC) suffered a serious setback when its first rocket failed to launch, caught fire on the pad, and fell to the ground. Since the failure, AMROC has cut its staff by half, and has begun seeking additional financing to carry its program forward.

The Soviet Program

During the year the Soviet Union began to market Earth photographs with sharper images than those of the U.S. Landsat and French Spot counterparts.

The Soviets suffered the loss of their 17th straight Mars mission on March 27, 1989. After receiving some data and a few photos of the tiny Martian moon Phobos, engineers lost contact with the internationally developed Phobos 2 spacecraft orbiting only 180 miles (290 kilometers) above the moon's surface.

Since Phobos 1 was lost seven months earlier, Phobos 2 was their only hope to land on the Martian moon and send back photos and data about its terrain, soil, and atmosphere, and to produce a temperature map of Mars.

Despite the failure, the Soviets announced their intent to try again in 1996. In addition, they plan to join France on a mission to the asteroid belt in 1997, return to Venus a year later, and send a probe to Mercury by 2003.

Voyager II:
NEPTUNE and BEYOND

by Dennis L. Mammana

They are part of a truly unique group on Earth: geologists, astronomers, physicists, chemists, engineers, meteorologists. Every few years these men and women leave their research and teaching jobs at their universities to gather at a very different world. They are members of the team that designed and operated the Voyager spacecraft mission during its remarkable odyssey through space to the outer worlds of our solar system.

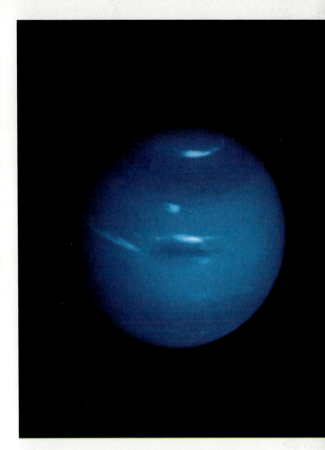

The Voyager II spacecraft (left) completed its grand tour of the outer solar system with spectacular views of Neptune (above) and a wealth of new information about the planet. Neptune was found to have furious weather systems, a strong magnetic field, a complex set of rings, and at least six new moons, for a total of eight.

Above: © Michael Carroll; upper right: NASA

A dozen years had passed since Voyager II left the launch pad from Cape Canaveral atop a Titan-Centaur rocket. Now it's the summer of 1989, and the scientists are assembling at the Jet Propulsion Laboratory (JPL) in Pasadena, California. Within weeks they will gain their first close-up look at mysterious Neptune.

Not the First Encounter

This would not be the first encounter Voyager ever had with a planet. Voyager II, along with its sister craft Voyager I, first flew by Jupiter in 1979. It was here they discovered active volcanoes on the Jovian moon Io, thin rings of dust encircling the planet, and three new moons. In 1981 both craft flew by Saturn. They found that the large Saturnian moon Titan has an atmosphere similar to that of the Earth billions of years ago when life was beginning to form. The craft found several new moons, and discovered that Saturn's famous rings contained thousands of tiny wavelike features caused by the gravity of small moons in and near the rings.

Because of the mission's overwhelming success at Saturn and Titan, scientists decided to send Voyager I up and out of the solar-system plane into deep space, where it could study the Sun's outer atmosphere like never before. And they set Voyager II on a course that would take it 1 billion miles (1.6 billion kilometers) farther from home—out to the distant world we know as Uranus.

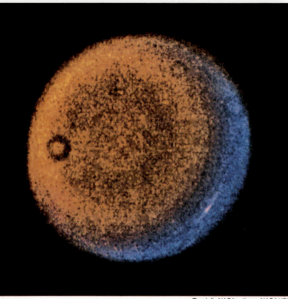

Top left: NASA; others: NASA/JPL

Voyager II flew by the three other giant planets before reaching Neptune. At Jupiter (above left), the probe examined the Jovian atmosphere and spotted volcanic eruptions on the moon Io. At Saturn (above), Voyager II provided an entirely new perspective on the planet's rings and moons. At Uranus (left), Voyager II discovered at least 10 new moons, many rings, a magnetic field, and a new type of atmospheric emission called electroglow.

It was in 1986 that Voyager II arrived at Uranus. Here it found a strange magnetic field with a corkscrew-shaped tail that extended for millions of miles into space. And it discovered that the small Uranian moon named Miranda had one of the most geologically diverse landscapes ever seen.

Though scientists were quite impressed with Voyager's discoveries, they had seen these planets before. Jupiter and Saturn had been visited by the Pioneer spacecraft in the early 1970s, and Uranus had been studied through large telescopes on the ground and in high-flying aircraft. Scientists had some ideas about what to expect when Voyager arrived.

But now it was the summer of 1989, and things were different. Voyager was heading toward an August 24 rendezvous with the planet Neptune. Even in the world's largest telescopes, Neptune appeared as but a tiny blue-green dot among the stars. Scientists had only the vaguest of notions about its nature, and they were understandably excited about the prospects.

First Photos

Neptune's great distance from the Sun (3.5 billion miles, or 5.6 billion kilometers) led scientists to believe that its atmosphere might be relatively bland, since weather systems cannot form where the Sun's heat is too weak to drive them. That's why the scientists were so stunned when they saw Voyager's first photos of the planet.

Even from millions of miles away, Voyager saw that Neptune was not the pale, blue-green world everyone had expected. Instead, photos revealed bright and dark bands and spots scattered around its disk. And as the spacecraft closed in on its planetary target by 1 million miles (1.6 million kilometers) each day, the excitement at JPL began to build.

Each new photo that came in revealed something new. By August 3, scientists had found four new moons orbiting the planet, and they temporarily named them 1989 N1, 1989 N2, 1989 N3, and 1989 N4. The moons orbited at distances ranging from 17,000 to 58,000 miles (27,300 to 93,300 kilometers) from the planet's cloud tops.

By August 11, faint, partial rings showed themselves to be encircling the planet at distances between 17,000 and 23,000 miles (27,300 to 37,000 kilometers) from the Neptunian cloud tops. Even though astronomers couldn't see these ring-arcs with ground-based telescopes, the existence of the rings had been suspected ever since the scientists watched a star's light blink off and on unexpectedly as Neptune passed nearby in the sky.

Getting Better and Better

During the weeks preceding the spacecraft's closest approach, it became clear that Neptune was not "just another planet." Every photo that came in was more detailed than the last.

Neptune itself appeared to be a spectacular world. With a diameter four times larger than the Earth's, Neptune looked like a "blue Jupiter." Its aquamarine color was caused by methane gas in its upper atmosphere that absorbed red light from the Sun and reflected back into space mostly blue.

Its largest feature was the Great Dark Spot (GDS), reminiscent of the Red Spot that has apppeared on Jupiter for more than three centuries. Nearly the size of the planet Mars, the GDS rotates counterclockwise, making one complete turn every 16 days. And just to its south appeared a group of feathery white clouds named by the scientists "S1." They believe that S1 exists because winds are forced to rise above the GDS, causing methane to condense into clouds.

There were other markings that had scientists intrigued. One of the most "popular" was a white system about the size of Mexico appearing in the planet's southern hemisphere. Perhaps a methane-ice plume deep within the Neptunian atmosphere, it was nicknamed the "Scooter" because it seemed to change its shape from day to day, and whip around the entire planet in just 16 hours.

Even farther south appeared a small, dark region with a central white spot known as DS2 (Dark Spot 2). Nicknamed the "eyeball," this feature often didn't follow its predicted path around the planet. Instead, it drifted unexpectedly northward into winds that slowed its movement around the globe.

And streaming across the blue-green face of Neptune were thin, wispy cirrus clouds of methane, some stretching as long as 125 miles (200 kilometers). These were unique because they cast shadows onto the lower-lying cloud decks, and enabled scientists to calculate their height at some 30 miles (48 kilometers).

Like the other planets visited by Voyager, Neptune was also found to have a set of rings revolving around it. The rings encircle the planet at distances between 17,000 and 23,000 miles.

INSIDE AN ICY GIANT

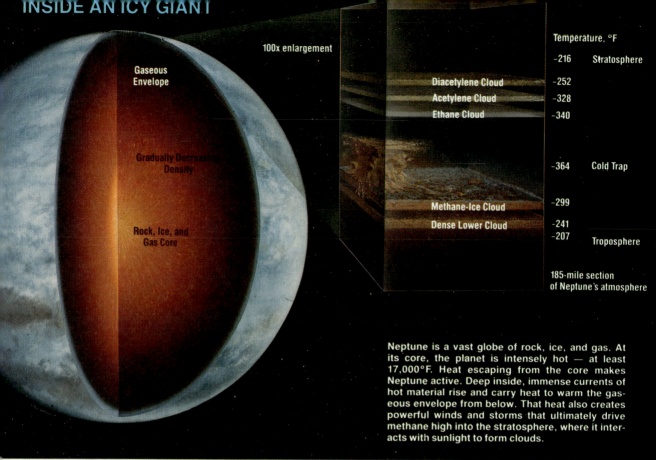

Neptune is a vast globe of rock, ice, and gas. At its core, the planet is intensely hot — at least 17,000°F. Heat escaping from the core makes Neptune active. Deep inside, immense currents of hot material rise and carry heat to warm the gaseous envelope from below. That heat also creates powerful winds and storms that ultimately drive methane high into the stratosphere, where it interacts with sunlight to form clouds.

© Paul DiMare

Close Encounter . . . and Beyond

And then, in the early-morning hours of August 25, Voyager whipped past the planet at 61,000 miles (98,000 kilometers) per hour, skimming only 3,000 miles (4,800 kilometers) above the Neptunian cloud tops. As it receded, the craft turned its full attention toward the bizarre moon known as Triton.

Since its discovery nearly a century and a half ago, Triton's strangeness had become legendary among planetary scientists. Not only did the moon orbit Neptune at a steep angle, it did so in a backward direction. Both of these facts made it difficult for astronomers to understand just where Triton might have come from. Was it born with the planet some 5 billion years ago? Or was it somehow captured by the gravitational pull of Neptune as it wandered innocently by?

As Voyager approached Triton from the south, the photos became more and more spectacular. Here was a world of exotic and complex geological features. Scientists were now seeing a brilliant white cap of nitrogen ice covering the moon's South Pole. Temperatures at the moon's surface were measured at an astonishingly low −400° F (−240° C)!

As the spacecraft moved northward, it revealed that the moon's surface darkened and changed in texture. Near the moon's equator was a terrain of ridges and valleys—"cantaloupe terrain," the scientists called it. And scattered about were broad, smooth plains of frozen ices of water, ammonia, and methane, broken by craters that were formed perhaps as long ago as 500 million years.

Some of the most fascinating features found by Voyager were dozens of dark, feathery markings that all seemed to streak toward the northeast. Scientists speculated that these might be the result of pockets of liquid nitrogen beneath the surface that boil and erupt into geysers, carrying sun-darkened methane ice crystals into the cold Tritonian air. Winds then blow these crystals northeastward until they fall back onto the surface, where they create dark streaks on the white ice.

As Voyager swung out beyond Triton, it looked back over its shoulder at Neptune once again. The planet's rings were now lit from behind, and Voyager's cameras showed that they were not partial rings, as scientists had once thought, but complete rings of clumpy dust and ice.

A Miraculous Program

While scientists and engineers on the ground were celebrating the latest success of Voyager II's mission, they realized that for the craft to have made it this far at all was nothing short of spectacular.

The Voyager program came about only after the cancellation of a more ambitious mission in the 1960s known as "The Grand Tour." The Grand Tour was designed to take advantage of a rare geometric arrangement of the planets Jupiter, Saturn, Uranus, and Neptune, to enable a properly aimed spacecraft to swing from one planet to the next without using tremendous amounts of fuel. But the mission was canceled to make money available for the development of the space-shuttle program.

Funding *was* available, however, for a more modest program. So, in place of The Grand Tour, scientists built two comparatively inexpensive space vehicles to conduct a more intensive flyby mission of only the giant planets Jupiter and Saturn. Their names were Voyager I and Voyager II, and they were designed to last only five years. Yet more than a dozen years later, both spacecraft are functioning well, and both are still returning important data to scientists back on Earth.

This is even more amazing when one considers that each of these robot craft is about the size of a subcompact car, and could fit nicely in the back of a pickup truck. Each contains some 65,000 parts—the equivalent of more than 30 color television sets. And if that isn't enough, all of the data that were sent billions of miles back to Earth by the Voyagers came by way of radio waves from transmitters with a power of only 23 watts—about the same as that of a refrigerator light bulb!

Problems

To carry out a successful mission, everything on board the craft had to be kept operating by remote control. When a piece of equipment misbehaved, controllers on Earth used a mock-up craft in the lab to determine the best way to fix the problem.

And there *were* problems. Soon after launch, the primary radio receiver of Voyager II malfunctioned, and the backup receiver automatically took over for the remainder of the mission. But even the backup began to experience tone deafness, and engineers had to learn to speak to it in a way that it could hear.

It was in 1981 that Voyager II suffered its most serious problem. While photographing one of Saturn's moons, the craft was struck by a chunk of ice from the Saturnian ring system. This knocked its camera off-target, and the spacecraft began to wobble dangerously. It wasn't until the vehicle was well on its way to Uranus that engineers could finally stabilize its motion, and begin to use the camera's pointing mechanism once again.

Despite their problems, both Voyagers have endured far longer than anyone had ever hoped. And when Voyager II encountered Neptune in August of 1989, it astounded even the experts. After a journey of some 4.5 billion miles (7.2 billion kilometers) through space, Voyager II arrived at Neptune within only 20 miles (32 kilometers) of its target, and within one second of schedule!

Farewell, Voyager

As Voyager II receded farther and farther into the distance, scientists and engineers on Earth were drinking champagne and celebrating the success of a most exciting mission. And Voyager II, 3.5 billion miles (5.6 billion kilometers) away in the darkness of space, was looking back as Neptune and Triton grew smaller in the distance. On October 2, 1989, engineers shut down Voyager's cameras for the last time, and its encounter with Neptune was officially over.

But its mission is far from complete, for its plutonium-based electrical generators may continue to function for another two or three decades. During that time the craft will venture farther and farther into space—a million miles (1.6 million kilometers) each day—and will join its sister craft Voyager I in exploring other, more exotic astronomical projects.

Together the Voyagers will examine the ultraviolet light of distant stars and galaxies. They will study the existence of magnetic fields, particles, and waves in interplanetary space. And sometime during the next 10 years, the spacecraft will cross an area known as the termination shock, the region where the million-mile-per-hour solar wind slams into the interstellar medium.

THE MOONS OF VOYAGER

Moons visited by Voyager ranged from tiny Iapetus (diameter: 900 miles), a moon of Saturn, to Jupiter's giant Ganymede (diameter: 3,270 miles). Earth moon (diameter: 2,160 miles) is considered average size.

Perhaps another decade or two later, the craft will cross the heliopause, the boundary between the solar wind and interstellar space. No spacecraft from Earth has ever entered this region—perhaps between 5 billion and 14 billion miles (8 billion to 22.5 billion kilometers) away. Voyager may be the first.

Voyager Interstellar Mission
When they do, the Voyagers will begin their last phase of research—the Voyager Interstellar Mission (VIM)—as both craft drift outward toward the stars.

Communications with the two spacecraft may continue until about the year 2020. Around that time the electrical power supplied by a radioactive plutonium generator will fall below the level needed to keep the craft alive. And Voyager will gradually lose contact with Earth.

For thousands of years, the craft will drift through the void of interstellar space—a million miles each day. In 20,000 years or so, the Voyagers will pass through the Oort Cloud, the swarm of trillions of comets held loosely to the Sun by gravity. If they survive, the Voyagers will continue outward into the galaxy.

Eventually, each will pass near other stars. In about 40,000 years, Voyager I will fly within 1.6 light-years (9.6 trillion miles, or 15 trillion kilometers) of the star AC +79 3888. Around the same time, Voyager II will pass within about 1.7 light-years (10.2 trillion miles, or 16.4 trillion kilometers) of the faint red star Ross 248. And after drifting silently for another quarter-million years, Voyager II will fly within 4.3 light-years (25.8 trillion miles, or 41.5 trillion kilometers) of the bright wintertime star Sirius. Both Voyagers are destined to wander the stars of our Milky Way galaxy—forever.

Another Encounter?
But someday, somewhere, the Voyagers may experience another encounter—not with a planet or its moons, but with an alien intelligence making its way through the galaxy. Just in case such an event should occur, the Voyagers have been prepared.

Scientists have mounted on the side of each craft a gold-plated phonograph record called

"The Sounds of Earth." Among its contents are: greetings in 55 human languages and one whale language; a 12-minute sound essay that includes a kiss, a baby's cry, and an EEG record of the meditations of a young woman in love; 118 digitally encoded images of our science, our civilization, and our culture; and 90 minutes of music ranging from a Navaho night chant to a Japanese "shakuhachi" piece to Beethoven to Chuck Berry.

From this record and its instructions, an advanced alien being—assuming there are any out there—might learn about the inquisitive creatures who created the Voyagers. They might wonder if we are still alive—whether we have annihilated ourselves from our newly developed technology, or whether we have gone on to bigger and better things.

There may even come a day when our science and technology have advanced so far that our descendants can venture out into the Milky Way and retrieve the Voyagers. Wouldn't it be exciting to visit a museum like the Smithsonian and see the craft that journeyed billions of miles through our solar system—the craft that was first to travel the space between the stars?

Our descendants might then look back with delight and wonder at the journeys of Voyager, much as we look back at those of Magellan and Columbus. And they may wonder what it must have been like to be alive at a time when humans sent robot craft on their first tentative journeys away from Earth.

A Remarkable Mission

The Voyager mission to the outer planets has indeed proven to be a remarkable achievement—perhaps the most stunning scientific success ever achieved. During their 12-year, $865-million odyssey through our solar system, the Voyagers traveled more than 4.5 billion miles (7.2 billion kilometers). They visited more than 60 different worlds—planets and moons—most of which had never been seen up close before. And together, the Voyager spacecrafts returned to Earth more than 100,000 photographs of their travels and adventures.

But more than just photos, the two spacecraft's combined 22 onboard scientific instruments provided researchers with answers to some of the most puzzling questions ever asked about the worlds of our solar system. They have uncovered mysteries that future scientists will ponder for decades to come.

And they have inspired an entire generation to look toward the stars—and wonder.

Scientists at the Jet Propulsion Laboratory tracked Voyager II every step of its journey through the solar system. They expect to stay in contact with the probe (and Voyager I, its sister craft) at least until the year 2020, when its electrical power will fall below the level needed to keep the craft alive. Once that happens, the Voyagers will likely drift through the galaxy for hundreds of years.

HARVESTING the NEAR-EARTHERS

by William Barton and
Michael Capobianco

When asteroid 1989 FC swept by the Earth earlier this year, missing our planet by a breathtaking 430,000 miles (691,870 kilometers), it generated considerable excitement. A "close call," the newspapers dubbed it, or a "near miss." No asteroid had ever been seen so close to the Earth. It wasn't the first time an asteroid scare had hit the headlines; some will remember the 1968 close approach of 1566 Icarus that sent people scurrying for higher ground.

Was the danger real? Yes and no. The Earth *has* collided with asteroids before, and the results have been horrendous, ranging from "minor" events such as the 1908 Tunguska explosion in Siberia to global catastrophes like the one that may have caused the extinction of the dinosaurs at the end of the Cretaceous period. But space is vast, and the Earth is a small target: 1989 FC missed by more than 50 times the diameter of the Earth, a trifle wide of the mark, to say the least. But can we justify sending humans and machines to these bits of asteroidal flotsam that litter the near reaches of interplanetary space? What will *we* gain?

Our space program must have explicit long-term objectives for our shorter-term objectives to make sense. We need to plan an evolutionary series of missions and goals that will turn the United States into a fully spacefaring nation over the next 30 to 50 years. And our ultimate goal must be the search for raw materials in space, not only for the space-based indus-

Asteroids may provide a source of valuable minerals to the next generation of space travelers. Astronomers already have discovered a number of "near-Earth" asteroids whose comparatively short distance from the Earth makes them prime candidates for future mining operations.

© Mark Maxwell

tries to come, but to supplement our dwindling resources here on Earth as well. Indeed, if we do not accomplish the transition from an Earth- to a space-based society in the 21st century, it seems unlikely that the impoverished and desperate civilization of the 22nd century will be able to do so. The near-Earth asteroids may represent humanity's best hope for making this bold dream a reality.

The Moon and Mars

The most often mentioned goals for our space program are flawed. The Moon is strongly depleted in the siderophile elements on which our industrial economy is based, and its geology is dominated by simple processes unlikely to concentrate useful minerals. Schemes to refine materials from the lunar soil have yet to be proven. Mars, which may have had a more "Earth-like" history, will probably have mineral resources. However, in many ways, Mars is the worst possible candidate for a deep-space mining facility. It is far away in terms of both mission energy and travel time and has a high surface gravity.

Enter the Asteroids

The most likely site of accessible resources in the inner solar system is the asteroids. A majority of asteroids are so small that they have negligible gravity, and therefore could be transported to a near-Earth industrial facility more or less intact, using a technology only modestly more advanced than what is now available.

More asteroids lie in elliptical orbits between Mars and Jupiter. In terms of both travel time and overall mission velocity—or "delta-vee," as it is sometimes called—this makes them considerably farther away than either the Moon or Mars. They are so remote that their potential will have to await the development of more advanced propulsion systems, perhaps for spacefaring generations of the future.

The near-Earth asteroids present an alternate target for our space program; they have all the advantages of the Main Belt asteroids with none of their drawbacks. These asteroids, also termed Triple-A's, are very close in terms of delta-vee. Six are closer than the Moon, and more than 50 are closer than Mars.

> **MISSIONS TO THE ASTEROIDS: THE NEXT DECADE**
>
> A number of U.S., Soviet, and European spacecraft may yield "up close and personal" looks at asteroids.
>
> • *Galileo* (1989), a joint American-European mission, will fly by Venus, Earth, and two asteroids, 951 Gaspra and 243 Ida, on its way to orbit Jupiter and drop a probe into the gas giant.
>
> • *CRAF* (1995) is another NASA/European Space Agency (ESA) venture. The as-yet-nameless Comet Rendezvous Asteroid Flyby mission will visit 449 Hamburga on its way to Comet Kopff.
>
> • *Vesta 1/2* (1996), a joint venture of the European, French, and Soviet space agencies, consists of two probes that will visit at least eight bodies, including a mix of asteroids and comets, during a five-year mission.
>
> • *Cassini/Huygens* (1996), a Saturn orbiter and Titan lander funded by NASA and ESA, is scheduled to visit asteroid 66 Maja on its way to the outer solar system.
>
> • *Rosetta* (2001), part of Europe's "Horizon 2000" program, was formerly known as the Comet Nucleus Sample Return mission. With NASA participation anticipated, it will have at least one asteroid flyby opportunity.

In March 1989, astronomer Henry Holt of Northern Arizona University discovered that a large asteroid had passed within 430,000 miles of the Earth. Widespread damage would have occurred had the asteroid struck our planet.

AP/Wide World Photos

The technology already exists to conduct reconnaissance and even prospecting expeditions to near-Earth asteroids.

More than 125 near-Earth asteroids have been discovered, most found in the past ten years. There may be more than 1,000 altogether. To determine how useful they may be, it is necessary to understand just what they are.

The solar system started out as a huge cloud of dust and gas, gradually condensing into the Sun and planets. But it was not an orderly process. Many small bodies were left over from the formation of the planets. In the region between Mars and Jupiter, a great ring of small planetoids formed, colliding, merging, and shattering to eventually produce the asteroid belt of today. Although many of these objects were ejected into interstellar space, there may be more than 10,000 bodies larger than 0.6 mile (1 kilometer) in diameter still wandering about the solar system in chaotic orbits. They range from "baked" asteroids, formed in the volatile-free high temperatures near the early Sun, to comets, primarily made of frozen ices and gases from the outer reaches of the solar system.

Near-Earther Specifics

Although no close-up observations of any asteroids have been made, we can make a few educated guesses about what these near-Earthers are like. They range in diameter from 25 miles (40 kilometers) to less than 100 yards (109 meters). Their physical structure is determined primarily by the results of high-speed impacts. The larger Triple-A's may have been reduced to "flying rubble piles," while the smaller ones are probably individual pieces that may still be reasonably intact.

The near-Earth asteroids display a wide variety of physical types, and may well represent a good sample population of the small left-over wanderers. They appear to have come from diverse places in the solar system, brought to their present orbits by the gravity of Jupiter. The spectra of many of the near-Earthers closely match those of meteorites, and they have been grouped on this basis. Some are made up of primitive material from the earliest history of the solar system. Some are stony minerals from the bedrock of partially or completely differentiated planetoids that shattered. Others appear to have a large metal content, primarily nickel-iron, from the cores of these planetoids.

Two types of near-Earthers stand out as particularly interesting. The carbonaceous chondrites contain carbon, primitive organics, and water and other volatiles; platinum-group metals and gold will almost certainly be by-products from refining these resources. The nickel-irons are obvious sources of industrial metals such as iron, nickel, and cobalt. Two nickel-iron near-Earth asteroids have been discovered so far.

Some Earth-grazing asteroids may be "burned-out" comets. Several of them are similar to the dark reddish Halley nucleus. Encke, a depleted periodic comet that follows an asteroidlike orbit, will be indistinguishable from an Apollo asteroid when its volatiles are depleted. If dead comets exist among the near-Earth asteroids, they would serve as excellent sources of water and other volatiles, as well as provide many useful hydrocarbons.

Why the Wait?

Except for some kind of hidebound planetary chauvinism, it is hard to understand why the near-Earthers have been ignored for so long. Launch vehicles capable of sending large and sophisticated probes to the Earth-grazers have been available since the dawn of the space age. Even piloted flights contain no formidable obstacles, because these asteroids are easy to reach and require no special equipment to "land." For the earliest missions, which would be Apollo-like "dashes" for prospecting and in situ planetological research, almost any off-the-shelf technology would do.

In the mid-1960s, a near-Earth asteroid mission was proposed using a two-man Gemini capsule propelled by Rover nuclear rocket stages. In the latter part of that decade, an expedition using unmodified Apollo/Saturn 5 hardware was given serious consideration as part of the Apollo Applications Program. Unfortunately, the plans were dropped when AAP, starved for funds, was reduced in scope to include only the Skylab space station.

The Soviet Union is in the best position to make a human near-Earth flight in the next decade. An Energia launch vehicle is capable of sending a Salyut-like crew module on a rendezvous-and-return expedition to any of the nearest asteroids with delta-vee to waste. The prinicipal hardship for such a mission would be its length. The Soviets, however, hold an edge: crews have been flight-tested for durations similar to those required for an asteroid trek.

The United States is in a less enviable position, if for no other reason than that most of its space-transportation system is as yet unbuilt. Even with the space shuttle back on-line, we still lack the means of reaching translunar trajectories with meaningful payloads.

An asteroid expedition would completely overshadow anything that any other spacefaring nation might achieve. However, it is reasonable to first send a series of small spacecraft to do reconnaissance of the near-Earthers. Several robotic flights have been planned that will take a look at the Main Belt asteroids, but, because of distance and delta-vee problems, such undertakings are expensive relative to the returns. Far more reasonable would be to design a multiple near-Earth asteroid flyby mission, which will more economically tell us what these asteroids are like, and allow us to make implications about their Main Belt relatives.

If these early and inexpensive flights show that the resources do indeed exist, then the nations of the Earth can safely make a commitment to more advanced missions, perhaps carrying automated fuel-processing plants, perhaps carrying mass drivers with which to move the asteroids themselves.

And perhaps sometime in the second quarter of the 21st century, a small near-Earth asteroid will be towed to an orbiting factory complex nurturing an industrial space-based economy.

AN ASTEROID USER'S GUIDE

Type	Description
C	Carbonaceous chondrites, dark bodies made of silicates and carbon. Similar to the Martian moon Deimos and comet nuclei.
S	Stony-irons, nodules of metal embedded in silicate rock.
M	Nickel irons, made primarily of metal.
E	Bright asteroids made of silicate rock and chondrites.
R	So-called "red" asteroids, probably with iron oxide surfaces.
U	Asteroids of unknown composition.
Aten	Asteroids with orbits lying primarily between Venus and Earth.
Apollo	Asteroids with orbits lying primarily between the Earth and Mars.
Amor	Asteroids with orbits lying entirely between the Earth and Mars.

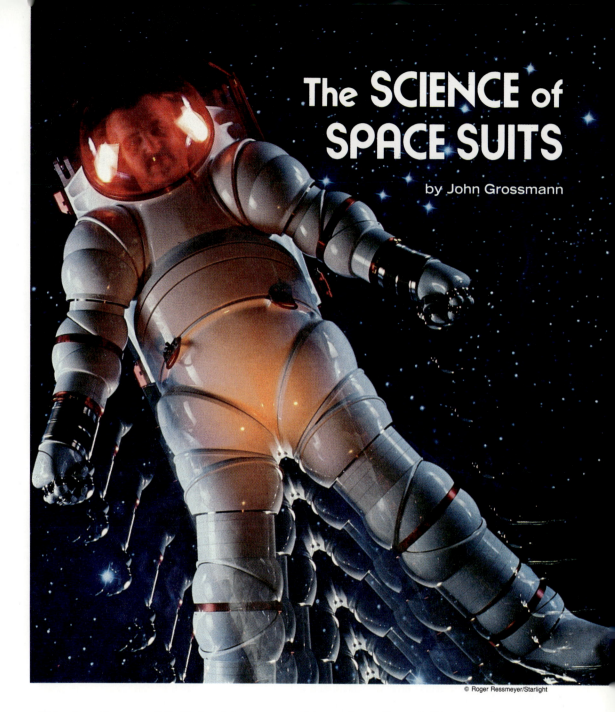

The SCIENCE of SPACE SUITS

by John Grossmann

S uch are the ways of the National Aeronautics and Space Administration (NASA) that before a new space suit can fly, it must first sink. "It's a bit like qualifying an F-16 to be a submarine," says Vic Vykukal.

The dunking, at NASA's Ames Research Center in Mountain View, California, will afford a last round of testing before Vykukal (pronounced "vick-a-call"), principal engineer in crew and human factors research, takes apart his AX-5 suit, packs it up, and sends it off to the Johnson Space Center in Houston. There it will compete head-to-head (as well as shoulder-to-shoulder, arm-to-arm, and leg-to-leg) in far more extensive testing with the Mark III, a space suit of considerably different design.

NASA subjects new space-suit designs to rigorous testing for endurance, radiation resistance, and comfort. The all-metal AX-5 suit above is competing to get aboard the space station, where space suits will have to withstand the most extreme conditions ever faced by astronauts.

ASTRONOMY AND SPACE SCIENCE 21

Shuttle astronauts have complained that their space suits lack flexibility. Makers of the Mark III suit (above) claim that their product's "soft" construction will allow space-station astronauts more freedom of movement.

Of the two competitors, the AX-5, nicknamed the Michelin Man because of its eye-stopping assemblage of segments, is more radical. There's not a stitch of fabric in it. Head to toe, it's harder than a turtle shell. The Mark III blends hard elements with more traditional soft goods. It hails from across the continent—specifically, the tiny town of Frederica, Delaware, home of a private company called ILC Dover, Inc., which has designed and built the Mark III in partnership with the Johnson Space Center.

Each design, years in development, is vying to be the astronauts' new clothes, a space suit that before the year 2000 will help build space station Freedom, and thereafter clock eight-hour days, week after week, on various projects in space. The Michelin Man is the underdog. Lockheed, the company ultimately responsible for supplying the station space suit, is planning to use a suit "very similar to the ILC hybrid" as the basis for the manufacturing contract, according to Joseph Kerwin, a Lockheed manager. Kerwin also says that Lockheed will feed the results of the Johnson review of both space suits into the contract requirements.

But first comes the water. At the Johnson Space Center, there is a huge tank, a 500,000-gallon (1.9-million-liter) swimming pool known as a Weightless Environment Training Facility—WETF (pronounced "wet if") for short. Here at Ames, what's sometimes referred to as "the hot tub" is more modest. The raised tank, walled with observation windows, is 9 feet (2.7 meters) deep, 11 feet (3.3 meters) in diameter, and holds 5,000 gallons (190,000 liters). Somebody has stuck on a sign: *Beware of sharks. Swim at your own risk.*

The idea, of course, is to have water stand in for space: buoyancy in the tank simulates microgravity. On this March morning, there's another stand-in. One of Vykukal's assistants, a suit designer named Phil Culbertson, is preparing to enter the tank. He's wearing astronaut long johns, otherwise known as a liquid cooling garment (plastic tubes circulate cool water throughout), as well as accessories that are not standard issue: skateboard kneepads and volleyball elbow pads. He climbs the stairs to the lip of the tank.

Until now, astronauts have pulled on their pants like firemen and then wormed their way up and into the top half, which is held in place by a frame. That's about to change.

Stepping up five more stairs, Culbertson reaches the AX-5, which hangs suspended from a gantry hook, toes just off the tank platform. The hinged suit back yawns open in a square hole that is tilted a helpful 9.5 degrees forward of vertical. Feetfirst, a bit like a child at the top of a slide, Culbertson slips in. The rear door closes like a refrigerator and is latched shut for him. Next comes his helmet. Life support switched on, communications working, Culbertson hangs like a marionette, flexing his arms and legs. All ready. The gantry groans, and soon, calling to mind Indiana Jones suspended over a pit of boiling lava, he's poised above the tank. Slowly, down he goes.

The Quest for a Range of Motion

Whether a hard suit or a semihard one emerges victorious from the big tank at Johnson Space Center, it will be a very different outfit from today's style, the shuttle suit. It must be. The shuttle suit is not the panoply for hundreds of eight-hour days in a life-threatening environment. Built for the relatively short periods of a shuttle flight, it requires too much maintenance to offer the astronauts long-term protection against micrometeoroids, radiation, and corro-

NASA evaluates the comfort and mobility of newly developed space suits in its Weightless Environment Training Facility, a 500,000-gallon swimming pool in which the water's buoyancy simulates the microgravity of space. Prevailing opinion seems to hold that neither the AX-5 "Michelin Man" all-metal suit (shown at right, about to be dunked) nor the Mark III "soft" suit (below) will be chosen for the space-station job. Most likely, NASA will opt for a hybrid space suit that combines the best elements of the two main contenders.

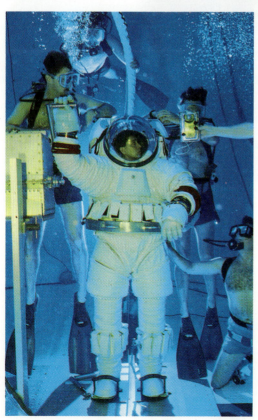

sive chemicals. And because station-based astronauts will be working near delicate external sensors, their suits, unlike previous models, will have to manage waste gases, such as carbon dioxide, internally. There will be no venting allowed. Finally, according to many of its wearers, the shuttle suit is a pain in the neck, as well as the shoulders, the arms, and so on.

"You're never unaware of being in it," says former astronaut Brigadier General Robert Stewart, now deputy commander of the U.S. Strategic Defense Command. Stewart experienced—or rather, endured—two periods of extravehicular activity (EVA) in the shuttle suit a day apart in February 1984. "You're constantly shifting positions to try and get a little extra motion. And if the suit is fit properly, you're kind of wedged into it. The pressure points for me were the collarbone and the fingertips of the glove. I bruised my shoulders and my fingertips, and my nails got a bit discolored. My first EVA, I worked so much with my hands—and at that time was using an earlier version of gloves—that my forearm muscles were just about in knots by the time I finished."

Above, left, and facing page: © Roger Ressmeyer/Starlight

THE EVOLUTION OF SPACE SUITS

Space suit precursor: when pressurized, the outfit worn by aviator Wiley Post (above) during his 1934 high-altitude flights kept him from walking or standing up straight.

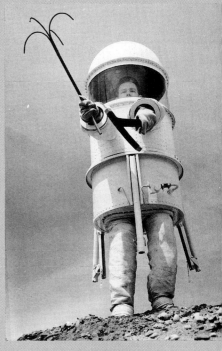

The evolution of space-suit technology has seen its share of reject designs. In the early 1960s, the man-in-a-can approach taken by the "tripod teepee" (above right) lost out, ostensibly because it required excessive space on already cramped spacecraft. In 1978, NASA turned down the "anthropomorphic rescue garment" (above), a space suit designed for the orbital transfer of an astronaut from a disabled shuttle.

NASA ultimately selected a modified version of a U.S. Navy pressurized flight suit as the space suit for the Mercury astronauts. Without its outer vapor-cover suit (left), the outfit had a distinctly robotic look. What the suit lacked in fashion savvy it made up for in freedom of movement: its flexibility even allowed for a game of baseball on the astronauts' training field (right).

"The shuttle suit has certain ways," explains mission specialist Kathryn Sullivan, speaking like a patient parent accounting for a child's occasional mulishness. "It does not have all the same ranges of motion that your body has. Its shoulder is not like yours, and its knee is not built like yours, so you learn, through several runs in the water tank, a whole set of lessons that have to do with suddenly being a person of greater mass and volume."

"But ours is not a recreational job," she adds. "This is not sailing the South Pacific for fun and a tan. The vacuum of outer space is a tough and unforgiving environment, and there's no escaping the reality: you have to be isolated from this environment to survive it."

Survival, Not Comfort

As even ground-bound mountain climbers know, the human body, which no doubt evolved at or pretty near sea level, starts sputtering when the air begins to thin. Above 38,000 feet (11,600 meters), most people lose consciousness in 30 seconds without a supply of oxygen.

Top three photos: National Air & Space Museum; bottom left: Litton Industries; bottom right: UPI/Bettmann Newsphotos

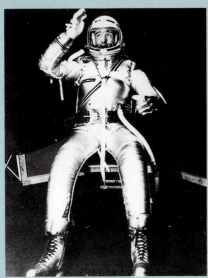

When fully pressurized to five pounds per square inch, the space suit worn by Mercury astronaut M. Scott Carpenter (left) allowed limited mobility.

After years of research and development, the Mercury astronauts emerged in space suits that were aluminized to reflect heat.

Long before any Apollo missions had flown, work had already begun on space suits that would protect astronauts on the upcoming lunar landings.

In the near vacuum of space, gas in the lungs would expand, preventing respiration. And without a device to supply pressure, body fluids would vaporize.

Survival, not comfort, has historically determined the design of these devices. For a series of high-altitude tests in 1934 and 1935, Wiley Post used a suit that was contoured in a sitting position. Once it was pressurized, he couldn't stand straight or walk. Accounts say "it afforded some degree of mobility in handling the airplane's controls." Alan Shepard's Mercury suit offered insurance, the ability to function as a backup in case of cabin depressurization. Its primary requirement sounds ironic today: it had to be comfortable *un*pressurized.

The difficulty of moving in a pressurized suit was dramatized in 1965 on the first spacewalk. Cosmonaut Aleksei Leonov's space suit expanded so much during his walk that it took him eight minutes to get back inside the airlock of his spacecraft—every movement was an effort. Three months later, when Ed White tumbled out of the *Gemini 4* spacecraft, he looked like a balloon in the Macy's Thanksgiving Day parade. "If you've ever seen any film of him, you know how immobile he was," says George Durney, ILC's chief engineer, a somewhat crusty veteran space-suit designer whose white hair is cut in a mid-century crew.

"God put the joints on the inside," says Durney, explaining that the quest is still to prove full range of motion through an external shoulder or knee joint, without undue effort from the wearer. Picture a cylindrical balloon, one of those frankfurter-shaped jobs. Inflated slightly, it's easy to bend, but with increasing inflation, it takes more and more effort to reshape it. On top of their liquid cooling garments, astronauts wear a kind of balloon—a rubber bodysuit, or gas bladder—that provides the desired pressure. But the rubber is highly elastic and, without a second skin to contain it, would merely expand under the pressure. Trying to duplicate the geometry of human bone structure externally in this second skin, which ILC has traditionally fashioned of specially patterned and reinforced polyester, has repeatedly sent Durney's fingers scratching his crew cut. "If you could just put pins through a guy's shoulder or elbow," he muses, "it would be a whole lot easier."

Durney has produced more humane solutions. One of the suits he is proudest of is the A7-LB, which improved neck mobility and thus visibility for Apollo astronauts driving the Lunar Rover. Another improvement was an "automatic" waist joint that took the yank out of sitting. Prior to the new joint, an astronaut had to tug and fasten a chest-mounted strap simply to sit down. To stand, he pulled again on the strap and released it.

The A7-LB proved nimble enough that ILC shot some 16-millimeter film of a systems engineer named Tom Sylvester wearing it on a high school football field. After a round of jumping jacks, toe touches, and push-ups, Sylvester tried his hand at making a number of tight-end-style catches, and his foot at punting and field goals. Hardly the stuff of NFL highlights, but his 20-yard (18-meter) kick through the uprights is still memorable.

The shuttle program, which opened the hatch to women, ushered in other changes in tailoring. Apollo astronauts and their predecessors wore custom-made suits. But for most of this decade, astronauts have been dressed like roommates getting ready for Saturday night: their suits were assembled off-the-rack—torso off this hanger; legs off another; arms, boots, gloves—from surely the most expensive clothes closet in the land.

Years back, simply out of curiosity, Durney tracked the cost of a production-model Apollo suit. "By the time you added in all the paperwork, it came to about 250 a copy." He nods. "Yes, $250,000." Estimates for the eventual cost of the space-station suit run several times that. Costs in general, and paperwork in particular, have risen through the years. But spending more on the suit may reduce costs in other operations.

A key requirement of the space-station suit is onboard maintenance and servicing. That's no small matter, considering the current protocol for a shuttle suit returning from a mission. By the time the suit has been inspected stem to stern—checked for pressure leaks, its backpack life-support system readied for the next flight—some 2,700 to 3,000 man-hours have elapsed. Aboard the space station, inspecting the suits will be the astronauts' responsibility, though they'll get a big hand from automatic checking systems and easily replaceable units.

Avoiding Space Bends

By far the biggest savings—and the biggest challenge—has to do with increased suit pressure. NASA has long batted around the notion of a "zero prebreathe suit," or ZPS. With good reason. The present discrepancy between the one-atmosphere, or sea-level, pressure of the shuttle cabin (14.7 pounds per square inch) and the pressure inside the shuttle suit (4.3 psi) currently squanders a precious commodity: an astronaut's time. Nobody's time is as costly as an astronaut's in space, but a shuttle crew member preparing for an EVA essentially idles, expensively, before getting dressed.

What's he doing? Breathing 100 percent oxygen during a so-called "washout" period to remove nitrogen gas dissolved in body tissues. As a scuba diver returning too quickly to the surface can suffer the painful and potentially lethal effects of the bends, so can an astronaut at the onset of EVA, if he changes pressure too drastically. One remedy is to decrease the cabin pressure from 14.7 to 10.2 psi for 12 hours before EVA, and then have the EVA crew members prebreathe 100 percent oxygen for 40 minutes. With the lungs acting as a gas separator, that procedure effectively rids the body of nitrogen. But lowering the overall cabin pressure for the EVA crew members is an option only when there are no pressure-sensitive experiments on board. Furthermore, the less dense air is more difficult to cool, raising some concern about temperature-sensitive instruments and controls. In order to prevent these problems, NASA prefers to keep the cabin pressure constant and have the EVA crew members prebreathe 100 percent oxygen—for three to four hours.

Wouldn't it be nice to have a higher-pressure space suit that would eliminate the need for prebreathing, make EVA missions more routine, and, not so incidentally, offer quicker response time in an emergency? Calculations show a suit pressurized to 8.3 psi—nearly twice the pressure of the current shuttle suit—would fit the bill. That, however, is like saying you'd have more office space in a mile-high building.

A Certain Robotic Look

Encased in the AX-5, Culbertson touches bottom in the Ames hot tub and soon works his boots into the awaiting foot restraints. In his bare feet, Culbertson stands 5 foot 10, about average astronaut height these days. To get a good fit from the AX-5, he has fixed three sizing rings above each knee and four below. These rings come in five sizes, ranging from 0.46 to 1.46 inches (1.16 to 3.70 centimeters) high, and, when inserted in all manner of combinations into the suit's legs and arms, should be able to accommodate about any size crew member, with the exception of a Kareem Abdul-Jabbar or a Mary Lou Retton.

Culbertson moves a bit like a physical-therapy patient relearning to use his limbs. There's a certain gawkiness, an appearance of programming some moves, but nothing like the jerkiness of the robots in science fiction movies. The AX-5 does, in fact, have a certain robotic look to it. But its human contents are apparent from the battery of high-tech computer-assisted equipment stationed outside the tank—equipment that measures the flow of oxygen into the suit and carbon dioxide out of it, and calculates metabolic rate.

"We're getting some work out of you," says Bruce Webbon, chief of Ames crew research and space human factors branch, through his headset to Culbertson, by now well into his exercises. "We're up to 1,727 BTUs [British thermal units, each equivalent to about 250 calories]." That's roughly half the heat put out by somebody playing racquetball.

Such measurements, taken in the Johnson WETF tank and also under weightless condi-

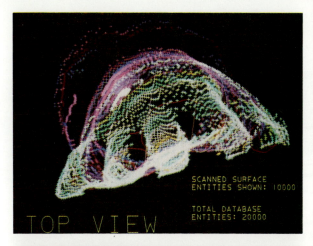

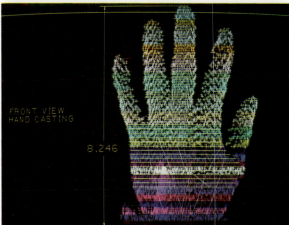

Custom-making an astronaut's gloves presents the greatest engineering challenge to the manufacturers of space suits. At ILC Dover, lasers scan the astronaut's hand (photos above), producing highly accurate models based on 20,000 bits of data. Engineers then fit pieces of fabric to the models. The finished gloves will have ribbed convolutes—ringlike ridges—at the finger joints, and a complex structure of diamonds and squares in the palm.

tions aboard a KC-135 aircraft, coupled with point-by-point recommendations made by the astronauts who test-wear the suits, will help NASA officials judge the performance of the AX-5 versus the Mark III. The two share certain design features: rear entry, a bubble helmet with better visibility, stainless-steel bearings, and Ortman rings for coupling and decoupling suit segments. But there are many elements to compare. Discounting agency politics, the competition boils down to a rivalry of engineering philosophies. To oversimplify, it's a contest of hard versus soft. ILC and the Johnson Center acknowledge that higher suit pressure requires more rigid construction, especially in large seg-

The space suits worn by the shuttle astronauts need only withstand the rigors of comparatively brief periods of extravehicular activity (left). For the astronauts building the space station, the suits will have to endure hundreds of eight-hour days in the hostile space environment.

ments such as the torso and brief, but they believe fabric can still do the job in the joints, theoretically enhancing mobility and comfort. Ames, which is chartered to explore innovative technologies, thinks it's time for an all-hard suit, and claims its design has numerous advantages over the Mark III.

For one thing, the suit's impervious skin would provide both gas bladder and containment structure, and possibly—perhaps in the form of a double-hulled, insulated construction—built-in micrometeoroid protection as well. Thermal protection, too, might be inherent, thus banishing the traditional overgarment, a cumbersome affair similar to the lead blanket a dentist lays over your torso before he departs to take X rays. The plan is to paint the suit with a thin coating that would reflect the sun's heat—in space, sometimes upward of 300° F (150° C). Gold is the coating of choice, but Fort Knox needn't worry over its stash: about 2½ ounces should do the trick.

The AX-5 prototype is made of aluminum and weighs about 185 pounds (84 kilograms), roughly 80 pounds (36 kilograms) more than a shuttle suit. Switching to a lighter metal such as titanium could subtract a few pounds. Aluminum or not, Ames says its suit will be durable, comfortable, easy to inspect after an EVA, and no mystery to manufacture.

"This new blue-collar suit should be an anvil, something you don't think about, you just use," says Webbon. "If you do the geometry of the suit properly and have bearings with low enough torque—and Vic has spent years on this—the manufacturing process is very cheap and very repeatable. It's all standard engineering practices. You just put it on a lathe. There's no black art in it."

"With an all-hard suit, you can use standard aerospace inspection techniques," Vykukal adds, "like pressurizing it to one and a half times operating pressure. You wouldn't have to minutely inspect it to see if any stitches are coming loose, or tear the suit down to see if the bladder inside is worn. Onboard inspection is going to have to be easier aboard the space station."

Even ILC, whose suit-assembly room contains rows of sewing machines, admits the amalgam of special patterns, tucks, and folds of its garment tailoring is an art, not a science. However, ILC engineers see merit, not demerits, in their creative use of soft goods. Soft, let's now acknowledge, is something of a misnomer, for by the time the Mark III is pressurized, it feels far more like a car tire than a pair of pajamas. Nonetheless, ILC claims a comfort advantage for the Mark III, pointing to the ad hoc padding—volleyball and skateboard pads—now sported by AX-5 wearers in an attempt to cushion the brunt of metal on bone. Ames researchers say these won't be necessary in microgravity.

Another touted advantage of fabric is the so-called "override" capacity. A hard suit, explains Joe Kosmo, Mark III project engineer at the Johnson Space Center, unavoidably provides a "hard stop." A metal joint moves only so far, then metal abuts metal.

With a prop in his ILC office, Philip Spampinato, ILC's advanced development manager, demonstrates the fabric advantage. The cylindrical model of a shuttle-suit knee is fashioned of accordionlike folds similar to those found at the midsection of articulated buses. A hand on each end, Spampinato compresses the knee into a giant elbow macaroni. Then, exerting more force—the override—he bends it still further. "You'll notice, however, that it's a one-way joint," he says. "There's not enough fabric on the underside for the knee to bend the other way."

When Vic Vykukal designed the very different joints on his AX-5 suit, he had in mind the wood-burning stove in the farmhouse where he grew up. "Think of a stovepipe joint," he says, explaining that he basically improved upon the hardware store thingamajig for joining straight sections of stovepipe in angles.

Play School 303?

Culbertson, who is about to move on to Phase II of his tasks—simulated EVA maneuvers such as "tube crimping in three orientations" and "ratchet wrench torquing"—owes his mobility to the bearings Vykukal carefully positioned between thingamajigs. It's the rotation of these bearings, through complementary planes, that moves the AX-5 joints.

"We're ready for Play School 303," says Webbon as Culbertson faces his task board, a panel that resembles nothing so much as a baby's crib toy. The board presents Culbertson with cranks to turn, aluminum tubes to crimp with a pair of shearlike pliers, and an assortment of holes through which he must pass similar-shaped boxes. Everybody's favorite task seems to be number six—peening aluminum. Slugging away with a hammer, Culbertson starts bending a sheet of aluminum over a mounting block. The hammering requires a good bit of arm movement, and progress is slow. "You realize," jokes Webbon, "that if you drop that hammer, you'll have to catch it on the next orbit."

The early feedback from those who *have* orbited and also spent some time underwater in the new space suits is a mixture of praise and concern. Astronaut Jerry Ross, a veteran of two shuttle missions, put both the AX-5 and the Mark III through some preliminary underwater paces. "Overall, they've done a pretty good job with both suits," he says, mentioning first that they're easier to get in and out of.

Ross also praises better mobility in both suits, which permitted him to nearly touch his toes, an improvement over the somewhat arthritic, shin-only touch afforded by the shuttle suit. On the downside, he reports some discomfort in the AX-5, especially centered on the kneecap, and an unexpected awkwardness common to both suits. "Unlike the shuttle suit, neither of these suits has a waist bearing for rotating around the vertical axis," he says. "You can rotate some, but the rotation comes from the bearings in the legs and upper thighs. You actually have to squat a bit, so it's a little awkward and definitely more work."

A Hybrid of Elements

It's his opinion, and the belief of almost everyone else—both Vykukal at Ames and Spampinato at ILC, for instance—that when all the astronauts' evaluations are considered and the official test data evaluated, probably sometime this fall, NASA will most likely *not* choose either the AX-5 or the Mark III. "I think we'll probably end up with a hybrid combining the best elements of both suits, and even then, I'm sure we'll see some additional tweaking of some of the details," says Ross.

When NASA finally has a suit for all sizes, each astronaut will be able to accessorize with his or her own wardrobe of gloves, the toughest engineering problem in the higher-pressure suit. A separate design competition, this one between ILC Dover and the David Clark company, makers of early NASA pressure suits and manufacturers of protective equipment and communications gear, is being run in search of a glove that fits like one. "We've all had hand troubles," astronaut Gerald Carr says.

"As envisioned now, each EVA crew member will probably carry up to as many as three pairs of gloves for his 90-day stint," says Johnson's Joe Kosmo. "Maybe one design that's better for manipulative tasks, another for gross handling of large trusses or payload packages."

Here, as with the suits themselves, test results for the competing sets of gloves aren't yet available. But at least one astronaut, Sonny Carter, has reported a pleasant surprise from his customized ILC glove. At the end of a long task-board session underwater, he not only tied knots in cord the size of a shoelace, but also threaded a nut on the end of a no. 6 screw, demonstrating the glove's dexterity on a screw shaft no wider than this capital *O*.

A new concept is at work in both suit and glove design. Initially protective, then flexible, they now must be durable. The direction is clear: astronauts are going to space to stay.

MOONSTRUCK

by Tom Waters

The landing of human beings on the Moon on July 20, 1969, was a momentous event by earthly standards, but it was still more so by lunar ones. Almost nothing ever happens on the Moon, so the slightest incidents are immortalized. With no atmosphere to erode them away, Neil Armstrong's and Buzz Aldrin's footprints will probably remain intact for hundreds of thousands of years. In fact, a detailed chronology of everything that ever happened on the Moon is inscribed on its surface—if you know how to read it.

For 360 years before the *Eagle* landed, ever since Galileo built his first spyglass, researchers had been trying to read that record. The moon was the nearest and most tantalizing otherworld, fatly filling a telescope's field of view; one could not help but wonder how it got there and how it came to look the way it does. What's more, as geologists came to appreciate just how long and complex Earth's history was, and how much of it had been expunged by erosion and a shifting crust, the Moon became more interesting. Its rarely cleaned slate, they hoped, might yield clues not only to its own distant past, but also to Earth's.

Such were the hopes that preceded Apollo. Centuries of merely looking at the Moon had brought only modest returns; now, at last, researchers would be able to hold Moon rock in their hands, compare it with Earth rock, analyze it chemically and mineralogically, measure its age, unravel its story. July 20, 1969, marked the beginning of a revolution in Moon studies.

Yet it was not quite the revolution that some had expected. "When I first went into the space program, I'd just got out of grad school," recalls planetary geologist James Head of Brown University. "This was 1968, and I was thinking, 'We're going to the Moon! And—they're going to land on the surface! And—they're going to let astronauts go out and collect

Above and right: NASA

Lunar Formation

A case in point concerns the answer to the biggest lunar question of all: that of how the Moon formed in the first place. According to a recent hypothesis, the Moon was born in violence, when a Mars-size piece of solar system debris—a planet that didn't make it—slammed into the young Earth, blasting off a cloud of rock that later self-gravitated into the Moon.

Computer simulations have shown that such a cataclysm could have produced a Moon with the right size and the right orbit—something earlier theories had trouble doing. And this proposed scenario has other charms as well. It explains why Moon rocks are depleted of volatile materials such as water and oxygen: they boiled away during the impact. And it explains why the Moon has at best a small iron core, unlike the other planets: most of the iron in the impactor would have sunk into Earth's core.

Neither the circumstantial evidence nor the computer simulations, however, are enough to verify the impact hypothesis. What lunar geologists really would like are more hard data. From

Manned exploration of the Moon ended with the 1972 flight of Apollo 17 *(left). Evidence gathered during the lunar landings has helped explain mysteries about the Moon that had haunted Earthbound observers for centuries. Scientists now think, for instance, that the Moon's craters (below) were created by meteorites bombarding the lunar surface, and not by Earth-style volcanoes.*

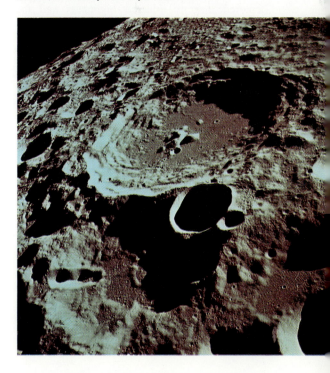

rocks! What more could you ask?' Well, my predecessors, who'd been working on this for 10 years, started out with the expectation that *they* were going to go, A; that, B, they were going to have a crew of geologists there, and one was going to land over here and another over there, and a big Rover was going to drive in between. It was going to be 15 landings or whatever—really large-scale stuff. Then, of course, the project got hammered down and down and down, until finally these two guys were going to drop down on the surface and get out for three hours."

The two guys and the 10 others who followed (including one geologist, Harrison Schmitt) did manage to cart back 841 pounds (381 kilograms) of rocks, and those rocks did answer some of the big questions about the Moon—but only some. After 20 years of hands-on lunar geology, there is much that remains to be learned about the Moon's history.

According to modern theory, an object the size of the planet Mars struck the Earth some 4.5 billion years ago. The collision hurled a cloud of vaporized rock into orbit. The rock eventually coalesced into the Moon.

At first, a layer of molten rock covered the Moon's outer surface. Lightweight solid rock gradually floated to the surface of the molten ocean, eventually forming the crust. Heavier rock settled into the Moon's lower layers.

their point of view, the Apollo astronauts left more than flags and footprints on the Moon. They also left a lot of unfinished business.

Meteorite Prints

The business of lunar geology—selenology—began in 1610, when Galileo noted that the Moon had two distinct types of terrain: light-colored, heavily cratered highlands and the darker, more sparsely cratered plains that came to be called *maria* (*mare*, in the singular), the Latin word for *seas*. But not until the middle of this century did researchers come to agree on how the Moon had acquired the craters Galileo had discovered. By systematically comparing lunar craters with ones created by volcanoes and bombs on Earth, astrophysicist Ralph Baldwin showed that no volcano (nor any bomb we've made so far) could have mustered enough energy to excavate the larger lunar cavities, known as basins, which are hundreds of miles across. Only the impact of a large meteorite or asteroid could have done the job. And in fact, nearly all lunar craters, large and small, are now thought to have been formed by impacts, with the heaviest bombardment occurring early in the Moon's 4.5-billion-year history, a time when the solar system was still thick with debris.

The realization that the craters were meteorite prints made them invaluable to Moon researchers. Each crater records the tick of a geologic clock: the more craters in a given region, the longer the crust there has been catching meteorites, and thus the older it is. By counting craters, one can make a rough guess of the relative ages of various regions on the Moon. The first thing selenologists noted when they made such tabulations is that the maria must be younger than the highlands.

The question of how the maria formed, however, was not answered until *Apollo 11* landed—but then it was answered almost immediately. There had been various theories of what Armstrong and Aldrin might find underfoot when they stepped out onto Mare Tranquillitatis. One of the most colorful was put forward by Thomas Gold of Cornell: he warned that the maria were seas of electrostatically charged dust, and that the *Eagle* might sink.

But Armstrong and Aldrin confirmed a more pedestrian view. Some researchers had long noted the similarity of the maria to the Columbia Plateau in Oregon and Washington. That plateau is made of basalt, a kind of rock that forms when lava flows out of a volcano and cools at the surface. Sure enough, the rocks Armstrong and Aldrin retrieved were basalts.

Lunar Lava

It is now agreed that all the maria were formed by a long series of lava flows during a period of intense volcanic eruptions that probably began around 3.9 billion years ago, after the worst of the meteoritic bombardment. The eruptions con-

Illustrations by Ian Worpole/© 1989 Discover Publications

The Moon underwent heavy meteorite bombardment during its first half billion years. One particularly strong impact may have caused dark, heavy minerals from the Moon's interior to push up through the light lunar crust.

The dark, cratered lunar plains called maria formed during a period of intense volcanic activity (inset) that began about 3.9 billion years ago. Meteorites continued to jolt the Moon long after the volcanoes died out.

tinued for at least 700 million years, and may not have ended until about a billion years ago. The lunar lava, fluid as motor oil, rushed out over the ancient crust, filling some of the impact basins. Then it froze into the dark seas of basalt we see today.

The maria, however, are not unadulterated basalt. Although the flux of meteorites began to taper off about 4 billion years ago, it continues to this day. The maria have relatively few craters visible from Earth, but they have been pounded by a steady rain of smaller meteorites, including many less than a millimeter across—the kind Earth is protected from by its atmosphere. These impacts have slowly bashed the basalt into a fine, powdery, charcoal-gray soil strewn with larger rocks. Many of the rocks and soil particles in the maria (as well as in the highlands) are breccias—composed of fragments of older rocks smashed by a meteorite.

Moon rocks in general are a jumble. And, in a sense, so is the entire Moon. A good-size impact can hurl rocks a considerable distance, as the bright rays emanating from the craters known as Copernicus and Tycho demonstrate: those rays, which consist of rock ejected from the craters and strewn across the surface, extend for thousands of miles. (The enormous impact that excavated Mare Orientalis may have flung material halfway around the Moon, where it collided with rock flying the other way; in this view, the large, bright smudge that lies exactly opposite Orientalis is where the colliding rock fell in a heap.) Some mare rocks contain bright fragments that are not derived from mare basalts at all. Those fragments came from the highlands.

Even before the highlands had been sampled directly by the later Apollo missions, Harvard geochemist John Wood had sifted highland pebbles out of mare soil and identified the bright material as a type of igneous rock called anorthosite. The landings in the highlands confirmed Wood's diagnosis. In fact, spectral measurements made by the orbiting command modules suggested that anorthosite, a notably light-colored and lightweight rock, was the dominant rock throughout the lunar highlands.

Finding anorthosite on the Moon was not surprising—it's not rare on Earth—but finding so much of it was. Unlike most igneous rocks, anorthosite is composed almost entirely of one mineral: plagioclase feldspar. It forms when a large body of magma cools slowly, allowing the lightweight, light-colored plagioclase to float toward the surface of the magma while heavier minerals (some of them dark as well) sink toward the bottom.

The implication of the Apollo findings was astonishing but unavoidable: since anorthosite is spread all over the surface of the Moon, the Moon must once have been covered with an ocean of magma. What's more, some of the Apollo anorthosites had been dug up from

ASTRONOMY AND SPACE SCIENCE

depths of 30 miles (48 kilometers) or more by large meteorite impacts. Thus, the primeval magma ocean must have been at least that deep. When it cooled and solidified, not long after the Moon's birth, it formed the crust that remains exposed in the highlands today.

The dark basaltic lava that covered this crust in the maria rose up, during a later period of melting, from below the anorthosite layer—that is, from the bottom of the original magma ocean. Mare basalt and highland anorthosite, in this view, are the yin and yang of lunar rock. Their unity was fractured in the primeval magma, and now they exist as separate layers.

Surprising as it was when Wood first proposed it, the magma-ocean concept is now generally accepted by geologists, and its implications extend well beyond lunar evolution. "At the time we went to the Moon," says Wood, "the conventional wisdom was that the Moon and planets had formed relatively cold, and only later heated up, due to radioactive decay inside the planets." But once it was admitted that the Moon had formed hot, it wasn't long before the principle was extended to planets: Earth, too, once had a magma ocean. "Conventional wisdom has swung the other way," says Wood.

Telescopic Analysis

In the years since the Moon landings, though, the magma-ocean theory has undergone a lot of refinement—or at least complication. "After Apollo, with samples in hand, we had a very simple picture of the Moon's evolution," says Carle Pieters, a geochemist at Brown University. "But things have gotten more complex since then." The magma ocean, Pieters has found, can't explain all the highland rocks, because not all of them are anorthositic.

Pieters is studying areas of the moon not sampled by Apollo, using a device that ought to seem quaint by now in lunar geology: a telescope. By attaching spectrometers to a powerful telescope on Mauna Kea in Hawaii, and by keeping the same spot of the Moon in view for hours at a time, she can analyze the mineralogy of the Moon's surface. And since craters bring material to the surface from various depths—the larger the crater, the greater the depth—observing a series of progressively larger craters is like descending into the Grand Canyon: you see a cross section of the crust.

It's in the large highland craters such as Copernicus and Tycho that Pieters and her colleagues have found their most interesting results. What they've found, in essence, is rock that has no business being in the highlands, because it contains heavier minerals—olivine and pyroxene—that should have been confined to the deepest levels of the magma ocean. Such discoveries do not disprove the theory. But they do suggest that after the anorthosite crust formed, and while it was being battered from above by meteorites, it was also suffering intrusions from below by material from the Moon's interior—perhaps the same sort of stuff that volcanoes later spewed into the maria. "The complexity of the lunar crust," concludes Pieters, "is far greater than we ever imagined."

Yet there is a curious pattern to the complexity—a pattern to the distribution of darkness and heaviness on the Moon. The pattern manifests itself in three ways. First, most of the intrusions of heavy minerals into the highlands are found in craters on the left (western) half of the Moon's near side (the side we always see from Earth); craters on the right (eastern) half tend to contain mostly anorthosite, like the Apollo samples. Second, the near side as a whole has many maria, whereas the far side has almost none. Third, the Moon's center of mass is closer to Earth than is its geometric center, indicating that the near side is made of heavier stuff than the far side.

Pieters and other researchers suspect that all three patterns might somehow have been produced by the impact of an immense object early in the Moon's history, while the crust was still crystallizing out of the magma ocean. The Procellarum Basin, a vast, apparently circular feature that is covered by the maria of the western near side, may be the disguised record of such an impact. But the idea is just speculation at this point. "We're data-limited now," says Pieters. "We have just enough data to know that ignorance isn't bliss."

More Data Needed

Where will new information come from? Twenty years after the first Moon landing, 65 groups of researchers are still working on the rocks brought back by Apollo, and much remains to be extracted from them. But the problem with those rocks is that they come from such a limited region of the Moon—roughly speaking, the central part of the near side. "To really answer the questions we have about the Moon, you need a global evaluation," says Pieters. "You can't just use a few pieces to extrapolate to the whole."

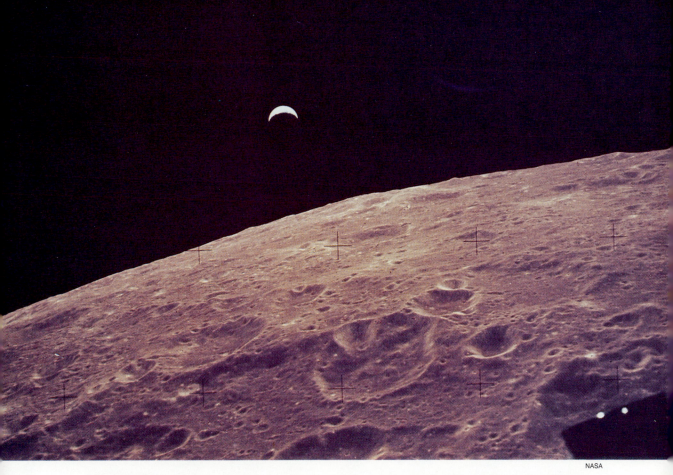

Scientists hope it won't be too long before an American astronaut again sees the Earth rise over the lunar horizon.

If you have good global maps, though, you can do some serious extrapolating, using the Apollo rocks as Rosetta stones. That's the premise underlying the National Aeronautics and Space Administration's (NASA's) modest plans, repeatedly shelved over the past decade or so, for a return to the moon. Although the entire Moon has been closely photographed in visible light, mostly by the pre-Apollo lunar orbiters, only about a fifth of the surface has been covered at the gamma-ray and X-ray wavelengths that are crucial for determining the composition of the surface. The *Lunar Observer*, a small probe now under consideration for launch in the mid-1990s, would extend those maps to the rest of the Moon. In so doing, it might help confirm or deny the impact theories of the Moon's origin and evolution.

The *Observer* could also do something else: it could identify good sites for another manned landing on the Moon. All lunar geologists agree that what they really need are more rocks. Of course, the needs of geologists are not what drives the space program, but the idea of a return to the Moon as a first step in the manned exploration of the solar system has undeniable romantic appeal. "Eventually we will go back to the Moon," says geochemist Jeffrey Taylor of the University of New Mexico. "The only question is when."

Taylor and Paul Spudis of the U.S. Geological Survey are ready now with an ambitious plan. They propose a permanent Moon base staffed by astronauts and robots. The robots would crisscross the Moon, doing fieldwork hundreds of miles from the base. Each one would be controlled by a geologist sitting back at the base, wearing a helmet that enabled him to see what the robot sees; an array of sensors on his body would transmit signals to the robot, causing it to mimic his motions. The vision may sound farfetched, but, as Taylor and Spudis point out, the technology is not. And if it is ever realized, lunar geologists will enjoy, if only vicariously, the full-scale field expedition they expected from Apollo, but never got.

BEHAVIORAL SCIENCES

Review of the Year

BEHAVIORAL SCIENCES

In 1989 studies shed new light on the myriad ways the mind works. Research elucidated the coping mechanisms of trauma victims, the trials of quitting smoking, the depths of depression, and the outcome of various child-rearing methods.

by Bruce Bower

Trauma Recovery

Many people assume that emotional recovery from a close brush with death or some other severe psychological trauma depends on confronting memories and feelings linked to the terrible experience. But a study of survivors of World War II's Holocaust suggested just the opposite: learning to forget, rather than reexperiencing, traumatic events may be the key to successful adjustment.

Israeli psychologists at the Technion-Israel Institute of Technology in Haifa found that well-adjusted Holocaust survivors had little or no recall of their dreams and remembered few details of their wartime ordeals. Survivors with serious mental, social, and work-related problems had better dream recall and remembered much more about the war.

Massive unconscious repression of Holocaust memories, such as imprisonment in a concentration camp, served well-adjusted survivors well, and spilled over into the repression of dreams.

However, another study suggested that talking candidly about past traumas boosts the mood and physical health of Holocaust survivors. Psychologist James W. Pennebaker and his colleagues at Southern Methodist University in Dallas found that people who openly and emotionally discussed their Holocaust ordeals with an interviewer reported that they felt better and had significantly fewer physical problems and physician visits one year later. Those who disclosed little of their past in the interviews reported the most subsequent health problems.

Some Holocaust survivors cope successfully with their ordeal by repressing memories of it; others fare better by confronting their experience.

Parenting Styles

A long-term study suggested supportive control, rather than permissiveness, is the key to raising a psychologically healthy teenager.

Parents who consistently set down clear standards for conduct and offer freedom within specific limits raise teenagers who perform better on academic tests, are more emotionally and socially stable, and use alcohol and illicit drugs much less than youngsters from other types of families, reported Diana Baumrind of the University of California, Berkeley.

Baumrind and her colleagues studied 124 children and their parents. Families were interviewed at their homes when the children were 3, 10, and 15 years old.

The researchers found that adolescents function best when parents are highly demanding as well as sensitive to a child's needs for independence. This parenting style is called "authoritative." Teenagers from "democratic" families, where parents stress permissiveness somewhat more than setting limits, also did well academically and socially, but they used drugs more than did the offspring of authoritative parents.

The outlook was less bright for teenagers from other types of families: "authoritarian" (extremely restrictive and demanding with little emotional support), "directive" (obedience-oriented and moderately supportive), "nondirective" (setting no limits and moderately supportive), "unengaged" (providing neither control nor support), and "good enough" (adequate, but not outstanding, in control and support).

Baumrind said single parents who were authoritative had teenagers who did as well as youngsters from intact authoritative families.

Children of demanding but sensitive parents grow into stable adolescents.

Infant Emotions

There is more to a baby's face than meets the eye. Psychologists at Wesleyan University in Middletown, Connecticut, documented greater emotional intensity on the right side of infants' faces while the youngsters smiled or frowned. Curiously, previous studies found emotional expressions are more intense on the left side of the face among right-handed adults.

Researchers think the adult pattern is influenced by the right side, or hemisphere, of the brain, which controls most muscles on the left side of the face and is critical in producing emotional displays. The left hemisphere controls much of the right side of the face, and is thought to inhibit emotional expressions, thus contributing to more intense expressions on the left side.

The examination of babies' faces suggests the right hemisphere matures more quickly during infancy, according to study director Catherine T. Best. At first, it apparently dampens the expression of spontaneous emotions, then gives up that function as the left hemisphere matures during childhood. Further study of facial expressions may clarify the ways in which infant and adult brains handle emotions, Best said.

© F. W. Binzen/Photo Researchers, Inc.

The right side of an infant's face expresses more emotion than the left.

Depression's Disabling Effects

Depression is a serious psychiatric disorder, usually involving a number of problems related to recurrent bouts of sadness and a loss of interest or pleasure in all activities. But aside from its mental symptoms, depression is as physically disabling as high blood pressure, diabetes, arthritis, and many other ailments, according to a study of medical patients in Boston, Chicago, and Los Angeles.

Depression can prove to be as physically disabling as diabetes, arthritis, and many other serious disorders.

Compared with patients suffering from full-blown depression or even a few symptoms of depression, only people with coronary artery disease or chest pains due to heart disease were more limited in routine daily activities—such as climbing stairs, walking, and dressing—said the study's director, Kenneth B. Wells of the University of California, Los Angeles.

Day-to-day functioning was most difficult for patients with symptoms of depression as well as a persistent medical condition.

How to Quit Smoking

Kicking the cigarette habit is notoriously difficult, but it is unclear how best to leave smoking behind. Several studies have concluded that smokers who quit on their own are two to three times more successful at staying away from cigarettes than are those who use various manuals describing strategies to stop smoking.

But that argument was challenged by evidence compiled from 10 new studies of smokers attempting to quit either on their own or with the help of "stop smoking" manuals. Researchers found that about 4 percent of both groups of smokers abstained from smoking for six months to one year after their initial attempt to give up cigarettes.

Previous studies did not check up on people for six months or more, the investigators said.

The new studies also found that smokers who use less than one pack of cigarettes a day were best able to quit smoking for a full year, whether or not they consulted a manual.

But the majority of people in the studies failed in their first attempt to quit smoking. Giving up cigarettes on one's own or with the help of a formal program usually requires several attempts, the researchers said.

Smokers rarely succeed in quitting the habit permanently in their first attempt to give up cigarettes.

Above left: © Louis Peihoyos/Matrix; below: © Peter Yates/Picture Group

Finding the ZONE

by Lawrence Shainberg

No one knows when it first came to be called "the zone," but if you want an example of the lofty, almost mystical, state to which sport can at times grant one entry, you couldn't do better than this one, recalled by the great Brazilian genius of soccer, Pelé, in his book, written with Robert L. Fish, *My Life and the Beautiful Game*. One day, Pelé said, he felt "a strange calmness" he hadn't experienced before. "It was a type of euphoria; I felt I could run all day without tiring, that I could dribble through any of their team or all of them, that I could almost pass through them physically. I felt I could not be hurt. It was a very strange feeling and one I had not felt before. Perhaps it was merely confidence, but I have felt confident many times without the strange feeling of invincibility."

Much as he excelled at his game, Pelé's experience was anything but unique. Athletes' reports of such phenomena are common. Basketball players say that when they play in the zone, the basket seems bigger, and they feel an almost mystical connection to it. Ted Williams, the legendary hitter for the Red Sox, has said that sometimes at bat he could see the seams on a pitched ball. Former collegiate gymnast Carol Johnson remembers that on good days the balance beam was actually wider for her, so that "any worry of falling off disappeared."

Olympic archer Denise Parker (facing page) and many other athletes report that their best performances occurred while in "the zone"—an elevated mental state during which the mind and instincts seem to function without any interference from thoughts or consciousness.

Many athletes have echoed John Brodie, the former quarterback of the San Francisco 49ers, who once told Michael Murphy, author of *The Psychic Side of Sports*, that there are moments in every game when "time seems to slow way down, in an uncanny way, as if everyone were moving in slow motion. It seems as if I had all the time in the world to watch the receivers run their patterns, and yet I know the defensive line is coming at me as fast as ever."

One mystery of the zone is that entry to it is not restricted to the elite. My own interest in the phenomenon—which has led me recently to speak with dozens of athletes, psychologists, and scientists—began on the basketball court years ago, on a day when my ordinary game suddenly (and, alas, temporarily) escalated so that it seemed I could not miss a shot.

Few among those I sought out were unfamiliar with the subject. Investigated psychologically, neurologically, and anthropologically, the zone has been related to hypnosis, spiritual or martial arts, practice, and parapsychology; and ascribed to genetics, environment, and motivation, not to mention skill. On the negative side, statisticians have offered studies that demonstrate, they say, that many "streaks" in sport are no more unusual than streaks in gambling, and some neuroscientists call the heightened perception reported by players like Ted Williams a sort of illusion or even a hallucination.

Liberating Athletes

For Keith Henschen, an applied-sports psychologist, the zone is a practical matter. From his office at the University of Utah, in Salt Lake City, Henschen conducts a sort of therapy, with a number of Olympic and college athletes, that aims to produce precisely the kind of concentration and energy that players like Pelé and Williams describe. With strategies gleaned from meditation practices and the martial arts, as well as psychological counseling and conventional athletic coaching, Henschen seeks to liberate athletes—the best of whom he calls "super-normals"—from their mental or emotional obstacles, give them an edge on the competition, and, he hopes, point them toward the state of grace that resides at the center of their games.

"No one can reach such levels by snapping his fingers," he says, "but the ultimate purpose of the exercises I use is to help an athlete get to the zone more frequently."

In 1985 Henschen set to work with one such super-normal as the first step on what he hoped would be the road to the 1992 Olympics, but which led instead to the 1988 Olympics in Seoul. She was Denise Parker, then a diminutive 12-year-old from South Jordan, Utah, and her sport was archery.

Two years earlier, her father, Earl, had introduced her to the bow and arrow, with the idea that she and her mother, Valerie, might join him in his passion for deer hunting. In fact, Denise did kill a deer last year, less than 30 minutes after obtaining her hunting license. But by then Valerie and Earl already knew they had a prodigy on their hands.

In 1990, Michael Jordan (far left) scored a career-high 69 points against the Cleveland Cavaliers. "Everything seemed to fall and I found myself in a great rhythm," Jordan said, a comment not unlike those heard from many other athletes after an excursion into the zone.

Right: © Carl Skalak/Sports Illustrated; facing page: © V. J. Lovero/Sports Illustrated

To many athletes, the zone has an almost mystical component. In his book My Life and the Beautiful Game, Brazilian soccer legend Pelé describes feeling "a strange calmness" when in the zone. "It was a type of euphoria," he says.

In 1983 Denise won the junior division at the Utah State Archery Championships, and a few months before her introduction to Henschen, she placed second in the juniors at the national indoor competition of the National Archery Association. Standing 4 feet 10 inches tall and weighing a little over 90 pounds, she had trouble stringing her bow as tightly as the longer target distances required; at 70 meters, the longest Olympic distance, she had to arc her arrows very high to reach the target at all. But no one who knew the sport could deny she had taken to it with an uncanny authority.

Henschen began by giving Denise a battery of psychological tests. Their results revealed a fairly typical American teenager, with an attention span that didn't go much beyond five seconds, and little of what psychologists call self-concept. What set her apart, in addition to her skill, was motivation—a determination to excel that was grounded in, among other things, a desire to please her parents and, more important perhaps, a tremendous competitive urgency. "She was always the sort of kid," says Valerie, "who could not stand to lose at anything. The sort of person who, if you were walking down the hall with her, would walk in the middle and not let you get by. In my opinion, it wasn't until she lost in a tournament—to a boy whom she beat the following week—that her interest in archery really took off."

Blanking the Mind

With the burgeoning of sports psychology, the exercises that Henschen suggested have become extremely popular in recent years at every level of competitive sports. In preparation for tournament pressure, he taught Denise to tense, then relax, each muscle in her body, and he directed her to wear earphones and listen to the radio while shooting. To strengthen her concentration, he assigned her seven different exercises, among them listening to her own heartbeat, reading a book while watching television and listening to the radio, and "blanking" the mind.

"I have found," says Henschen, "that athletes who are best at this are the ones who go into the zone most easily."

Finally, he asked her to create for herself what he called a "happiness room," a place to which she could withdraw in her imagination in order to visualize an upcoming meet. Of all the exercises, this was the one to which Denise brought the most enthusiasm. The room she created was primarily a replica of her bedroom, but it had its magical dimension, and it was anything but austere. "There's stairs leading up to it, and these big doors you go through," she explains. "It has brown wall-to-wall carpet, a king-sized water bed, stack stereo, a big-screen TV, and a VCR, posters of Tom Cruise and Kirk Cameron on the wall, and a fireplace that's always blazing. That's where I go when a meet's coming up. I drive up to it in a Porsche, go inside, lie down on the water bed, and watch a tape of myself shooting perfect arrows. Later, when I get to the tournament, everything seems familiar. Even at the Olympics, I was calm as soon as I began to shoot."

Like archery itself, which the Parkers practiced together in a cornfield behind their house, Henschen's exercises quickly became a family affair. Since Denise found it difficult to do them unless her mother and father joined in, all three repaired to the couch after dinner in the evening for visualization and relaxation exercises. Even today, at the age of 15, she cannot imagine doing such things alone.

In addition to Henschen, Earl Parker hired a new coach for Denise. He was Tim Strickland, a professional archer from Pine Bluff, Arkansas. Like Henschen, Strickland found her to be a remarkably willing student. "Her mind was open; that's what was unusual about her. She had a cleaner attitude than most anyone I'd worked with, an ability to take instruction and put it at once into practical application."

Strickland stiffened her practice routine. On a typical day, she ran three or four miles, did aerobics before school in the morning, shot for two to three hours, lifted weights to strengthen her shoulder, and, of course, practiced her mental exercises. Like any coach, he was a fanatic about technique as well as the state of mind his sport required, but he may have been a little more precise about the connection between the two. "The better your technique," he says, "the more you can anchor your mind in it. And the more you anchor your mind, the better your technique will be."

Avoiding Conscious Intervention

In archery, as in any sport where aiming is involved, an athlete deals with the subtlest kind of interaction between active and passive instincts. Among other things, one has to face the terrible fact that one's sight is never still. "If you don't believe that," Strickland says, "try pointing your finger at an object and see how much it moves."

One of the most difficult skills to acquire is the ability to acquiesce in such movement. Because the eyes dilate when the level of adrenaline is high, an archer's sight always seems to be moving more in pressure situations, when one wants it most to be still. Great archers, says Strickland, have the paradoxical ability to welcome such pressure and can, as incredible as it sounds, even draw strength from a drifting sight. "If you let your sight move," he says, "you'll shoot within the arc of its movement. But if you try to hold it, say, within a one-inch arc, you'll be lucky to hit within six or eight inches of where you're aiming."

Then, too, there is the matter of releasing the arrow, which, Strickland says, must not be an act of decision or will. The great enemy for an archer, as perhaps for any athlete, is conscious intervention. "Your conscious mind always wants to help you, but usually it messes you up. But you can't just set it aside. You've got to get it involved. The thing you have to do

The heightened concentration and sense of invincibility that Steffi Graf and other tennis players feel during tournaments may well be a manifestation of the zone.

is anchor it in technique. Then your unconscious mind, working with your motor memory, will take over the shooting for you."

Another thing Strickland rails against is also a function of consciousness: concern with the target or the score. Nothing, he declares, interferes with performance like concentrating on the goal rather than on the process of one's game. "An archer who worries about his score will try to *make* his arrows go in instead of *letting* them go in," Strickland says. "If your technique is correct, the target never enters your mind. It's just there to catch your arrows. Oriental archers"—the Korean women are the best in the world—"learn this way, but Americans are not trained with the same thought pattern. They're always thinking about the score."

A New Order of Existence?

If there's any theme that dominates the reports of zone experience, it is this subtle freedom from intervention—from volition and thought and finally consciousness itself.

"The mind's a great thing as long as you don't have to use it," says Tim McCarver, currently a broadcaster for the New York Mets and ABC Sports, who not only knew the zone himself as a catcher for the St. Louis Cardinals and the Philadelphia Phillies, but also had the privilege, when he caught Bob Gibson in 1968, of participating in what many believe to be the greatest year a pitcher ever had: Gibson's earned run average that season—1.2 earned runs allowed per game—was the best in the history of modern baseball.

To hear McCarver talk about Gibson is to hear of a zone that endured for an entire season. "Gibson was on a mission. You could see it in his eyes. He had tremendous energy and animation, a confidence that he could do anything . . . and I mean anything . . . put the ball anywhere he wanted. I'd just put my mitt out, and he'd put the ball in it. Look, people still talk about Gooden in his great year," says McCarver, referring to New York Mets pitcher Dwight Gooden in 1985. "He was good, all right—24-4 with a 1.53 ERA. But here's a guy that gave up half a run less per game!"

Even statisticians acknowledge that such performances as Gibson's—or Joe DiMaggio's 56-game hitting streak in 1941, or Orel Hershiser's string in 1988 of 59 scoreless innings—rise so far above the norm that there is no room for them on probability curves. So much do they exceed common levels of mastery that they seem, like a mutation in evolution, to define a new order of existence.

Says McCarver: "Many ballplayers think too much. Players like Gibson and Hershiser seem to have a sort of paradoxical intelligence—one that allows them not to do anything to hinder themselves. It's a sort of intelligence you use almost paranormally. It allows people to do phenomenal things. People who really use their mind—they free it from impeding their activity."

Japanese hitter Sadaharu Oh describes a similar intelligence when he says that, of all the qualities he had to master, none was more important than the ability to wait—learned from Ueshiba Morihei Sensei, the master of the quintessentially fluid martial art known as akido.

Waiting, says Oh, in his autobiographical book *A Zen Way of Baseball*, co-written with David Falkner, "was the most active state of all. In its secret heart lay the beginning of all action . . . the exact moment to strike . . . With the ability I had acquired to wait, I now could move my contact point somewhat farther back. This in turn gave me slightly more time before I had to commit myself."

Brain and Environment Interact

Most ballplayers—out of confusion, or perhaps superstition—maintain silence on the subject of the zone. Neuro-scientists are no less baffled, but new research may shed some light on the brain's participation in such experiences. At the University of California at Irvine, Dr. Monte S. Buchsbaum, a professor of psychiatry, uses new computer and brain-scanning technology (known as PET scan, for positron emission tomography) to image the brain's metabolism during performance. His particular interest is in the so-called "arousal state," an escalated level of energy and concentration, which is an obvious component of the zone.

Buchsbaum has yet to study an athlete, but working with subjects involved in problem-solving, he's found that during periods of intense concentration, there is a marked *decrease* in the overall metabolic rate of the brain. The research indicates that the more skill one brings to a task, the more efficient the brain becomes.

"If I were pitching a baseball, my whole brain would be active," Buchsbaum says, "but if we could see Hershiser's brain while he's pitching, my guess is that he'd be using only particular areas. In all likelihood, his skill re-

In 1988, Orel Hershiser (left) pitched a string of 59 scoreless innings, an achievement ascribed to "a sort of intelligence" that frees the mind from impeding the body's actions. Kevin Mitchell (above) and others who enjoy hitting streaks may have had sustained access to the zone.

sults, not from an overdevelopment of different areas, but from the ability to use certain areas more efficiently. We have found that higher levels of metabolism correlate with worse performance."

In work recently begun with a colleague, Dr. Richard Haier, Buchsbaum has studied the brain of a subject learning a video game. Predictably, they have found that the metabolic rate decreases as the game is mastered, but there is one interesting exception. In the visual cortex, the part of the brain that processes visual imagery, the metabolic rate increases. Buchsbaum suggests that this is because the subject is able to process more visual information as his skill increases.

It is interesting to correlate this observation with that of researchers in the area of time perception, who have often posited an inverse relationship between information processing and the speed with which time seems to pass. As the psychologist Robert E. Ornstein writes in *On the Experience of Time*: "When an attempt is made to increase the amount of information processing in a given interval, the experience of that interval lengthens."

Consider the amount of information processed by a quarterback like John Brodie as he dropped back to pass. If his overall brain metabolism were lowered and his visual cortex highly activated by the level of skill, concentration, and excitement he brought to his game, would this account for the degree to which time slowed for him when he was in the zone?

The same sort of process may make it more credible than many neuro-scientists believe that Ted Williams could see the seams on a pitched ball—or, for that matter, that, for Williams, a 95-mile (153-kilometer)-per-hour fastball actually took longer to reach home plate than it did for lesser hitters. This idea may seem far-out, or even occult, but it is a truism in modern physics—a direct inference from the theory of relativity—that speed and time are relative phenomena, functions of the point of view through which they are perceived.

Accurate though they may be, however, such explanations are slightly misleading, because they portray the zone as a linear phenomenon, issuing directly from the brain. In fact, the zone experience represents an interaction of brain, body, and environment. While it is true that Denise Parker's brain affects her perception and therefore her shooting, it is no less true that archery itself transforms her brain fundamentally from moment to moment.

For Denise, as for any archer, tournament pressure increases adrenaline, which makes her process more visual information; slows time for her; increases metabolism in her visual cortex and decreases it elsewhere; and finally—if she's fortunate—decreases the activity of her consciousness and permits the more intuitive, primitive, sensorimotor systems to hold her bow and release her arrows. As Strickland said to her, "If you can set aside your consciousness, your motor memory will take care of your shots."

Buchsbaum's speculations certainly do not contradict the coach's intuition. "When one experiences the mind turning off," he says, "it may be that those primitive regions of the brain we call the basal ganglia take over."

Certainly, the basal ganglia, a cluster of nerve cells concerned with modulating motor behavior, would seem to be a crucial component in any zone experience. Situated beneath the outer layer of the brain, the source of many of the symptoms found in diseases such as Parkinsonism, they are believed to have evolved millions of years before the so-called "cortical brain," which is considered the source of "higher" consciousness. Biologists tell us that because of their very primitivism, the basal ganglia contain enormous quantities of preconscious experience about the nature of the world and how to survive in it.

A Paradox of Evolution

If skill can be correlated with efficient metabolism—and thus less activity—in the conscious brain, it cannot be accidental that many players, remembering their zones, become inarticulate or

In the zone? The precision passing of Joe Montana led the San Francisco 49ers to victory in the 1990 Super Bowl.

© Richard Mackson/Sports Illustrated

mystical, frequently using words like "automatic" or "unconscious" or, as football players like to say, "playing out of my mind." They are describing a process that offers them access to an ancient wiring system. The euphoric state that Pelé reported is a moment in which ordinary conscious authority is effectively silenced.

Our fascination with the zone—and, indeed, with sport in general—may be due—in part, at least—to the possibilities it reveals, the energy and strength and flexibility of the organism when liberated from its ordinary neurological and psychological constraints. Nor does one have to be an athlete to experience such phenomena. Learn to type, and you know what it means to acquire complex knowledge and then "forget" it. For all the skill and cerebration that may be involved in such tasks, the sense when they go well is that the body is doing them on its own.

What deepens the plot—or, more precisely, makes it circular—is the fact that such tasks as typing and baseball are not themselves wired into the sensorimotor system. Complex and sophisticated, they are products of the very reasoning process that is turned off when the primitive wiring system takes over. In fact, as anthropologists point out, the complexity of the games we play has increased with the complexity of our brains. The paradox is that the higher—which is to say, most recently evolved—cerebral regions that create games are precisely those circumvented when the game is played at its highest level. When you wait, as Sadaharu Oh learned to wait, you may be exploiting a talent shared with cats and frogs, but cats and frogs do not set themselves the task of hitting a small white sphere with a wooden stick.

It cannot be incidental that Oh not only set himself this task, but also mastered it absolutely. Wasn't it skill, in Buchsbaum's experiments, that lowered brain metabolism? Time and again we see that the athletes for whom the zone is most accessible are those who are simply best at what they do. The operative—and merciless—fact is familiar to us all: the better you are at what you do, the more you can forget it; the more you forget it, the better you do it.

As Oh says, "All this training, all this minute attention to detail, rather than complicating hitting, seemed to make it simpler." Or, as Buchsbaum and his colleagues might put it, when you're hitting well, credit your basal ganglia, the "primitive" regions of your brain; when you're in a slump, credit those "higher" regions of your brain that offer you thought and reason. Better yet, listen to that great neuroscientist Yogi Berra, who once, when offered pointers about his hitting, responded: "How can I hit and think at the same time?"

We cannot doubt that Berra spoke from zone experience, but let us not forget that he was a very good hitter indeed. Had this not been the case, he might have found himself hitting and thinking a good deal of the time, and doing neither very well.

Hard to Understand

When attacked by his master for being too willful, Eugen Herrigel, author of *Zen in the Art of Archery*, said: "How can the shot be loosed if 'I' do not do it?"

" 'It' shoots."

"And who or what is this 'it'?"

"Once you have understood that you will have no further need of me."

We expect such conundrums from books on Zen, but not from a 15-year-old whose idea of meditation is to lie on a water bed and watch herself shoot arrows on a VCR.

By the time I met Denise in Utah, it was four years since she had begun work with Strickland and Henschen, and she was pointing toward her second Olympics. In 1987 she had become the youngest gold-medal winner in the history of the Pan American Games. Though she'd done poorly in the individual competition at the 1988 Olympics (which was won by a 17-year-old Korean girl whose training, according to Henschen, included two hours of meditation every day), she had taken a bronze medal in team shooting. And along the way she had become the first female archer in the country to break the 1,300-point barrier in the sport's standard international scoring system.

She is still just a teenager, and her mind is not inclined toward the paradoxes that interest the likes of McCarver and Buchsbaum, but when she speaks of her record-breaking tournament, she is no different from any other athlete talking about the zone.

"I don't know what happened that day," Denise explains. "I wasn't concentrating on anything. I didn't feel like I was shooting my shots, but like they were shooting themselves. I try to remember what happened so I could get back to that place, but when I try to understand it, I only get confused. It's like thinking how the world began."

BEHAVIORAL SCIENCES 47

WOMAN in a CAVE

by Alcestis Oberg

Top: © Tom Ives; above: AP/Wide World Photos

Inside Lost Cave, New Mexico, there was no sunlight. No wind blew the window curtains, cut from blue-colored paper and taped onto the wall of Stefania Follini's small home, a 9- by 12-foot (2.7- by 3.6-meter) Plexiglas module. A paper sun was taped up; so was a paper cat that looks like it's sitting on a ledge. Paper grass "grew"; a bowl of paper fruit and paper bread added some cheer to the small place. Outside the window a paper crescent moon and a few paper stars kept unmoving vigil over the solitary occupant. The only forms of life were crickets, tiger salamanders, red-spotted toads, and an occasional mouse. It was too cold for snakes.

Small in build, quiet, almost serene, the 27-year-old Italian volunteered to imprison herself in Lost Cave for nearly five months. By profession a designer who advises people on how to arrange interior spaces, Follini began her solitary venture on January 13, 1989. When Pioneer-Frontier Research and Explorations, a nonprofit organization based in Italy, sought a place for Follini's experiment, Lost Cave, one of approximately 600 caves in southeastern New Mexico, seemed perfect. Twenty-five feet (7.6 meters) deep, it is dry, with a relatively constant temperature of about 68° F (20° C). The "great hall" in the cave is 150 feet (45 meters) long, with an average ceiling height of about 13 feet (4 meters). The Italian foundation had already conducted a number of cave experiments in Europe. In one experiment, 15 people lived together in a cave for 45 days.

Maurizio Montalbini, the director of the organization, holds the *Guinness* world record for staying in a cave—210 days. He hopes to interest the National Aeronautics and Space Administration (NASA) in a cooperative experiment involving three to eight people dwelling in a cave for 12 months—the time it would take to fly to Mars. According to Montalbini, the purpose of Follini's experiment—called Frontera

Stefania Follini (above) lived alone in a cave (top) for 131 days as the subject of an experiment that studied how a human would react when deprived of all time-measuring devices and such cues as sunrises and sunsets.

Dona (Frontier Woman)—was to study the psychological and physiological effects on a human deprived of all time measurements and cues—clocks, sunrise, sunset. Artificial lights in the cave could be turned down but not off, so she couldn't create her own time cues.

Constant Monitoring

Follini wore a blood-pressure monitor, providing data that are analyzed at the University of Minnesota by Franz Halberg, the father of chronobiology, and the man who coined the term *circadian rhythm*. This was the first continuously run blood-pressure test on a person in isolation. Follini took her temperature four times a day; collected her urine, an important indicator of calcium levels; and drew her blood every couple of weeks, sending the urine and blood vials through a hose connected to a trailer aboveground. The team sent the vials of blood, Express Mail, to laboratories to check her hormone levels, stress indicators, and red- and white-blood-cell counts. Some of her blood was immediately frozen; after the experiment was completed, her calcium levels as well as her urine samples were analyzed. Periodically she hooked herself up to an electroencephalograph (EEG) to measure the electrical activity of her brain, its alpha and beta waves. Sometimes when she slept, the crew tested her REM (rapid eye movement) sleep patterns. Occasionally she was asked to take tests that NASA, the Air Force, and the Navy use on pilots to determine reaction times, pattern recognition, and fine motor functions, such as finger coordination.

Follini fed this and other information by computer to the crew aboveground—Montalbini, his wife, and team physician Andrea Galvagno—who monitored her 24 hours a day. When the crew wished to communicate with her, they beeped her. They could hear her talk and sing, but she could not hear them. A black-and-white camera, affixed to the wall of the cave, provided a full view of her module. One camera situated inside the module caught Follini's every move. The inside camera, positioned a few feet from the computer, surveyed Follini as she typed out answers to questions from the crew. When reporters or researchers interviewed her in the cave, she could turn and look directly into this camera, laughing, gesturing, or smiling at people she could not see.

Follini's home was designed like a small spacecraft. There was an entrance area, similar to an air lock, where Follini cleaned off dirt when she returned to her module from the cave. Outside the module a tiny bathroom had a toilet and a mirror but no bath—only towels for washing. The hose through which she sent her blood was in the cave; so were bottles of water for drinking and bathing and replacement cans for the toilet. She stored the toilet's filled containers in the cave. The team figured out how much food and water she would need for the five months. Nothing was resupplied. Inside the module were a small worktable, medical equipment for blood tests, the computer, and a wall of shelves. Follini took 400 books with her, a guitar, a diary, and, being a vegetarian, plenty of cereals, legumes, dried fruits, and seaweed.

Fragile Environment

The Carlsbad area—where Lost Cave is located—had been an underwater reef 250 million years ago, near the coastline of the North American continent. After about 60 million years, the limestone compacted, then cracked, and water began to seep underground. The acid content of the water ate away at the stone, carving out embryonic caves. The steady drip of water—over thousands and thousands of years—formed the dramatic stalagmites for which Carlsbad Caverns is famous. Then, about 12,000 years ago, the area turned into desert. The caves became dormant, awaiting the rains and fresh water that would create more eerie underground sculptures.

By the time she emerged from the cave, Follini had developed much altered sleep/wake cycles, staying awake for 35 hours straight, then sleeping for 21 hours. She underestimated her time spent in isolation by two months.

Prehistoric animals—saber-toothed tigers and ground sloths—lived in the entrances to the caves. A tribe of prehistoric Native Americans, the Guadalupe Basket Makers, inhabited the area as early as 4,000 years ago. The Mescalero Apache, who lived on a reservation in the Sacramento Mountains, displaced the Basket Makers around A.D. 1400.

Although the rocks always project a feeling of power and strength, cavers and park rangers alike are quick to point out that caves are among the most fragile of earth environments. The carbon dioxide people exhale, the dust turned up by footsteps, can change the underground environment. Despite their appearance, caves are delicate, as fragile as tiles on the space shuttle.

The Isolation Experience

Follini entered Lost Cave, New Mexico, at three in the morning on January 13, 1989. Montalbini decided that going in and coming out of the cave in darkness would be less traumatic. The research team accompanied her into the cave. The occasion was festive. They left at 5:15 A.M.

When she was first locked in the cave, she experienced some vertigo. Montalbini—and scientists at NASA's Space Biomedical Research Institute—are particularly interested in what happens to the bones of a person completely cut off from all sources of vitamin D, a vitamin essential to absorption of calcium by the body for bone renewal. Follini received a complete calcium workup and skeletal survey before she entered the cave. Even her dried milk is vitamin-D-free. These tests were rerun when she finished the experiment. (According to Joseph Degioanni, a medical officer at the space agency, NASA did not officially support Montalbini's experiment. The Biomedical Research Institute scientists worked independently.)

Within a few days after entering the cave, Follini's biorhythms began to drift, moving from a 24-hour day to a 28-hour day. She tended to go to bed later and later. Throughout January she awoke between six and nine in the morning. By early February she awoke anywhere between noon and two in the afternoon, and by the end of February, she was waking up around eight in the evening.

On February 27, she went into a sudden, drastic biorhythm shift. She awoke at one in the morning and went to sleep at one in the morning the following day. On March 4, she woke up at 6:30 in the morning and went to sleep March 5 at noon. She had shifted from a 28-hour cycle to a 44-hour one. Galvagno noted that the proportion of sleep was the same: she still slept one-third of the time. She was losing weight, however, because she was eating three meals a day for a 44- rather than 24-hour cycle. Another researcher, John DeFrance, a neuropsychologist at the University of Texas Medical School, noticed that her REM sleep had begun to occur earlier in her sleep cycle, a symptom sometimes associated with depression. She had had only one menstrual period in 53 days. Through the camera, one could see Follini's range of emotions—anger, irritation, sadness, happiness, laughter—the picture of human vulnerability, laying bare the psyche's own brutal rhythms.

As the weeks wore on, Follini's sleep/wake patterns took an even more dramatic turn. She regularly remained awake for 35 hours consecutively, only to sleep for 21 hours. She began to perceive 14-hour sleeps as 2-hour "naps."

Finally, on May 22 at 6:40 P.M., Follini received an unusual message from Montalbini on her computer screen: "You have returned from your journey of solitude." Follini was confused. She had estimated the date to be about March 15—a difference of over two months. The next day, May 23, 1989, she crawled from the cool, damp cave into the arid heat of the New Mexico desert.

"I love the sun; it is beautiful," she said, putting sunglasses on to reduce the glare.

Follini's experience supplied researchers with reams of data. Her 131 days of sunless isolation dramatized the essential role that daylight plays in keeping our body's biological clocks in tune with the earth's cycle of day and night. Scientists have already gained a better understanding of how the immune system, bones, and muscles—not to mention behavior and problem-solving abilities—will react to long periods of solitude. This information can now be applied to astronauts and other people destined to spend extended periods of time alone, with no visual cues to mark the passage of hours and days.

As for the test subject herself, Stefania Follini carried from the cave a deeper sense of herself and her relationship to others.

"I know better now who Stefania is," she explained a few days after surfacing. "I love myself more, and I think I will love others and the world more as a result. I am more honest now and more happy because I faced myself.

"I have seen solitude, and it does not frighten me," she said.

DANGEROUS DELUSIONS

by Daniel Goleman

Intense public attention on the lives of the famous is leading an increasing number of mentally disordered people to place celebrities at the center of delusions that can lead to harassment and occasionally to violence, psychiatrists say.

The delusion most frequently takes the form of the celebrity having a romantic interest in the disturbed person, a condition called "erotomania." But it can also take other forms, like the belief that only the celebrity can provide protection from some imagined persecution or injustice.

The former Beatle John Lennon was shot to death in December 1980 by Mark David Chapman, a delusional fan. John W. Hinckley, Jr., stalked the actress Jodie Foster before he tried to kill President Ronald Reagan in 1981; David Letterman, the talk-show host, obtained a court injunction last year to keep a woman who falsely claimed to be his wife out of his house; Rebecca Schaeffer, an actress who starred in the television series "My Sister Sam," was killed in Los Angeles in July 1989 by an apparently delusional fan who the police said had been writing her letters for two years.

Another actress, Theresa Saldana, survived a knife attack by a man who had traveled from Scotland under the belief that the two of them shared a common destiny and had to be "united." When he was about to be paroled, Ms. Saldana objected; prosecutors kept him in custody by filing new charges that he had threatened Ms. Saldana from jail.

No More Napoleons

Psychiatrists say the increasing public attention to the lives of celebrities has led them to play the central role in people's delusions that used to be filled by figures like Jesus or Napoleon.

One security consultant to entertainers has, in the past eight years, collected 140,000 letters

Mentally disturbed people who develop fixations on celebrities occasionally express their obsessions violently. In one extreme case, John Hinckley, Jr. (upper left) attempted to assassinate President Ronald Reagan (center) as a means of impressing actress Jodie Foster.

Top left and center: Wide World Photos; right: Photofest

written to celebrities by people apparently suffering from delusions, and the Capitol police say a similar problem with letters to members of Congress is also growing.

"Since 1968 there have been as many injurious attacks on public figures by mentally disordered people who gave some sort of warning as there were in the preceding 175 years," said Dr. Park Dietz, a psychiatrist in Newport Beach, California, who has completed the most thorough study to date of the delusions that lead people to seek out prominent people. Searching public records for the past two centuries, Dr. Dietz has so far found 16 such well-documented cases, nine of which were in the past 20 years.

Perhaps the most startling finding from Dr. Dietz's study is that those who write letters containing outright threats, obscenity, or sheer vituperation are among the least likely to try to see a celebrity in person. Those most likely to try to track down a celebrity, he found, are people with romantic delusions or other fantasies of personal intimacy, and who express a desire for a meeting.

"Nearly everyone makes the mistake of assuming that unless there's a threat, you can safely ignore 'nut mail,' 'kook calls,' and 'weird visitors,' " Dr. Dietz reported in an October 1989 summary of his findings to a convention of personal managers in the entertainment industry. The full report was recently submitted to the National Institute of Justice, which financed the research.

Mostly Men

In another surprise, Dr. Dietz found that a large majority of those who write romantic or otherwise inappropriate letters to celebrities are men, often writing to famous women. It had long been thought that the problem was most prevalent among women. Of the 300 letter writers studied, just over 80 percent were men.

One need not be among the very famous to be the target of such delusions, other studies are

In 1980, Mark David Chapman (left), a deluded fan, murdered former Beatle John Lennon (right) on the steps of his New York City apartment building. Just hours before the crime, Chapman had obtained Lennon's autograph.

Robert John Bardo (above), described by police as "an obsessive fan," stands accused of killing actress Rebecca Schaeffer (right) in July 1989. He owned a video collection of her television series My Sister Sam.

Above and facing page: Wide World Photos

finding. Local prominence, or even simply being of a higher social status than the person with the delusion, is enough to put one at the center of such mental scenarios. But the attacks on public figures are easier to track.

In his study, Dr. Dietz was able to draw on 5,000 letters written to celebrities by people with delusions, 800 of whom were known to have also tried to approach the celebrity in person. On average, about one in eight people who write such letters to celebrities also tries to make a personal approach, Dr. Dietz said.

The letters were among 140,000 collected since 1981 by Gavin de Becker, a security consultant to 23 entertainers whose assistants routinely forward odd, inappropriate, or threatening letters to him. Mr. de Becker's archive is currently growing by 50,000 such letters a year, according to Dr. Dietz. The most prolific writer, a woman, sent over 10,000 letters, the vast majority to a single entertainer, in six years, and is still going strong.

Dr. Dietz was also able to study more than 200 similar letters to Congressmen, letters that were forwarded to the Capitol police by congressional staff members. Half of the 86 letter writers studied were known to have tried to see the member of Congress in person.

The Capitol police's intelligence division has added new files on mentally disturbed people who write or try to meet Congressmen. While in 1986 there were 330 new cases added, by 1988 that rate had reached 415.

The letters in the archives would strike most any reader as alarming, Dr. Dietz said. For instance, one man sent an entertainer a Valentine's card, writing on it, "We've shared many fine years together. See you next week."

"I'm So Lonely"

Another wrote, "I am afraid I made a mistake when I told you I was your father. The reason I thought so was I used to date someone who looks like you. I asked your manager to borrow $10,000. I hope she lets me."

Still another person wrote, "I liked your last movie so much and I hope to find someone like you who will love me. I'm so lonely. Love, Bill."

Those who have delusions about one public figure frequently have them about others, too, Dr. Dietz found. "When I walked into the

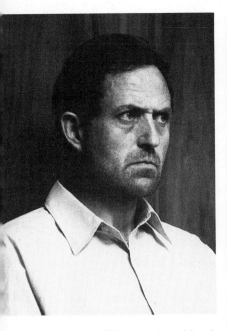
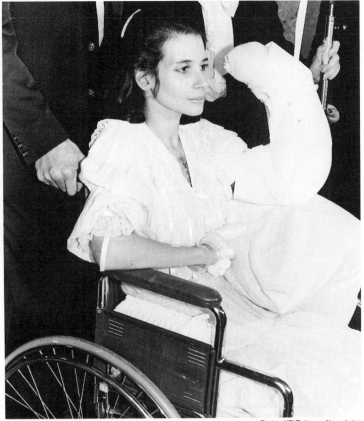

Arthur Richard Jackson (above), convicted of brutally knifing actress Theresa Saldana (shown at right, recovering from her wounds), traveled from Scotland to commit his crime. Amid much publicity, authorities stayed his parole when he continued to harass his victim from jail.

Photos: UPI/Bettmann Newsphotos

offices of the Capitol police, I found names familiar from those who wrote to the entertainers," Dr. Dietz said. "At least one-fourth of those who pursue one public figure pursue another public figure."

Dr. Dietz pointed to the incident in October 1989 in which a man struck Senator John Glenn of Ohio. Earlier in the day, the same man approached Vice President Dan Quayle's motorcade and tried to hand him a letter.

In his study, Dr. Dietz used a technique that allows a tentative psychiatric diagnosis to be made from a careful analysis of the letters. Dr. Dietz found 95 percent of the letter writers were mentally ill. Of these, one-half to two-thirds were schizophrenic. The technique searches a given letter for signs of specific symptoms, and then analyzes the pattern of such symptoms to arrive at a probable diagnosis.

Four of five letters to entertainers gave evidence of romantic feelings. When romantic delusions occur without any other sign of mental disturbance, the diagnosis is erotomania. While 16 percent of those who wrote entertainers qualified for the diagnosis, only 5 percent of those writing to members of Congress did.

A romantic delusion was also common among people whose additional psychiatric symptoms qualified them for diagnoses ranging from manic to schizophrenic.

Though romantic delusions may seem benign, they can easily turn to violence. "The initial delusion is typically that the figure involved is in love with oneself from afar," said Dr. Jonathan H. Segal, a psychiatrist at the Palo Alto Medical Clinic in California. "The delusional person then falls in love and wants to approach the figure. When the other person denies being in love, the delusional person typically persists, often seeing the denial as some sort of test."

Anger at Rebuffs

For some who are rebuffed, however, the reaction is anger and violence. Dr. Dietz tells of one man who sought out the actress who was the object of his romantic delusion. When she refused to see him, the man, who was armed, threatened suicide.

DELUSIONS OF INTIMACY

• *CASE 1.* "Mr. C.," 31, met a well-known lawyer over a business matter. Although his meeting with the woman was casual, Mr. C. "knew" she was in love with him and, to declare his love, began besieging her with calls, letters, and flowers. Although she denied any interest in him and finally filed charges of harassment, he saw her reactions as a "test of love." He left his wife and abandoned his business to pursue the lawyer. When she continued to rebuff his advances, he sent threatening letters and was finally committed to a psychiatric hospital.

• *CASE 2.* "Ms. B.," a 65-year-old retired clerk, had been forced into early retirement because of her conviction that the president of her company was in love with her. She then became convinced that an elderly bank manager who lived in her apartment house had a romantic interest in her, although he had never spoken a word to her. She knew of his interest "by the way he looked at her," and because for months, he had been "contacting her telepathically" with romantic messages. Outraged by his "advances," she accosted him several times to protest, and when he got a restraining order barring contact between them, she followed him to his office one day, leaped across his desk, and apparently tried to strangle him, though later she told a therapist she had only been trying to touch his tie.

• *CASE 3.* A Frenchwoman, 53, became convinced in the early years of the century that King George V of England was in love with her. English sailors and tourists in France, she was certain, were messengers sent by the king to proclaim his love. On several trips she made to England, she waited patiently outside the gates of Buckingham Palace, watching intently for a sign from the king. Once, she saw a curtain moving in a window, and she "knew" this was a signal to her from the king. When she lost her luggage, she insisted that it was the king's doing, and that it was clearly a sign of the intensity of his feelings for her: "I could never be indifferent to him, nor him to me."

A man shouting "the earthquakes are coming" shocked the Washington press corps when he suddenly slugged Senator John Glenn in the jaw. The day before, the same man alarmed Secret Service agents when he tried to approach Vice President Dan Quayle's motorcade to hand him a written message.

In an article on erotomania in the October issue of *The American Journal of Psychiatry*, Dr. Segal observes that while romantic delusions can be about almost anyone, they typically fix on a person of prominence. Apart from celebrities, though, that can mean someone prominent in the person's immediate circle, like a priest, a boss, or a physician.

The majority of people with such delusions are withdrawn and lonely, Dr. Segal said; many have had few sexual encounters, and some are "notably unattractive." In contrast, the people they fixate on are of superior status, looks, or authority. The delusions allow the person to feel "plucked from her obscurity." Such problems are notoriously difficult to treat.

Most of the cases of erotomania that have been reported since the syndrome was first identified in 1921 have been women. But Dr. Dietz said that may be because women are more easily cajoled into treatment than are men. The greater threat of violence, according to Dr. Dietz, comes from men with the delusion.

Often relatives or acquaintances of those with delusions know of the fixation, though few report the problem to the authorities, Dr. Dietz said in his report. "The public needs to know that the persons most likely to be hurt when these subjects turn to violence are the subjects' own family members and neighbors."

Panic Disorder

by Dina Van Pelt

Your palms get sweaty. The blood drains from your face. You feel a pounding in your ears. Your heart thumps in your chest. You feel dizzy and faint, and you get scared—so scared that you fear for your life.

These discomforting symptoms are all signals that your body is reacting to a situation by shifting into a state of panic, and life is fraught with situations that trigger it: losing your wallet, struggling to meet a deadline, catching a flight about to take off, a near accident on the road.

But what about going out to lunch? Or taking a walk? Or just leaving the house? Mundane situations that contain no real threat can trigger intense, overblown emotional and physical reactions in people with panic disorder.

A woman may clutch her chest in fear and dash from a grocery store checkout line. A man may become physically immobilized as he is about to cross the street. A person with panic disorder may have attacks several times a day, or once a week, or may simply be so afraid of having one that he or she withdraws from certain activities and places.

Since the time and place for attacks are usually unpredictable, a person with panic disorder plays it safe by routinely picking aisle seats, avoiding certain thoroughfares, or venturing from home as little as possible.

"Imagine your heart racing rapidly as if you're on a roller-coaster ride. . . . Say you're sitting in a restaurant where it's dark, and for no reason your heart starts to race, your palms become clammy, your brow perspires, you hyperventilate and get dizzy," says Glen, a 37-year-old who has panic disorder. "You then become focused on yourself to the point that you can't stay, or else you will pass out because the more you dwell on it, the worse it gets."

People who experience this kind of spontaneous attack four or more times in a four-week period have panic disorder, medical scientists

Millions of Americans suffer from panic disorder, a condition in which everyday situations containing no real threat generate intense physical and emotional reactions. The pounding heart and other characteristic symptoms of panic disorder often lead to an incorrect diagnosis.

say. Glen reached the point where he was having them once or twice a day. He now controls his panic through medication and counseling, but for 10 years he either avoided seeking treatment or had tests for maladies such as blood-glucose abnormalities and inner-ear infections.

Misunderstood and Misdiagnosed

Although 11 million to 13 million Americans suffer from panic disorder, and as many as twice that number can have panic attacks at some point in their lives, American Psychiatric Association researchers estimate that it is one of the most misunderstood and grossly misdiagnosed conditions around, largely because it is a great imitator. Panic disorder's typical symptoms could fill the examination chart for a number of conditions, ranging from heart attack to hypoglycemia to epilepsy. Panic disorder also brings on tremendous feelings of impending death, loss of control, and demoralization.

The sensations can be so strong, in fact, that many people seek emergency care during panic attacks because they believe they are having a heart attack. At the hospital, though, a patient is often told not to worry about it, that "it's just nerves," and is sent home, says Jerilyn Ross, president of the Phobia Society of America, and counselor at the Roundhouse Phobia Treatment Center in Alexandria, Virginia.

A study by Jeffrey H. Boyd, director of ambulatory psychiatric services at Waterbury Hospital in Connecticut, found that people with panic disorder use ambulatory-health-care systems more often than those with any other psychiatric disorder. But as is often the case with patients with chronic fatigue syndrome, an anomaly for years, those with panic disorder are often shuffled from one specialist to another. The average panic-disorder sufferer, say researchers, will see a minimum of 10 health professionals over an eight- to 12-year span before being correctly diagnosed.

Doctors "were always trying to link my condition to something they could treat, like the flu, so they could give me something to minimize the symptoms," says Glen. "But nothing like that ever helped."

What could possibly bring on a feeling of terror and reaction so powerful that it is mistaken for cardiac arrest? Although more neurological evidence points toward a biological cause, considerable debate remains. Some health-care professionals believe that the condi-

Below: © The Washington Times; facing page: © Van Bucher/Photo Researchers, Inc.

A restaurant (right), shopping mall, or countless other ordinary settings can trigger overblown reactions in people with panic disorder.

People with panic disorder may mistake their symptoms for a heart attack, stroke, or other dire medical emergency (left). Many physicians now think that panic disorder has a biological cause: detailed PET scans of the brains of panic-disorder patients (below) indicate abnormal patterns of activity in the section of the right hemisphere associated with emotion.

tion is much more psychological, while others contend that certain people with other seemingly unrelated conditions are somehow more prone to panic attacks.

Abnormal Patterns

A biological predisposition certainly seems to be a component, since medical scientists have documented the condition's tendency to run in families. New research indicates that a localized portion of the brain's right hemisphere is responsible for the bursts of activity coinciding with attacks.

With detailed brain scanning made possible by positron emission tomography (PET), Marcus E. Raichle, professor of neurology and radiology at Washington University School of Medicine in St. Louis, Missouri, has pinpointed abnormal patterns of activity in the limbic section of the right hemisphere, an area associated with emotion and motivation. He found that those prone to panic had greater blood flow, glucose metabolic activity, and oxygen in this section than in the left when they were calm and not having an attack. During an attack, however, little difference in brain activity was observed between people with panic disorder and those experiencing normal anxiety.

"It now appears that those prone to panic disorder do in fact have a predisposing abnormality in the brain," says Raichle. "This may lead to more effective treatment in terms of drugs which act on the chemical message system in the brain."

Assuming that the biological foundation already exists in certain people, a traumatic event

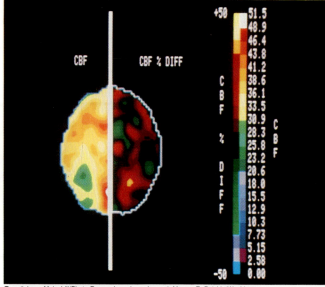

Top: © Larry Mulvehill/Photo Researchers, Inc.; above: © Marcus E. Raichle/Washington University School of Medicine

such as an illness, a personal loss, the birth of a child, or another major life change then usually brings on an attack. "The first attack usually occurs during a particularly stressful period, but not always," says Bernard J. Vittone, director of the National Center for the Treatment of Phobias, Anxiety and Depression. "Any one of a number of events may trigger the circuitry in the brain predisposed to panic disorder, and an attack then occurs."

As described in the report of a major new study by Myrna M. Weissman and colleagues at the Columbia University School of Medicine, the presence of panic disorder usually guarantees that a person will be beset by other personal difficulties such as secondary depression (which results from another illness), marital problems, and attempts at self-medication with drugs and alcohol. The incidence of alcohol abuse among patients with panic disorder is estimated to be as high as 50 percent.

In their study of 18,011 adults, Weissman and her colleagues found that people with panic disorder have a high propensity for suicide, brought on by secondary depression and demoralization. Twenty percent of the adults with panic disorder in the study had attempted suicide, and those with panic disorder were three times as likely to have attempted it or to have thought about killing themselves than were people with other psychiatric illnesses. Panic-disorder patients were 18 times more likely to attempt suicide than were people with no psychiatric disorders, they found.

"This is a serious illness, which can be very disabling, possibly leading those affected to get very desperate, especially if the condition is not diagnosed," says Weissman.

Ross, though, calling the finding of 20 percent "extraordinarily high," says she does not see the tendency for or actuality of suicide attempts among the panic-disorder patients she counsels. But, she notes, "the fear [of harming themselves or others] is there."

Some people report experiencing panic attacks during their sleep, and many say these attacks are worse than those during waking hours because they are so much more surprised by it, says Harold Koenigsberg, an associate professor of clinical psychiatry at Cornell University Medical College.

Apparently, these attacks are not a product of REM sleep during dreams, he says, suggesting that a biological component and not an emotional state influenced by a dream sets off the attack.

Panic versus Phobia

Although phobia is in a category of its own—and, like panic disorder, considered in the general area of anxiety disorders—symptoms of panic disorder and phobia are often linked, even though someone suffering from panic disorder does not necessarily have a phobia.

However, several panic attacks, or even one that occurs in a public place, could create a phobia. And years of having panic attacks in various places and situations often evolve into agoraphobia—the fear of being in open or public places, which researchers consider the most complex and potentially debilitating phobia.

"Phobia attacks are not the same as panic attacks, mainly because phobics avoid those situations which they already know make them uncomfortable," says Koenigsberg. "Phobics slowly develop anxiety as they approach the uncomfortable situation. Panic attack is much more sudden, and involves a wider range and intensity of physiological symptoms."

Once panic disorder is diagnosed, treatment is available and is highly effective. With a combination of medication, usually involving antidepressants, and behavioral and cognitive therapy—in which patients are purposely exposed to situations in which they usually panic, and are taught to tolerate the stimuli that usually lead to panic—patients can dramatically reduce the number and severity of their attacks, and in some cases can eliminate them altogether after a few months. Perhaps now that panic disorder is more widely recognized for the distinct condition that it is, those getting help can save themselves from years of agony.

Most panic-disorder sufferers realize that their fears have no logical basis. Some can control their symptoms through a combination of drugs and counseling. Others work with therapists to confront their fears (below).

© Brig Cabe/Insight

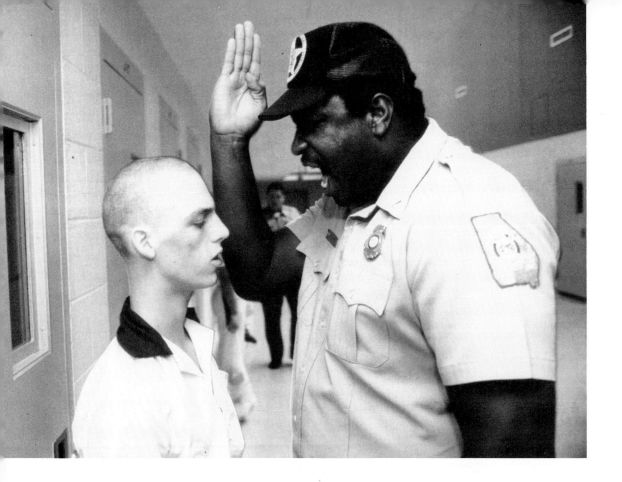

SHOCK Incarceration

by Daniel Kagan

Several voices have lately been sounding a new note in the chorus of prescriptions being offered for ways to attack America's drug problem: to combat the sale of crack—and the violence and social ills it breeds—send the dealers to camp.

Prison camp, that is.

The logic is that crack sales spiral out of control largely because the justice system, short of jail and prison space, cannot incarcerate those arrested for peddling the intensely addictive drug. Instead, the system is a revolving door. Police can arrest enough street dealers to greatly reduce the street presence of crack, or at least curtail its open sale, but the chronic shortage of space keeps them from holding most of those convicted or who plead guilty.

Limited jail space also means that arrests for much more serious drug felonies are plea-bargained down to lesser charges so those individuals will serve less time to free space sooner for others.

Chronic prison overcrowding has led corrections officials in some states to set up special boot-camp-like institutions where first-time, nonviolent drug offenders serve out their sentences performing hard labor in an atmosphere of strict, military-style discipline.

A similar problem exists in overcrowded prisons. Even though many serious drug offenders draw long terms in these institutions, many more could be incarcerated if the space were available.

"Most criminals play the odds, and the odds are clearly in their favor" against being incarcerated, says Norman Carlson, former director of the Federal Bureau of Prisons.

"Most people are not getting any meaningful punishment because there's no room," agrees Francis C. Hall, the recently retired chief of the narcotics division of the New York City Police Department.

More jails and prisons could be built. But this is prohibitively expensive—$50,000 or more to construct cell space for one inmate—too much for state budgets already overburdened with the costs of corrections systems.

Consequently, some law-enforcement and policy experts are considering camp-style institutions to provide the needed space. In contrast to expensive, fortified, high-security walled prisons, these would be cheap, quickly built, low- to medium-security, relatively Spartan facilities outside urban areas, where inmates would live in barracks, probably do hard physical labor such as clearing land, and might be subject to military-type discipline. Like jails, they would hold those convicted of lesser to moderate drug offenses and serving time from 30 days up to several years.

Incapacitating Criminals

Unlike prisons, the camps would not offer a range of programs for education or self-improvement. Their purpose would not be rehabilitation, but what corrections professionals call "incapacitation": getting criminals off the street and keeping them there so they cannot commit crimes.

Camps, it is hoped, would create something that does not exist now: a clear-cut risk of punishment for even minor crack sales. Some believe that by ensuring incarceration, camps would also act as a deterrent for some crack sellers. Others think that juveniles, now enticed by the easy money in crack, and emboldened by the sight of arrested dealers appearing back on the streets in a few days, might choose not to

Above, right, and facing page:
© Mike Greenlar/Picture Group

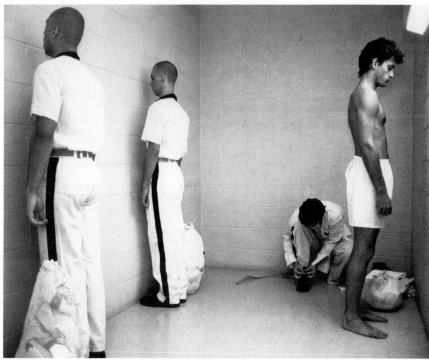

Some sociologists feel that the harsh dose of prison life provided by shock incarceration programs might prove enough to set young drug offenders straight. The rigid discipline starts immediately: upon arriving at the facility, new inmates must strip, undergo delousing, have their heads shaved, and put on uniforms.

sell the drug once they see that there is a high probability of doing time.

Variations on this basic model have also been suggested. These include boosting the deterrent effect by making life in such camps especially uncomfortable, keeping amenities to the minimum allowed by law. It has also been suggested that the camps might make up a sort of American Gulag—one expert calls it "the Devil's Island approach"—located in harsh, remote climates such as in Alaska, where the distance from crack sales centers is so great that no wives, girlfriends, mothers, or cronies could easily visit.

A Harsh Dose of Prison Life
Camps might also be run like the popular new "shock incarceration" programs. These programs, begun in Georgia in 1984, and rapidly being instituted in many other states, are aimed at shocking first-time inmates age 17 to 25 out of committing future crimes by dealing them an intensely harsh dose of prison experience. They are also aimed at instilling order, discipline, pride, and a positive self-image in young offenders whose lives lack all those features.

Limited to first-time inmates guilty of nonviolent felonies such as burglary and drug offenses, these programs offer an early release to intensively supervised probation in exchange for participation. The course lasts 90 days to six months and is a version of military boot camp. It is run by former drill instructors, who shout at the inmates, bully them, and run them ragged through 16- to 18-hour days of grueling calisthenics, standing at attention, marching, and heavy physical labor. The most recent advocate of these programs is William J. Bennett, the director of the Bush administration's drug-control policy.

The words "prison camp" evoke a bad image, unfortunately, even though American camp prisons are far more comfortable places than actual prisons. The term conjures up World War II visions of the internment camps for Japanese-Americans, and of squalid Nazi and Japanese POW camps full of emaciated American G.I.s. Indeed, some have misunderstood camps for drug offenders to mean detention camps that would be filled with inner-city residents indiscriminately swept from the streets by the National Guard. One black leader recently confronted Bennett and demanded reassurance that blacks would not be rounded up and herded into drug "concentration camps."

Despite the range of possible applications, there is much disagreement about the value of such a system. Some dismiss its workability; others view it as a panicky notion born out of government leaders' frustration with the immensity and intractability of the drug problem.

"A lot of governors and legislators mounting a war on drugs are looking desperately for something," says Dale G. Parent of Abt Associates Inc., a sociological-research firm in Cambridge, Massachusetts.

Romantic Idea?
Arguments for camp-style prisons for drug offenders are "romantic ideas by people who don't know the problem," says Norval Morris, a former dean of the University of Chicago Law School and an influential architect of current correctional theory. Advocating camps "is a way of not looking at the problem," which he says is best addressed via more support for treatment to help addicts stop using crack.

"I'm not saying there's not merit in locking people up. But the real thing that matters is the prevention of the continuing use of drugs. We can do this [with strict probation] involving regular weekly drug testing," says Morris. "Jail should only be used as a backup for failing to become drug-free."

John Irwin, a sociology professor at San Francisco State University, calls the interest in prison camps an "extension of a policy that hasn't worked and an extension of silliness." He believes that there is no guarantee of incarceration strong enough to deter crack sellers. A recovered drug addict who served time in California's Soledad Prison in the 1950s for burglary, Irwin sees young crack dealers as hopelessly disenfranchised from the mainstream of society.

"The fact of being sent to a prison camp won't stop them," he says. "These kids just don't think about consequences. . . . These kids have nothing to lose. Given the chance to make a profit, they'll take it. They don't care. Most have had some jail anyway, so it's not so frightening to them."

Mark D. Corrigan, director of the National Institute for Sentencing Alternatives in Waltham, Massachusetts, says, "Deterrence via prison camps is a nonissue. A kid 17, 18, 19, minority, poor—do you think he goes through a rational calculation of consequences before committing a crime? That is a calculation engaged in by people who live on campuses, not

The military-style regimen aims to instill order and discipline in the previously unstructured lives of the inmates. At a shock incarceration unit in New York state, inmates must stand at attention before meals (above). Early-morning push-up drills begin the day at a similar facility in Florida (right).

Top: Jack Manning/NYT Pictures; above: © Steve Starr/Picture Group

the streets." He thinks an expanded system of prison work camps could be put to better use housing nonviolent, nondrug inmates to free up room in prisons for drug offenders.

And Irwin cites research from several states indicating that locking up more offenders does not decrease crime rates.

The Disorientation Effect

Nevertheless, such far-flung camps might have possibilities. For urban street people who both use and sell drugs, says Parent, "there may be some value in putting them in an environment that is 1,000 percent different from what they are used to." This could have a "disorientation effect that will strip away their defenses and make them vulnerable to change. Then bring them back into society with a heavy emphasis on drug treatment."

This is similar to the principle behind shock incarceration. The tricky part is not the camp program but the close supervision afterward, when the inmate is back in the environment that helped shape his criminal behavior. This follow-

BEHAVIORAL SCIENCES 63

During frequent group therapy sessions, inmates discuss their shortcomings and come to terms with their mistakes.

up phase usually fails. Indeed, quite a few corrections experts believe that shock incarceration has no lasting impact partly for this reason. Boot camp works because it is followed by years of military service that reinforces the principles laid down in training.

In fact, crack's rapid rise notwithstanding, some law-enforcement officials note that street-drug-related crime and related problems began to soar only after the end of the draft, which plucked idle young men out of the world of drugs and gangs, trained them, taught them some discipline and how to survive in totally different environments, and removed them from the cycle that led to drugs and the street. Part of the appeal of prison camps may be that they appear to replace, at least in part, this function once served by the draft.

By providing cheap, additional space and guaranteeing that all drug offenders do time, camps can also restore to the justice system some of the credibility that the overcrowded prisons have taken away.

"I think any plan for cheaper incarceration is attractive," says Paul H. Robinson, acting dean at Rutgers Law School and a former member of the U.S. Sentencing Commission. "The idea of having a system where we make threats about punishment and then say, 'We will let it pass'—not doing what we promise hurts the system."

The lack of space also reduces the effectiveness of probation. "One-third to one-half of the guys coming into prison are probation or parole violators," says one corrections expert. Probation works, he explains, only when the threat of a return to jail for a violation hangs over the probationer's head "like a sword. Now it is an empty threat."

The more creative variations on such camps, however, carry built-in obstacles. The Gulag model presents a gigantic problem in terms of recruiting staff, since they would also have to live in those godforsaken places. Making the camps as uncomfortable as possible raises problems as well. "One worries that 'Spartan' can be used as a code word for skating close to the edge of unconstitutionality," says Tom Gilmore of the Wharton Center for Applied Research. Asks one corrections expert: "Just how far can you go before you cross the line to cruel and unusual?"

A modified-boot-camp approach drawn from shock incarceration presents difficulties because of staff burnout, a problem even in programs that run for less than six months.

Out of Circulation

There does seem to be one scenario for the use of such camps, however, that is so simple and straightforward that it appears almost too good not to work. It is the one suggested by former narcotics officer Hall, in which camps would be used as a tactical weapon, enabling the state to take street-level drug dealers out of circulation on a conveyor-belt basis. Hall says the idea would be to build medium-security camps in upstate New York on state-owned land and create a stripped-down system of high-turnover courts, prosecutors, legal assistants, and public defenders to service them. Then set the machinery in motion and see how long the dealers can take being incarcerated at every turn.

"You don't need draconian punishment. Imagine that each time a street dealer gets arrested on Friday, by Monday he's on a bus upstate for 30 days," says Hall, who has an excellent record of success with street-level antidrug operations, such as the legendary Operation Pressure Point in the Lower East Side during the early 1980s. "Each successive time [a drug dealer is convicted] maybe you add 15 days to the sentence. But the point is, he goes every time.

"If the guy is back out for a week after doing 30 days, and he gets busted for selling again," Hall adds, "he goes right up again for 45 days—and so on until he accrues perhaps a 90-day sentence. Then he might simply do 90 days every time he is convicted for crack sales."

In this way the camps become a formidable mechanism for actively harassing and disrupting the rhythm of street drug dealing, breaking up the organized structures built by the sellers, and preventing dealers from gaining a foothold and establishing an economic power base in a neighborhood.

"We have space for 45,000 inmates now," says Hall. "With cheap camps we could easily add space for 45,000 more.

"People will replace these guys, but not one-for-one," Hall continues. "We have identified 10,000 career street dealers in New York City. Twenty percent of the career dealers account for 70 percent of the drug sales. If you can neutralize them permanently" by holding many at once in such a system, "it would have some impact."

Most crack sellers will not quit, because it is too good a living, he acknowledges. But "if you bang on a guy often enough, hit him on the head every 30 days—if he knows he'll spend six months a year upstate—you're not only cutting his income; the streets will be quieter. Let's focus on things that are doable."

Most important, says Hall, is not to look at such a camp system—or any other law-enforcement effort, such as large seizures of drugs—as a solution to the drug problem. "It won't eliminate drug trafficking," he says. "The problem with drugs is that everyone wants an immediate solution."

This, he says, is unrealistic. "I am thinking long-term—20, 25 years," he says, comparing the war against drugs to efforts against smoking. "It took us 25 years [from the first surgeon general's report that smoking is a health hazard] to reduce smoking 11 percent. The drug problem will be in evidence into the far future. It's long-term, like the battle against cancer."

At "graduation," a model inmate calls marching orders to his group. The programs claim a high success rate.

Jack Manning/NYT Pictures

Review of the Year

BIOLOGY

Scientists continue to learn more about how plants and animals evolved, how and why they behave the ways they do, and how to improve the measurements of various biological processes in the laboratory.

by Joanne Silberner

New Tools and Techniques

While scientists can win Nobel prizes, molecules garner no awards. At least, that is, until 1989, when *Science* magazine gave "Molecule of the Year" honors to a type of DNA polymerase, the key player in a chemical reaction that allows scientists to produce millions of copies of a single gene. Polymerase chain reactions can be used to more accurately diagnose sickle-cell anemia, detect minute levels of infectious agents in tissue, pinpoint active genes, and determine if tissue found at a crime scene belongs to a particular suspect.

University of California researchers designed a colorful way to check for genetic transfer. They cloned the genes that allow click beetles to glow in four different colors: green, yellow-green, yellow, and orange. These genes can now be saddled onto other genes being transferred in other experiments. To check for the success of the transfer, scientists will need only to detect cells displaying the signal color.

Italian scientists announced a major advance in genetic engineering, although other scientists claimed the method, at least as described, doesn't work. The method presents a new, simpler way to create animals that bear foreign genes. Scientists currently make such transgenic species by injecting foreign genes into the nuclei of eggs, a procedure that requires tiny pipettes, a steady hand, and, most of all, a good deal of luck. According to the Italian researchers, however, sperm soak up free

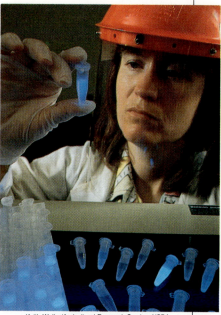

Keith Weller/Agricultural Research Service, USDA

To check the success of genetic transfer experiments, scientists now use a technique that makes cells into which the genes have been transferred glow.

DNA from a solution. Therefore, to make a transgenic animal, all they need to do is add the DNA to the sperm, then use the sperm for *in vitro* fertilization. To date, U.S. researchers have been unable to produce the feat.

Genetic Engineering

Scientists began planning for the 1990 start of a vast project to map all the genes on human chromosomes. Similarly, the U.S. Department of Agriculture (USDA) announced plans for an analogous project of its own, mapping the genes in food and forest products. Meanwhile, the Environmental Protection Agency (EPA) approved the first release of a genetically altered plant virus that destroys the cabbage looper. In Australia an antibiotic-producing bacterium that battles another strain of bacterium went on the market. A USDA committee gave the go-ahead for a field test of a genetically altered fish, a carp that carries a growth-hormone gene extracted from trout. It also approved field tests of alfalfa seeds that carry a bacterial gene that imparts herbicide resistance.

Belgian researchers devised a method to keep plants from fertilizing themselves. Self-fertilizing plants generally resist tampering by agronomists, who wish to crossbreed them to yield new strains that produce abundantly while being less prone to disease. To make a hybrid of a self-fertilizing plant, the pollen-producing tissue must be removed. In the case of corn, this is relatively easy: seed companies simply snip off the male tassel and expose the female ear to pollen from another line of corn. But other species present more of a challenge. The male and female parts of rice, for example, occur so close together that workers in China must remove the male structures using tweezers. The Belgian scientists, however, successfully inserted into self-pollinating rapeseed a gene that stops pollen production. They claim the procedure could also work for other species.

Bacteria and yeast may have their own form of genetic exchange. Two University of Oregon biologists observed *Escherichia coli* bacteria trading genes with *Saccharomyces cerevisiae* yeast. The trick took some laboratory manipulation; whether bacteria and yeast trade genes in nature remains to be seen.

Insects

The African honeybees, buzzing inexorably toward the U.S. from South America, are due here in the early 1990s. Experts had hoped that by the time of the bees' arrival, their genetically programmed aggressiveness would be at least somewhat diluted through cross-mating with calmer local honeybees. Unfortunately, the bees don't seem to be getting any more tractable. University researchers now think they know why: the current swarms came from a purely African line, meaning that little or no crossbreeding has occurred that might help dilute their aggressiveness.

Colorado State University researchers may have discovered a way to protect mosquitoes from infection. Although the new procedure won't protect humans from annoying bites, it could someday offer protection from malaria, yellow fever, dengue, and other mosquito-borne scourges. The researchers have already successfully inserted test genes into the mosquitoes. Their goal now is to insert genes that will help mosquitoes fight off infection. If the mosquitoes can avoid infection themselves, they can most likely be genetically altered to avoid transmitting disease to humans.

A Michigan zoologist announced that a minute insect, a type of thrips, can lay eggs or bear live young, a trick not generally seen in nature. In laboratory-bred thrips, about a third of the mothers use both strategies. Thrips hatched from eggs are always female; those born as larvae were male. In a related discovery, a University of California entomologist found that within a few generations, colonies of thrips genetically adapt themselves to the specific plant on which they are living. This adaptation greatly increases the available spots for thrips colonies.

Researchers have developed a microphone that differentiates the wing-beat frequency of a common honeybee from that of an African "killer" bee.

Oak Ridge National Laboratory

Evolution

Angiosperms—flowering plants—have thus far done a good job of hiding their age. Fossil records of these plants go back about 150 million years, but what happened before that is anybody's guess. Three West German scientists have now made a stab at it. Their conclusion: angiosperms have been around some 300 million years. The scientists studied the genetic material of current species of flowering plants and calculated the rate of genetic mutations. By assuming these plants arose from a common ancestor, they were able to calculate how many generations ago their forerunners must have lived.

In another evolutionary find, a British researcher made a startling discovery: ears may well have started out as lungs. While studying the evolution of a bone in the middle ear, J. C. Clack of Cambridge University determined that in early amphibians, this bone was used not to transfer sound, but to pump air in and out.

Colorful or extravagant bird plumage has long been thought to have evolved at least partially as a means of attracting mates. Male swallows, for instance, attract female swallows with their long tails. But in 1989 a Swedish scientist found that male swallows can actually become too attractive. When the scientist clipped the tails of some male swallows, female swallows shunned them. When he glued extra-long tail feathers onto other male swallows, female swallows grew quite attentive. But beauty has its price: the well-feathered male swallows apparently caught the attention of potential prey, who were then able to elude capture. Even worse, after the cosmetically altered males shed their long tails, they didn't do as well winning females in the next mating season as did their unadulterated brethren.

J. Roch

Scientists now estimate that flowering plants evolved 300 million years ago.

New York Zoological Society

Colorful displays of feathers, while helping to attract mates, do not work so well at dinnertime: potential prey who notice the extravagant plumage can often elude capture easily.

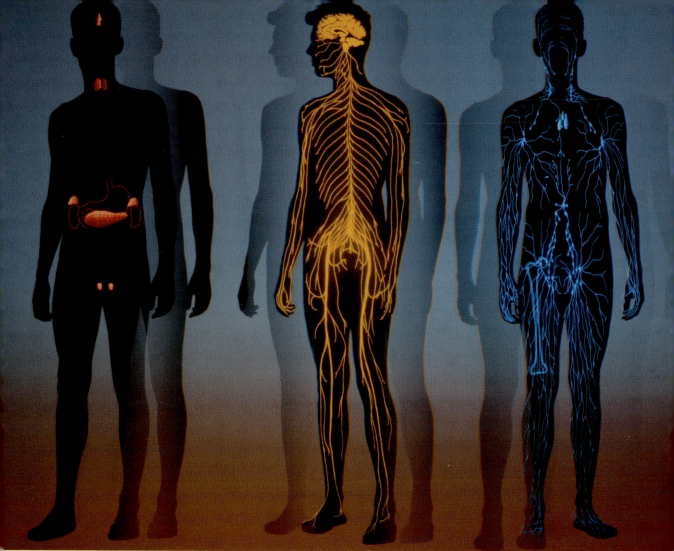

All photos: © Michael Freeman

BIOCHEMICAL CODES:
The Language of Life?

by Stephen S. Hall

Scientists tend to remember the day they arrive at a conclusion they are certain no one else is going to believe. For J. Edwin Blalock, that day came in 1981, when his research group at the University of Texas Medical Branch at Galveston found a biological molecule where it wasn't supposed to be. The molecule was a hormone called ACTH (short for adrenocorticotropic hormone); according to every medical textbook of the time, this hormone was made exclusively by the pituitary gland in the brain, and "belonged" to the endocrine system. Blalock, working with his collaborator Eric Smith, kept finding it in the wrong place. They kept finding it in a laboratory flask filled only with human cells belonging to the immune system. In fact, they began to make the unbelievable suggestion that ACTH, a hormone, was made by white blood cells.

The scientific community politely cleared its collective throat in public, and not so politely cast aspersions on the work in private. A com-

70 BIOLOGY

mentary in the British journal *Nature* harrumphed about "radical psychoimmunologists," and went on to imply that Blalock's work lay beyond even that suspect fringe. "That was not a whole lot of fun," Blalock says now, in retrospect.

Before she joined the ranks of radical psychoimmunologists, Candace Pert was a doubter, too. When researchers suggested around the same time that white blood cells had qualities in common with brain cells, Pert—then chief of the Brain Biochemistry section at the National Institute of Mental Health (NIMH) in Bethesda, Maryland—reacted the way scientists often do when things don't fit: they question the rigor of other people's science. "I thought perhaps the work was sloppy, shoddy, didn't have the right controls," she admits. "When you can't fit it into your cosmos, it's a lot easier to think someone else is a slob." Then Pert and her NIMH collaborator, Michael Ruff, began to find the same thing: certain white blood cells were equipped with the molecular equivalent of antennas tuned specifically to receive messages from the brain.

Tearing Down the Barricades

As more and more of these "misfit" molecules turn up in unexpected places, some biologists believe we need to rethink some long-cherished principles, beginning with medicine's traditional separation of the central nervous system (the seat of thought, memory, and emotion) from the endocrine system (which secretes powerful hormones) and the immune system (which defends the body from microbial invasions). To be sure, many other scientists argue that these misfit molecules are, at best, not well understood, and, at worst, the product of unfit science. But scientists like Blalock, Pert, and Ruff are prepared to declare a revolution. The barricades they seek to tear down separate nothing less than the mind and the body.

Bit by bit, these and other scientists are assembling a mosaic of data suggesting that our anatomical systems, separated by 19th-century tradition, routinely communicate with one another. Carrying the messages back and forth, moreover, are small go-between molecules. There may be from 60 to 100 of these powerful biochemicals; traditionally known as neurotransmitters, hormones, neuropeptides, growth factors, and lymphokines, these chemicals are now seen by some as a related family of biochemical words, essential turns of phrase in the vocabulary of life. Francis O. Schmitt, a neuroscientist at the Massachusetts Institute of Technology (MIT), refers to them as "informational substances." As they are better understood, some researchers believe, these chemicals may assemble into the intercellular equivalent of the genetic code, and may influence medicine in the

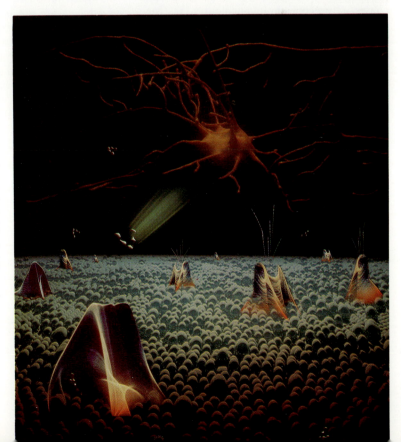

The new science of psychoneuroimmunology holds that "messenger molecules" closely link the body's endocrine (in red, facing page), central nervous (yellow), and immune (in blue) systems. Such messengers include peptides (the flying V-shaped structures at right). One particular peptide, CCK, released by the stomach after a meal, has a specific chemical structure that matches certain opiate receptors that stud the brain's surface, much like a key matches a lock. Once the peptide connects with the right receptor, it "tells" the brain to shut down the appetite, while also contributing to a general feeling of well-being.

Nerve threads run like wires between the central nervous system and key immune organs like the thymus. In a landmark experiment, Dr. Karen Bulloch traced the regeneration of nerve fibers in a mouse thymus transplanted into a mouse without one. Her work helped demonstrate the importance of nerves to the immune system. At left, Dr. Bulloch points to the sheath of the newly generated nerve fiber in a superimposed cross section of the transplanted mouse thymus.

21st century the way research on DNA has influenced 20th-century biology.

Research on informational substances forms the molecular avant-garde of a relatively young scientific discipline that holds enormous promise. The discipline travels under various aliases—"psychoneuroimmunology" is perhaps the best known of several tongue-tying alternatives. The informational substances, many of which are known to have a powerful effect on mood and emotion, provide a molecular way to understand the long-suspected connection between state of mind and state of health.

How interconnected are mind and body? Ed Blalock, now at the University of Alabama at Birmingham, believes the immune system functions as a sensory organ, just like the eyes or nose; white blood cells recognize what he calls "noncognitive stimuli," such as bacteria and viruses, and these immune-system cells influence behavior by unleashing a gush of powerful biochemicals. In a similar vein, Pert, Ruff, and their colleagues, then of NIMH, theorized about a "neuropeptide and psychosomatic network," where the mind and body constantly chatter back and forth using a vocabulary of biochemicals, the detectable result of which is the full range of human emotions. In their scenario, white blood cells are "bits of the brain floating around the body." They make and discharge hormones. They receive messages directly from the brain. They may even send messages to the brain that affect behavior, too.

What all this means is unclear, but some researchers are not bashful about speculating. "Just as there are only a finite number [of nucleic bases] that code for all forms of life, whether it be viral or human," says Pert, "there's going to be a finite number of information molecules that code for intercellular communication, whether we're talking about communication between two separate organisms or we're talking about intercommunication between the cells of your gut and the immune system and your glands and your brain."

It is not an easy scientific leap to make, as Pert herself acknowledges. Last year at Lake Arrowhead, California, at a meeting sponsored by the UCLA School of Medicine, she described a recurring dream about a giant chasm that separated the "old medicine" from the "new medicine." "And there's a guy on the other side who is telling me to jump," she told her colleagues, "and I'm scared to jump."

In truth, Pert's group has helped define the field in theoretical leaps, and is taking a running start at the far side of that chasm, where emotions play an important role in disease and health. That world lies somewhere in the twenty-first century, but Pert is confident we are going to get there. "It's going to be so quickly forgotten," she predicts, "how controversial this all was."

Metaphysical Journey

It was another dream—possibly the most important and most debated dream in the history of philosophy—that helped build the walls that have so recently been set atremble by molecular biology. More than three centuries ago, on the morning of November 10, 1619, the French philosopher René Descartes awoke from a dream that, he revealed later, inspired him to embark on a lifelong philosophical mission. That metaphysical journey culminated in his most famous assertion—that the mind and the body are segregated into two separate branches of worldly existence.

Not that everyone subscribed to the idea, before or since. Aristotle was among the first to suggest the connection between mood and health. ("Soul and body, I suggest, react sympathetically upon each other," he once noted.) Charles Darwin, too, believed the connection was important; it was a major premise of his largely overlooked book *The Expression of the Emotions in Man and Animals*. And no less revered a practitioner than Sir William Osler, the turn-of-the-century physician described as the "father of modern medicine," once remarked, "The care of tuberculosis depends more on what the patient has in his head than what he has in his chest."

Aristotle, Darwin, and Osler form an impressive trio of observers; all were struck by the apparent connection between mind and body, emotions and health. Yet all, too, were essentially powerless to sketch out that connection in anything but the broadest and fuzziest of strokes. The tools they brought to the task—dissection, microscopes, X rays, and the like—simply were not powerful enough to discern the links. Only with the convergence of molecular biology, immunology, and neuroscience have scientists begun to span the huge gap among emotions, mental processes, and molecules.

One remarkable pioneer was Russian émigré S. I. Metal'nikov. Working at the Pasteur Institute in Paris with V. Chorine in the 1920s and 1930s, Metal'nikov showed that classical (Pavlovian) conditioning could both suppress and enhance the immune response. At the turn of the century, Pavlov had shown that if you rang a bell shortly before presenting dogs with meat powder, the dogs mentally associated the bell with food and, after many repetitions, began to salivate solely at the sound of the bell. Metal'nikov paired such cues as heat, hand scratches, and trumpet blasts with injections of bacteria that stimulated the immune systems of guinea pigs and rabbits. After repeated trials, the animals in effect "learned" to rev up their immune systems with a horn, suggesting that the central nervous system communicated with the immune system.

Two Americans stand out both as pioneer researchers in the field and as major figures in establishing its credibility. Psychiatrist George F. Solomon and colleague Rudolf Moos of Stanford University, in a landmark study of women published in 1964, demonstrated a link between emotional conflict and the onset and course of rheumatoid arthritis. Despite a genetic predisposition, certain women remained free of disease. "They were all emotionally healthy," Solomon recalls of the disease-free group. "They were not depressed. They were not alienated. They had good marriages. We felt that emotional health protected them from rheumatoid arthritis." The study, not surprisingly, provoked great skepticism in the scientific community.

The other historical giant in the field is psychologist Robert Ader of the University of Rochester School of Medicine. Ader is credited by some with putting psychoneuroimmunology (he coined the term) on the map in the 1970s in a series of groundbreaking experiments with Nicholas Cohen. Ader's rats were trained to associate an initial stimulus with a subsequent event. In this case they "learned" to depress their immune system when given sweetened water. In other words, mere mental association could put a damper on the immune system—supporting Metal'nikov's precept that the central nervous system could influence the vigor of the immune system.

Anatomical Answers

Connections, however, demanded mechanisms—demanded, literally, to be fleshed out, and, at about the same time, links between the central nervous and immune systems began to provide some anatomical answers. Karen Bulloch, now of the University of California, San Diego, and David Felten of the University of Rochester medical school both have done pioneering work tracing nerve threads that run, like wires, from the central nervous system to two key immune organs, the thymus (Bulloch) and the spleen (Felten).

So the early research provided a rough anatomical sketch, suggesting areas where mind and body intersected. But all these links begged several fundamental questions. How could the central nervous system influence the immune system? Moreover, shouldn't there be two-way communication, with the immune system talking back, as it were, to the central nervous system as well? And by what means did they communicate with each other?

At this point the questions of psychoneuroimmunology began to overlap some long-standing puzzles of brain research, and that is where Candace Pert comes in. Although she set out to be an English-literature major when she went to college in the mid-1960s, it was the precise science of psychopharmacology—studying the effect of drugs on the brain and behavior—that ultimately hooked Pert on the neurosciences. She went on to the Johns Hopkins School of Medicine for graduate studies. Trained there in the powerful techniques of modern biology, she began research that led to a bold new theory—that a relatively small family of molecules could be identified as the biochemical basis of emotions.

Working with Solomon Snyder, Pert in 1973 discovered an important biological landmark called the opiate receptor. Psychopharmacologists had assumed that in order to have an effect on the brain, natural painkilling substances such as morphine and other opium derivatives (known collectively as opiates) needed some specific way of interacting with brain cells. What Pert found was a molecule on the surface of the cell that accommodated the drug like a keyhole accepts the notches of a particular key. That cellular keyhole became known as a "receptor."

As Pert later put it, "God presumably did not put an opiate receptor in our brains so that we could eventually discover how to get high with opium." If the body naturally made opiate receptors, the inescapable implication was that the body made opiates, too—natural "in-house" painkilling molecules that resembled "foreign" molecules like opium, morphine, and heroin. Implication became reality in 1975 with the discovery of the first in-house, or endogenous, opiate.

"The key thing about the opiate receptor discovery," Pert says now, "is that it pointed the way." As evidence accumulated, it indeed became clear that the human body is an impressive pharmacological factory. Since the early 1970s, beginning with the isolation of a pain-inducing peptide called substance P, about 50 similar molecules have been discovered. These peptides are formed by strings of amino acids—joined "like poppet beads," Pert says—and, like all proteins, each of these molecules folds into a unique three-dimensional shape that can be likened to the unique grooves and notches on a key. It interacts with a specific type of receptor just as a particular key, for example, engages only the tumblers of a particular lock. As with a lock, a small event such as the insertion and turning of the key leads to big changes—the biochemical equivalent of throwing a dead bolt—inside the cell. Here, then, were the elements of Pert's theory: these peptides and their receptors, she proposed, are the biochemistry of emotions.

The hormone insulin provides a good example of how these molecules work. Secreted by the pancreas into the bloodstream, insulin affects metabolism by allowing cells throughout the body to use glucose as a fuel. In the early 1980s, a group at the National Institutes of Health (NIH) headed by Jesse Roth discovered insulin receptors on brain cells. Later they found that some brain cells *make* insulin, and that, in the brain, insulin performs different tasks: by promoting cell growth, it acts as a growth factor instead of as a hormone, and by suppressing appetite, it also functions as a neuropeptide.

A Molecular Map of Emotions

Peptides, like keys, are portable and they can move around; receptors, like door locks, are stationary, rooted to one spot. Candace Pert and her NIMH colleagues, principally Miles Herkenham and Joanna Hill, began surveying neighborhoods of the brain for specific receptors. They took very thin slices of rat brains and, using radioactively tagged molecules, created maps of specific receptors localized in the brain. Pert began to see these maps as a kind of molecular atlas of emotions, because certain dense clusters of receptors appeared in parts of the brain long associated with emotional processes.

Pert refers to these "hot spots" as nodal points that seem to correlate with emotions. Three key neighborhoods of the brain—the frontal cortex, the hippocampus, and the amygdala—are especially dense with many peptide

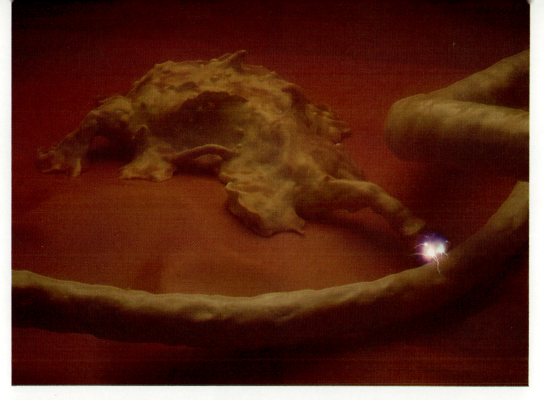
Macrophages can act as mobile nerve cells because their surface contains virtually every peptide receptor known.

receptors, as is a part of the spinal cord known as the dorsal horn, which runs from the neck to the tailbone and is a major point of convergence for sensory information. The nodal point that lies in the frontal cortex is not only rich in peptide receptors, but is heavily threaded with specialized neurons associated with touch, smell, sight, sound, and taste. In Pert's view, nodal points mark junctions where information molecules commingle and influence behavior. Indeed, she argues that some of the figures of speech we use out of habit have molecular underpinnings.

"Take a deep breath" is an injunction to calm down. Breathing is controlled in the brain stem, and research shows this region is saturated with opiate receptors; Pert even argues that changes in breathing—like those experienced by athletes and yogis—can alter the flow and concentration of peptides. Or consider "gut feelings." The stomach and gut are heavily wired with nerves and with receptors to many peptides. A painkilling peptide, CCK, is released in the gut after a meal. Not only does it tell the brain to shut down appetite, but it also contributes to a feeling of well-being. "No wonder we feel so good after a great meal," Pert observes.

Even more surprisingly, researchers have found the same—or very similar—substances in other animals, in higher plants, and even in single-celled organisms. Investigators were astounded recently to find that a substance virtually identical to insulin is made by a primitive unicellular protozoan called tetrahymena. The peptide that tickles yeast cells into mating seems to be a cousin of human gonadotropin releasing factor, which controls the release of human sex hormones. Prolactin, the hormone that stimulates the development of mammary glands in mammals, allowed fish to adapt to a saline environment in an earlier stage of evolution. The same powerful neurochemicals that alter mood in the human brain have been found on the skin of frogs.

What does this mean? It means, according to Pert, that evolution has plucked certain serviceable molecules and used them again and again—not just in humans, but in all living creatures—to do what they do well. And what these molecules seem to do well is regulate, modulate, and convey information. Whether it is a hormone that influences sexual urges, a neuropeptide that elevates mood, or even an immune-cell-manufactured chemical like interleukin-1 that can spur fevers, these substances clearly

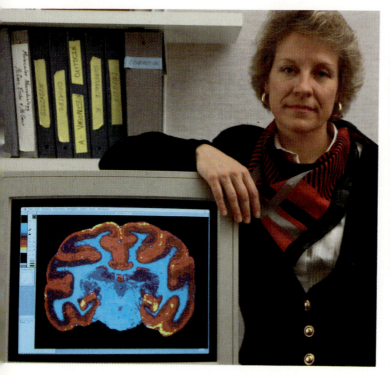
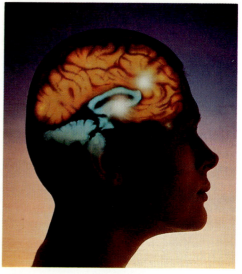

Concentrations of peptide receptors, the white "hot spots" in the brain image above, occur in those regions where human emotions originate. Researcher Joanna Hill used "tagged" molecules to show that the AIDS virus attaches at similar sites (in yellow, left) in a monkey's brain.

allow cells to talk to one another, and the cumulative din of those chemical conversations can make us feel frisky or euphoric or feverishly ill. "Emotions are so important in terms of regulating behavior, and regulating survival, that the very first successful evolution in that direction would be preserved," Pert believes. "It would tend to be used over and over again. After all, the great pleasures of life—eating and sex—are both necessary for survival, and there's tons of emotional wiring around those behaviors in humans."

The Immune System

Perhaps the most important mechanism for survival is the way the body preserves health and fends off disease, which is the province of the immune system. And that is where the misfit molecules really began to confound conventional thought.

The immune system is extraordinarily complex and fluid. Like branches of the armed services, it maintains a varied corps of task-specific cells in a perpetual state of vigilance and readiness. Some cells (monocytes) act as sentinels, others (antibodies, and natural killer and cytotoxic T cells) as infantry and artillery; some (helper and suppressor T cells) rouse or mute the general alarm, and others (macrophages) come around to clean up the battlefield and cart off the results of the carnage. Humans can generate up to 1 billion different antibody molecules to attack foreign invaders. The system is sophisticated enough to "remember" any invader, which is why vaccination works—it gives the immune system something to lodge in its memory cells. And it is amazingly self-preservative—when an individual suffers a severe, hemorrhaging injury leading to massive blood loss, these crucial memory cells know enough to seek refuge in the bone marrow, so they won't bleed out onto the street.

How do these cells coordinate this flurry of activity? The answer is that they send messages to one another. A cell will send a message by squirting one of those very small, short-lived molecules known as peptides. And in order to receive those messages, immune cells also need to have receptors. Back in 1976 Nicholas Plotnikoff of Oral Roberts University first reported that immune cells possessed, of all things, the opiate receptor (in other words, immune cells theoretically received messages from the brain).

In his work at the University of Alabama, Blalock, coming from the other direction, has reported that immune cells can make the mood-altering brain peptides known as endorphins.

One of the most intriguing networks has been worked out by Michael Ruff at NIMH. His work has concentrated on macrophages, once derided as the "garbage collectors" of the immune system. When first formed in the bone marrow, macrophages enter the bloodstream as nomadic adolescent cells called monocytes. Monocytes are among the first cells to spot a foreign intruder and raise the alarm. Later on, monocytes can settle down in the skin, lungs, or brain and take up permanent residence as macrophages.

Ruff's research shows that these cells, in motion or at rest, have enormous potential for sending and receiving biochemical messages. Macrophagelike cells in the brain (called glia) possess receptors to powerful molecules made by the immune system; both monocytes and macrophages have receptors on their surface that accommodate virtually every neuropeptide known. Because of these cells' movement and informational dexterity, Pert and Ruff call them "mobile synapses."

Molecular Communication

What all this work has established is that there are the molecular equivalents of telephone lines between the brain and immune system. "What we have done, with the help of others, is maybe pointed out the molecular basis for the communication between these systems, but that's all," says Joanna Hill of the Pert lab. "You don't know if it is being used, and you don't know what's being said." One obvious—and difficult—next step in research is to eavesdrop on those conversations; a more distant goal is to enter into the dialogue pharmacologically to aid healing.

The mere existence of those lines of communication, however, convinces Ed Blalock that the immune system is a sensory organ as much as the eyes or ears. "The immune system serves to recognize and sense those things you can't hear or taste or touch or smell or feel, things that are not recognized by any of the other senses," he says. "If you came into contact with a grizzly bear, for example, the visual recognition would evoke the production of stress hormones, which would change your physiol-

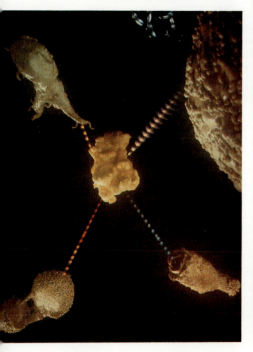

Michael Ruff (right) helped show how macrophages relay messages between the brain and the immune system. Helper T cells (below, center) coordinate other immune cells, including macrophages.

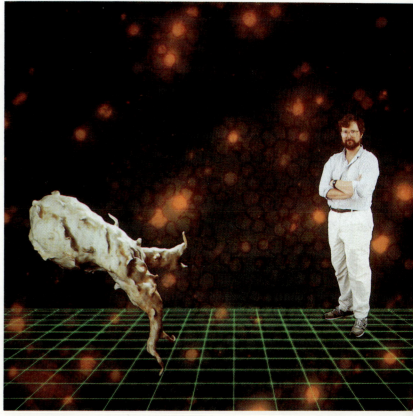

ogy in a way that would help you deal with the situation. If your body comes into contact with a virus, you can't see it or smell it or touch it. But your physiology needs to change nonetheless. Well, our argument is that your physiology changes when your immune system senses it and produces the hormone that changes physiology.''

It bears repeating at this point that many—perhaps the majority—in the scientific and medical communities view these developments with wariness, if not outright disbelief. Part of this skepticism is the legacy of overblown, unsubstantiated claims connecting stress with illness or certain personality traits with cancer. In a 1985 editorial in the *New England Journal of Medicine,* Marcia Angell called the connection between emotions and health ''largely folklore,'' and argued that the ''literature contains very few scientifically sound studies of the relation, if there is one, between mental state and disease.''

Nonetheless, as data accumulate, that vast gulf between molecules and state of mind is being gradually closed. ''I used to say we were at square one,'' says Robert Ader. ''Now I think we're at about square three.''

Ronald Glaser and Janice Kiecolt-Glaser, both of Ohio State University, have done studies of first-year medical students showing that during periods of academic stress (just before or during exam week), the immune function suffers both in numbers of cells and in the ability of cells to perform. ''It suggests,'' says Kiecolt-Glaser, ''that even commonplace events that we associate with emotional arousal or discomfort can be associated with [decreases] in immune functions.'' When the students return after summer vacation, normal vigor has returned to the immune system.

Trying to Track the Signals
Robert Ader's recent work with rats suggests that behavior may somehow be influenced by the immune system—more evidence pointing toward the long-sought feedback mechanism where the immune system affects the central nervous system. He has shown that a special breed of laboratory rat defies the normal aversion to water laced with the nausea-inducing drug cyclophosphamide. When the drug is paired with sweetened water, rats with systemic lupus erythematosus—a disease caused by an overly aggressive immune system—ignore the unpleasant side effects and continue to drink.

Cyclophosphamide also depresses immune function, but that is beneficial for lupus-prone animals and their hyperactive immune systems. ''The animals know it's 'good' for them, as it were,'' says Ader. ''Signals generated by the immune system are being read by the central nervous system. What that signal is and where it is going—that I don't know.''

Even AIDS can be viewed through the prism of psychoneuroimmunology. Candace Pert and her NIMH colleagues have advanced a hypothesis that AIDS is a disease rooted in the disruption of peptide communication. Specifically, her group argues that a portion of the AIDS virus's external spike, used to gain entry to cells through a particular receptor, actually mimics a ubiquitous neuropeptide known as vasoactive intestinal peptide (VIP). This molecule is active not only in the gut, but also in the brain, where it apparently promotes the growth and health of neurons. When the AIDS virus attaches to the VIP receptor, it preempts the activity of VIP, and neurons die. Pert believes this is the cause of the dementia that afflicts many AIDS patients.

Viewing the disease as a peptide disorder, the NIMH group has gone on to design a protective peptide that covers the receptors of neurons like a cap and prevents the AIDS virus from attaching. Called peptide T, in small-scale clinical trials, its use has been accompanied by significant physical and neurological improvement. Whether peptide T works or not—and there is considerable skepticism in the scientific community about it—it represents a novel departure from traditional disease treatments, such as vaccines and antibiotics. So optimistic about the possibilities is Pert, she and Ruff have formed a biotechnology company called Peptide Design L.P.

There is a sign in Pert's new office that reads: ''If you are being run out of town, get out in front of the crowd and make it look like you're leading a parade.'' There are still a lot of people raining on this particular scientific parade, but Pert and her allies are convinced they are on the right track. ''I've got the goods, the little nuts and bolts, the peptides,'' she likes to say. ''Our work is just beginning to prove some of this. A lot of this has been my interpretation of the significance of my lab's work. But I know it's right.'' Scientists don't often talk that way, at least not for public consumption. Revolutionaries do, however, and we're in the midst of a revolution.

Farmers Learn New Tricks from MOTHER NATURE

by Emily T. Smith

There came a time when Fred L. Kirshenmann didn't like what he saw in the fields of his 3,100-acre (1,250-hectare) farm. The soil was getting harder to plow, and he was applying more expensive pesticides and fertilizers to get the same yields. So Kirshenmann, whose father was the first in his township to adopt agrichemicals during the 1950s, became the first to abandon them 12 years ago.

The Medina, North Dakota, farmer substituted compost and legumes for synthetic fertilizers. To control insects and weeds, he began rotating crops, adopted special plowing and planting techniques, and turned to mechanical weed control. In the process, Kirshenmann had to forfeit 40 percent of his wheat and soybean acreage eligible for federal subsidy payments. "Our neighbors thought we would lose our shirts," he recalls. Yet today Kirshenmann maintains an average harvest of 33 bushels of wheat per acre—equal to the best yields of his neighbors who still use conventional farming techniques. He also brings in a profit from a panoply of other farm products, from cattle to sunflowers. The 54-year-old farmer hasn't had to borrow operating funds for nine years.

Fred Kirshenmann (right) and many other farmers turn a profit without using polluting chemicals and expensive technology. Rather than using herbicides, for instance, such farmers can control weeds by combining old methods like ridge tilling with modern mechanical means (left).

Left: © Ed Landrock/Rodale Stock Images; right: © Greg Booth

Chemical-free agriculture gained widespread consumer support in the wake of the 1989 scare over Alar, a possible cancer-causing agent that helps ripen apples (top). The cyanide that tainted grapes from Chile (above) appears to have come from tampering, not farm chemicals.

Farmers like Kirshenmann, advocates of so-called alternative agriculture, used to be oddballs. Today, though they total only about 100,000 of the nation's 2.17 million farmers, their ranks are growing—and they are gaining political clout.

Fear of Food

In Iowa, California, and Minnesota, farmers are uniting to experiment, support one another, and exchange information. The Practical Farmers of Iowa, founded in 1985, has 300 members. State legislatures in Minnesota, Iowa, Texas, and California have dedicated funds for research and ag-extension programs. When lawmakers debate the 1990 farm bill this year, alternative ag will be on the agenda. "We're taking a whole fresh look at how we farm," says Paul E. O'Connell, administrator of the Agriculture Department's Low-Input Sustainable Agriculture (LISA) program. Indeed, its advocates believe that alternative agriculture "could transform American farming—make it profitable and save the family farm," insists Jim Hightower, head of the Texas Agriculture Department.

Alternatives to conventional agriculture are winning broader support. Concerns over the safety of food are one reason: recently there have been scares over Alar, a possible cancer-causing chemical made by Uniroyal Chemical Company to help ripen apples, and cyanide-contaminated grapes from Chile. Also, farmers concerned with their health, rising production costs, and the long-term productivity of lands seek more options. Meanwhile, water pollution, erosion, and other environmental damage caused by farming get more attention.

Evidence of harm from traditional farming techniques is not lacking. Today farmers use an estimated 10.5 million tons (9.5 million metric tons) of fertilizer a year, and some 55,000 tons (49,900 metric tons) of pesticides. In recent years the Environmental Protection Agency (EPA) has withdrawn 27 agricultural chemicals from the market because of evidence of health hazards. Moreover, the EPA considers agriculture the largest nonlocalized source of surface-water pollution. The EPA and states such as Iowa, Florida, and Minnesota have detected agricultural chemicals in groundwater, which supplies 50 percent of all drinking water.

At the same time, some pests continue to best chemical pesticides by developing resistance to them. Irrigation is depleting some aquifers and causing a buildup of salt in some areas, such as California's Imperial Valley. Intensive cultivation, wind, and water erode as much as 3.1 billion tons (2.8 billion metric tons) of topsoil from U.S. cropland annually.

The price of such damage and waste is staggering. David Pimentel, an entomologist at Cornell, estimates that costs from conventional agriculture—including health effects, damage to the environment, and wasted resources—total $75 billion to $100 billion a year.

The purists of the alternative-agriculture movement prefer the term "sustainable," because they advocate practices that will not deplete resources or harm the environment. They want to adopt techniques that require less of such nonrenewable resources as fossil fuels, do not pollute or degrade the environment, rely less on increasingly expensive technology, and that will allow rural economies to prosper. "In the future, agriculture is going to have to be economically viable and environmentally sound," says William C. Liebhardt, director of the sustainable-ag program at the University of California at Davis. "Current practices aren't sustainable."

Field Practice

While still small and controversial, alternative agriculture is no longer being dismissed by agrichemical makers as a fringe movement that will go away. In 1988, 20 chemical companies—including Ciba-Geigy, Monsanto, Dow, FMC, and Du Pont—created the Alliance for a Clean Rural Environment (ACRE) to counter criticism of chemicals and instruct farmers on safer use of pesticides. Industry groups, including the Fertilizer Institute and the National Agricultural Chemicals Association (NACA), have lobbied and contributed money to oppose alternative-agriculture initiatives in Iowa and Minnesota.

The chemical industry argues that many problems with conventional farming can be solved by careful management. Chemicals in groundwater come from "accidents or abuse," not judicious application, says Jay J. Vroom, the new president of NACA. Adds James L. Searcy, manager of global agricultural products development at Du Pont: "Once we learn of a problem, we can solve it with new technology that will not degrade the environment."

Industry groups are pushing what they call Best Management Practices, which promote such methods as integrated pest management—inspecting fields for pests before applying chemicals—to reduce use. A new generation of narrowly targeted, low-dose chemicals, such as a group of Du Pont herbicides that block an enzyme found only in specific plants, is coming to market. Each chemical acts on a certain weed, and less than an ounce per acre is needed. At the same time, companies are working on safer pesticides based on naturally occurring chemicals and microbes. Agriculture and biotech companies also are enthusiastic about crops genetically engineered to resist pesticides, disease, and insects.

Sustainable-ag advocates, however, remain wary of a technological fix. "It doesn't bring deeper changes for a fundamentally better agriculture," says Ronald Kroese, executive director of the Land Stewardship Project, a Minnesota sustainable-ag group. Adds Jack Doyle, project director at the Environmental Policy Institute: "We want agricultural production techniques that supply adequate returns without locking farmers into an expensive and environmentally damaging array of new inputs."

Alternative agriculture entails exploiting naturally occurring biological relationships on the farm. "We're trying to alter the environment so it's not favorable to pests, and create healthy soils," says Charles H. Hassebrook, program leader at the Center for Rural Affairs, an agricultural think tank in Nebraska. But it isn't easy. Making a complete transition can take from five to 15 years. Alternative-ag farmers use multiple crop rotations rather than sow large acreages in just a few crops, such as corn or wheat, year

The "salad vac" machine sucks insects up from lettuce plants, saving the farmer from using chemical insecticides.

© Carl Doney/Rodale Stock Images

BUG-EATING BUGS

It may sound like a giant step backward at first. But not to Stahmann Farms Inc. in Las Cruces, New Mexico. The nation's largest pecan ranch gave up using chemical pesticides entirely on its 3,600-acre (3,300-hectare) spread. Instead, it gets the same result with the tiny green lacewing, a natural predator that feasts on insect pests. By turning to nature, the farm cut the costs of controlling pests by well over 50 percent.

The creepy-crawly business is booming, and not just among backyard gardeners and onetime hippies who drifted into organic farming. "We're encouraged when not only the progressive farmers but also the largest in the business start coming to us," says Jake T. Blehm, marketing vice-president at Rincon-Vitova Insectaries Inc. in Ventura, California, which is also testing lacewings at the vineyards of E. & J. Gallo Winery.

The green lacewing feasts on farm pests, not plants.

Weevil Empire

Insects aren't the only "bugs" that can eliminate pests. A host of companies—biotech and otherwise—are scrambling to commercialize microbes and other natural enemies of agricultural blights. Biosys Inc., for example, has devised a way to grow, package, and ship several strains of nematodes that live in the soil and feed on immature insects. It sells the tiny worms to cranberry farmers and citrus producers. "The nematode is designed by nature as an insecticide," says Venkat S. Sohoni, chairman of the Palo Alto, California, venture.

More and more, the new bugs will be designed by man. "We take naturally occurring toxins and engineer them into a delivery vehicle," says Jerry D. Caulder, president of Mycogen Corporation in San Diego. For Mycogen, the vehicle is bacteria; its first product, called M-ONE, produces a toxin that knocks out beetles and weevils. So far, it has been set up against the Colorado potato beetle and sprayed on the Washington Mall to control elm leaf beetles. Others hope to engineer a biopesticide into the host plant itself. Monsanto is developing a tomato with leaves that are toxic to the tomato pinworm. Crop Genetics International Corporation is testing corn with stalks that kill corn borers.

Broad use of these so-called biopesticides is still several years away. But they are finding markets where chemical pesticides must be abandoned because of groundwater contamination or pests that have become resistant. That leaves a bright future for bugs of the garden variety. After all, it's a technology that's already here.

LARRY ARMSTRONG

---

after year. Livestock operations, which many farmers abandoned because they weren't profitable, and the cultivation of legumes become sources of fertilizer. Mechanical weeding and planting techniques, such as ridge tilling (in which some residue of the previous year's crops is not plowed under) are also mainstays.

When used in combination, these practices can cut pesticide use and erosion by more than 50 percent. But they call for more labor and management. "Sustainable-agriculture approaches are extremely scientific," says Ken Tull, a director at the Rodale Institute, one of the pioneers in alternative agriculture. "It requires a better farmer."

In addition to the experiences of farmers such as Kirshenmann, there is increasing evidence that alternative-ag yields can match those of conventional practices and that farmers can make money at it. In September 1989, a report by the National Research Council cited 14 farmers who got comparable yields for a variety of crops from apples to wheat to milk. It concluded that alternative ag was so promising that federal research should be boosted from its current $4.5 million to $40 million. Other studies indicate that even without comparable yields, profits can still be made, because cash outlays are lower.

Even so, the case for alternative ag is not strong enough to persuade most farmers to convert. "There is a great deal of skepticism," concedes Richard Gauger, director of alternative-ag programs for the state of Minnesota. James C. Wiers, who spends $300,000 a year on chemicals for his 3,000-acre (2,700-hectare) produce farm in Willard, Ohio, has done all he can to cut his chemical use. But he still sprays—because, he says, "it's the only way to maintain yields and quality. I'd love an alternative, but it must be economically feasible."

To critics, U.S. farm policy is one reason that alternative agriculture may not be economically feasible on a larger scale. Commodities programs, which cover 70 percent of U.S. cropland, offer subsidies for such crops as corn, wheat, soybeans, rice, and cotton. Farmers are paid based on a so-called base acreage and their yields, so they strive for maximum yields—even if that means using more chemicals for marginal gains. If they don't plant their base acres, farmers lose subsidies. That discourages crop rotation along with various other alternative techniques.

Even in Washington, D.C., however, alternative agriculture is making some headway.

Richard G. Lugar of Indiana, ranking Republican on the Senate Agriculture Committee, has introduced legislation that urges changes in the 1990 farm bill to give farmers flexibility to try alternative-agriculture techniques without completely losing their subsidies. Whatever the fate of that proposal, "the sustainable debate will influence the 1990 farm bill," says Neville Clarke, director of the Texas Agricultural Experiment Station in College Station, Texas. Even Clarke and other mainstream agriculture experts, along with some Washington policymakers, expect the final farm bill to reflect greater concern for the environment and make it easier for farmers to experiment with alternative techniques.

Underfunded

Still, more research is needed to prove that the techniques will work on a large scale. Yet today only a fraction of the Agriculture Department's $600 million research and ag-extension budget is allocated for sustainable ag. LISA, its key program, was swamped with more than 800 proposals that it considered worthy of funding over the past two years—but it could support only 74. Without more research, "we don't know how far we can go," says I. Garth Youngberg, executive director of the Institute for Alternative Agriculture Inc. in Washington, D.C.

Advocates remain optimistic that alternative ag will win more support. That's progress: when farmers like Kirshenmann walked away from the conventional techniques of the Green Revolution a decade ago, the powers that be wouldn't have given alternative agriculture a second thought.

Preying on predators: ladybird beetles released into an orchard will feed on mealybugs, a destructive pest of fruit trees.

© Nigel Smith/Animals Animals

The WONDER OF BIOMECHANICS by Bruce Fellman

"I think that I shall never see, a perambulator tree. . . ."

Cornell University botanist Karl Niklas, with no apologies to Joyce Kilmer, laughs at the rhyme. "A walking tree is an impossible organism," he says. "The energy necessary to create motion in a structure as big as a tree exceeds its ability to trap sunlight and convert it into energy."

Nature cannot violate physical laws. Delights such as walking trees or winged horses are just not possible, no matter how much we might like them to exist. But the natural world does take advantage of those same physical laws and appropriates them for its own uses in surprising and, until quite recently, underappreciated ways. The discoveries of an emerging scientific discipline called comparative biomechanics have shown that nature has a better imagination than any fantasy writer—and many biologists. The central insight is how the design and construction of plants and animals reflect the oppor-

Pinecone Aerodynamics

But when Niklas looked at a pinecone, he saw a very different kind of structure. "Anything that's wind-pollinated is subject to the same laws of aerodynamics as a plane fuselage or the flapping of a sea gull's wings," he explains. "But no one had ever looked at it that way."

To determine how a pinecone really works, Niklas used a wind tunnel. After working first with a scale-model papier-mâché cone and helium bubbles to show the currents on which pollen traveled, he turned to real cones and pollen, photographing the patterns of currents the pollen wove around the cone. Here was no baseball glove, succeeding by simply being in the way. Rather, what Niklas observed was a coniferous air-traffic controller deliberately bending the flight paths of the pollen grains to its aerodynamic will.

The pollen grains flew over the top of the cone, and once on the lee side, a secondary, corkscrew current carried the fliers in a spiral around the scales. These set up eddy currents of their own that swept the pollen between the scales, where it could complete fertilization.

"I see the cone as a turbine that can't rotate on its axis," says Niklas. "Instead, this collection of little airfoils—wings, if you will—causes the air to rotate around it."

Biomechanics studies how the design and construction of plants and animals obey and even capitalize on the laws of physics. The shape of a pinecone, for example, makes the most of air currents passing by it. In a wind tunnel (left), helium bubbles swirl around the scales of a papier-mâché pinecone in much the same way that air circulates around a real pinecone. Such eddies increase the chances of airborne pollen grains landing on the scales and fertilizing the waiting ovules. The superimposed formula relates the scale of the model to life-size cones. Evaluating animal biomechanics presents a far greater challenge. To study the muscle forces of large mammals, for instance, scientists had a llama work out on a treadmill (right).

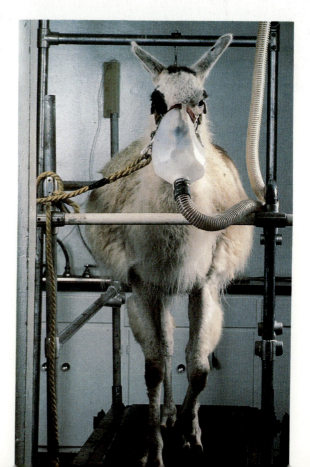

tunities as well as the constraints of physics. Comparative biomechanics is the study of kangaroo hopping, the fluid dynamics of seawater around limpets, the catching mechanics of spiderwebs, the reason trees don't fall down and bones hold us up, and why a tetrakaidecahedron (a 14-sided polyhedron) is the best shape for the storage cells of plants—to mention just a handful of research projects in this wide-ranging merger of biology and engineering.

Consider the pinecone. For the past 200 million years, this intricate, scaly creation has sat out on a limb and dumped or received pollen, depending on the sex of the cone. The standard textbook line on female cones is that they are catcher's mitts, haphazardly snagging whatever pollen happens to be on the breezes.

All photos: © Peter Angelo Simon/Phototake

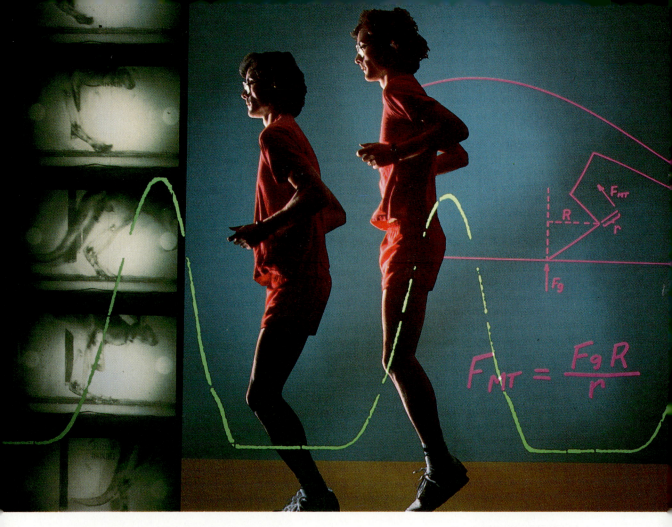

Forces exerted by a kangaroo hopping (left) and those of a person jogging are governed by the same laws of physics. When the jogger hits the ground, the green line peaks, indicating that maximum energy is stored for his next step.

Engineering Implications

This explanation seems intuitively obvious. So why was the obvious missed for so long?

"Biologists don't know the everyday, practical kind of physics. They may know what a quark is, but it's more important that they know how to build a bridge," declares Steven Vogel of Duke University.

Vogel is one half—the other is Duke colleague Stephen Wainwright—of this country's prime moving force in biological (as opposed to orthopedic) biomechanics. The two are singularly responsible for pushing principles of mechanical engineering into the mainstream of biological thought.

Of course, a working knowledge of biomechanics long predates Vogel and Wainwright. As they readily admit, the earliest human technologies were based on a practical understanding of plant and animal engineering. Roman weapons designers scored a very large coup when they came up with the ballista, a fearsome device that could hurl a 90-pound (40-kilogram) stone one-quarter of a mile. The ICBM of its day stored energy by twisting a bundle of animal tendons. When these twisted tendons unwound, it was, "Bombs away!"

One of the earliest written considerations of biomechanics comes from Galileo Galilei, the Italian astronomer. In myth, fairy tale, and science fiction, creatures from beanstalks to capillary-cruising submariners are forever changing size, but, more than 300 years ago, Galileo demonstrated, in his treatise *Two New Sciences,* a clear feel for the consequences and limitations of the changes.

"Thus it is impossible to build enormous ships, palaces, or temples, for which oars, masts, beamwork, iron chains, and in sum all parts shall hold together; nor could nature make trees of immeasurable size, because their branches would eventually fail of their own weight . . . ,'' he wrote. Galileo knew that for every alteration, for every decrease and increase, there were biological and engineering implications. But what were the rules of the game?

Biomechanics and the Mystery of Life

The best—and certainly the most exhaustive—exploration of this question can be found in D'Arcy Wentworth Thompson's classic book *On Growth and Form*. First published in 1917 and still in print, the exquisitely written and illustrated volume is responsible for many seductions into the biomechanics field. The Scottish biologist exhorts his readers (and assails those disinclined to follow) to ''explain by geometry or by mechanics the things which have their part in the mystery of life.''

But Thompson never trained a cadre of students. The task of channeling the enthusiasm generated by the book fell to a Cambridge professor, Sir James Gray. As a researcher, Gray was fascinated by how animals move, and his 1968 book *Animal Locomotion* remains a standard reference on the subject. He was also a diligent teacher and an employment counselor, as well as the organizer and editor of the discipline's preeminent publication, the *Journal of Experimental Biology*. In essence, Gray engineered the field.

''Of course, what we do is the reverse of the engineer's approach,'' says Vogel. ''An engineer says, 'I want to do the following; what's the best design I can come up with?' We say, 'Given this design, what was it designed to do?' We've got the answer—the organism—and we have to figure out the problem.''

A biologist's traditional education is not much help in framing the right questions. ''There's nothing in a basic physics course that would teach you why, for example, the portholes on a ship or a plane should be round,'' Vogel notes. ''And if you don't know that, you can't appreciate the skeletons of echinoderms that have all kinds of little round perforations, or the shapes of holes in bones. These, to the best of our knowledge today, are the best designs to minimize crack propagation.''

The coquina clam evolved a wedge shape so that water washing over it (demonstrated by the green fluid) rotates the clam (upper right) until it is oriented to the flow (left, side view). The clam can then quickly dig into the sand.

Helically wound fibers, as demonstrated below, allow cylindrical structures like an elephant's trunk to bend without kinking. Leaves follow the path of least resistance by folding over one another in a strong wind (left).

Seeing the natural world from an engineer's viewpoint has enabled biologists to explain the unexplainable. Take the case of the seemingly suicidal coquina clam. These brightly colored, fingernail-size mollusks live in wet sand, just below the surface, and they're common on wave-swept beaches throughout the warmer parts of the world. In 1983 they made quite an impression on Olaf Ellers, who had just come to Duke as a graduate student of Vogel's. The soft-spoken biologist, now a professor at the University of California, Davis, had taken his parents sight-seeing to the North Carolina coast, and as they watched the breakers rolling in and out, Ellers noticed something very strange. "There were little clams jumping out of the sand," he says. "It was a really dramatic thing. You'd look out at the beach, and there'd be no animals on it at all. Then a wave would come in, and all of a sudden, the sand would be paved with clams. This wasn't anything I remembered learning about in invertebrate zoology class."

Mollusks leaping into the surf: to Ellers, such behavior seemed, well, stupid. At first he thought the clams came out of the sand so that the waves would carry them from one place to another, the wave power harnessed for their own locomotion. But, he wondered, would not the outwash carry them out to sea?

"I like to think of animals as being 'clever,'" says Ellers, and when he put the coquinas in a flow tank that mimicked the surf, he learned just how clever they were. There is, however, no molluscan "intelligence" involved. The clam's ability to avoid being washed away lies almost entirely in its shape.

Coquinas are built along the lines of wood splitter's wedges, and because of this unusual (for a clam) shape and density, these mollusks are not at the mercy of waves. When they make

88 BIOLOGY

what seems to be a fatal leap from the sand into the surf, the water washing over them automatically reorients them with their thin end facing upstream. There they can briefly remain—suspended as if by wires—until the wave subsides, and the clam drops anchor with its foot and burrows back under the sand.

The interaction between the current and the animal's design sets up what Vogel terms "flow-induced pressures." The notion that flow patterns around objects can cause subtle and not-so-subtle variations in pressure is old hat to engineers. Indeed, Swiss mathematician Daniel Bernoulli mapped out the concept in 1738, and the Bernoulli principle is the foundation of aviation. An airplane flies because the shape of its wings creates more pressure below them than above them. Without that lift, a flight remains grounded.

Getting into Flow with Fruit Flies

The tricks plants and animals might play with pressure variations have interested Vogel since his student days in the 1950s and '60s. "I got into the flow business when I started looking at how insects control their wingbeats," he explains. "I was trying to fly fruit flies in a wind tunnel. It sounds preposterous, but it's not as ridiculous as you think. You learn how to solder little wires onto the miserable insects."

For his efforts the brash biologist demonstrated that the physics of air moving over the wings was as important to drosophila (fruit flies) as it was to a 747. The branch of science that describes this kind of physical behavior is called fluid mechanics, and it is not for the fainthearted. When applied to airplanes, fluid mechanics involves hairy mathematics, and still the results must be tested in wind tunnels. When applied to the infinitely variable world of animals, however, the mathematics is not powerful enough to give answers. Vogel found he nevertheless could get the answers he wanted by going directly to the wind tunnel. The work has opened up a new way of seeing the world.

As an example of how fluid mechanics can help the observer appreciate natural design, consider a fast-swimming, streamlined fish. "The mouth is at the point of highest pressure so the water goes in," Vogel explains. "The gill covers are located at the point of lowest pressure so the water is forced out. Wherever pressure is positive—that's inward—you have a bony skull so the fish won't collapse; and where the pressure is outward and negative, you have a tension-resisting skin. The eyes of the fish are located precisely where the pressure goes from positive to negative so the focus of the eye won't change with swimming speed. The whole fish is beautifully designed to make these pressure distributions work for it rather than against it."

At least some creatures are able to tap into the pressure fields that are created by wind and water currents flowing around their bodies. They can then take advantage of the Bernoulli principle to perform a remarkable variety of tasks. The squid, for instance, uses flow-induced pressures to assist its jet propulsion. The way Vogel went about demonstrating this is typical of his approach to biology. "I don't especially like animals—they bite me," he quips, at least half-serious. "So I make them on a lathe. It's just as well."

When these streamlined aluminum squid were monitored in a wind tunnel under conditions that approximated swimming, Vogel discovered that squid have Bernoulli on their side. Forward movement reduces the pressure alongside a squid's body, and this pressure reduction makes it easier for the animal to fill up with the water that will be squeezed out its downstream end to shoot it ahead. The faster it swims, the faster it fills.

Vogel and his students have documented many other Bernoulli principle profiteers. Sponges use flow-induced pressures as a "low-cost" way of pumping water through their bodies. Scallops rely on pressure fields to help reopen their flapping shells. The great baleen whales are aided by water pressure in expanding their massive throat pouches.

Art, Sculpture, and Engineering

Bernoulli also figures into Steve Wainwright's work, but not as directly. When the Wright brothers took off from Kitty Hawk, the specially modified bicycle chain that linked propeller to engine came from the Wainwright family's factory in Indianapolis. Engineering was in his blood, but so was art, sculpture in particular.

Biomechanics provided him with a way to combine these endeavors. "I look at the various structures that exist, and I try to figure out what they're there for," Wainwright explains. For example, he and his student Lisa Orten showed that whale blubber is more than mere fat. "It's a highly complex material that's actually part of the whale's motor," he says. To be sure, blubber is insulation, but the fatty material is also interlaced with a network of tendons. Like rubber bands, these absorb and release energy as the whale swims. "Roughly 20 percent of the power in the swimming stroke comes from the elastic energy stored in the blubber."

Wainwright is perhaps best known for the work he's done on the helically wound fiber systems that seem to pervade the natural world. The helix, Wainwright believes, is a key to understanding how life evolved beyond a population of essentially two-dimensional creatures. "Even today there are some very successful sheetlike organisms, like sea lettuce. And if that's all there was, eventually the whole world would be covered by a mat," he says.

But anything that could grow up or down would have all this new territory to exploit. Of course, the organism would also have to deal with gravity and a number of other physical forces that limit possible body plans. The more the sculptor looked at the shapes nature had crafted, the more he realized that they were based on cylinders. However, a simple long, thin balloon would not do the trick.

To demonstrate why, Wainwright reaches into his desk and pulls out a package of balloons. He blows into one, enough to make it fill out, but not enough to expand any of the material. He blows again, and now a bulge appears, a kind of aneurysm. Of course, aneurysms are counterproductive, since they have an annoying habit of bursting. Fortunately for the evolution of life on earth, these bulging instabilities are easily corrected. Wrap the cylinder with a couple of sets of fibers, and organisms can become three-dimensional.

As I watch, Wainwright gleefully demonstrates how well this solution works by clothing the balloon in a sheath made from one of those red-and-white checkered tablecloths found in Italian restaurants. "Sure enough, this solves the aneurysm problem, but it's a terrible worm. When you bend it, it kinks," he continues. Such a creature, made with a sheet of longitudinal fibers wound over a sheet of circumferential fibers, would also have trouble moving.

The way around this is to cut the tablecloth on the bias, like a necktie. "It gives you a cross-helically wound pattern that allows bending without kinking, which in turn allows movement," Wainwright says.

Here was Wainwright's organizing principle. He and his students have found it at work throughout the plant and animal kingdoms, in organisms as widely diverse as palm trees and octopuses.

William Kier, a biologist at the University of North Carolina, and his wife, Duke University anatomist Kathleen Smith, have figured out the modus operandi of lizard tongues, elephant trunks, and octopus tentacles. All, it seems, depend on muscles—including muscles that are helically wrapped around the limb or organ. These structures are alike in that they lack visible means of support. Most organisms depend on skeletons that are internal (mammals) or external (beetles and Dungeness crabs, for example). Others, such as worms and sea anemones, use "hydrostatic" skeletons in which connective tissue surrounds a cavity filled with liquid. The liquid is incompressible; when muscles tighten around it, the organism is supported.

The Logic of Limbs

The legs and arms of some animals, however, don't appear to use either system. Kier was particularly puzzled by the tentacles of octopuses and squid: "If you've ever eaten them, you know that when you saw across the tentacles with a knife, they're solid muscle." Muscles have to pull on something, normally a skeleton of some sort. There are no bones in a tentacle, and no fluid-filled cavities to compress. Kier and Smith have found that such body parts work as muscular hydrostats, a novel arrangement in which muscle itself takes the place of the fluid. A squid, for example, can play one group of muscles in a tentacle off against another. Contracting one group while keeping another stiff (so that in effect they act like bone) allows the limb to bend. And because there are no joints, the tentacle has the capacity to bend at any point and in any direction.

The dossier compiled by biomechanists is full of similar examples of elegant design that, to become obvious, required an observer with a different perspective, not to mention a healthy dose of engineering. Now chicks carry weighted backpacks and run while wearing strain gauges on their legs to determine the impact of stress on bones. Labrador retrievers, horses, kangaroos, rats, and humans jog on a device called a force plate to help scientists to understand muscle use. There are swingers of oaks, horsetails, and birches who are searching for the elastic similarities among plants. Researchers peer through microscopes to learn the mechanics of helical swimming among microorganisms and prey capture by copepod arms. Paleontologists take the engineering approach to figure out how fossils lived and developed.

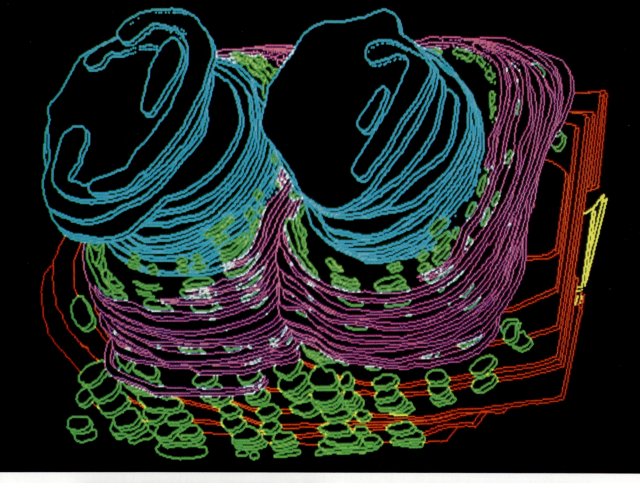

Sections cut through the suckers on octopus tentacles reveal the various muscle groups (indicated by different colors above) involved in movement. The interplay between the muscles gives the tentacles their strength and allows them to bend at any point in any direction (left).

Don't assume, however, that the underlying goal of comparative biomechanics is to demonstrate a kind of perfection in nature. "I don't like to use the word 'optimum,'" Vogel admits, "for there may be a better solution that an organism hasn't discovered, or can't easily discover. We make the assumption that if a plant or an animal is here, it must be doing something right. If it looks badly tuned, it's a clue that the observer is just barking up the wrong tree. It may be doing things that are laughably crude, but remember, natural selection can't try every possibility. It hasn't had unlimited time."

And time, as Woody Allen, surely the most unlikely of evolutionary theorists, noted, is "nature's way of preventing everything from happening at once." In the world of natural engineering, there can be neither walking forests nor gridlock.

BIOLOGY 91

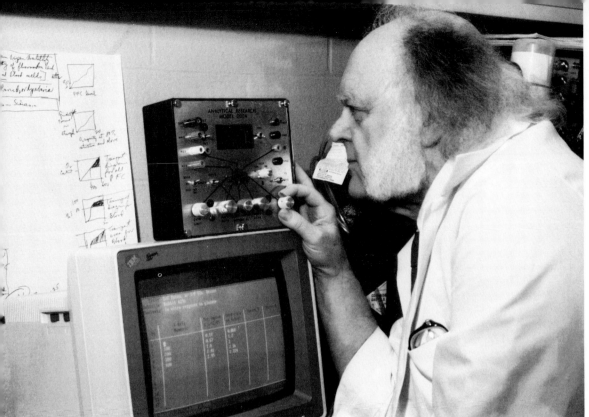

© Robert Spokane/Children's Hospital Research Foundation

BIOSENSORS by David Holzman

It is not as if cocaine were the most volatile stuff in the world. Wrap it up several times in plastic and stick it in a briefcase, and who would think it could be sniffed out from the pile of luggage on a conveyor belt? But a cocaine detector did just that recently at the airport in Munich, West Germany.

In this instance, no one was arrested. The purpose of the exercise was merely to demonstrate the power of a new technology. But smugglers beware: the device may be coming to an airport near you.

The biosensor, as this technology is called, is really just a highly specialized artificial nose. "What the biosensor does is simply to mimic nature," says George G. Guilbault, the inventor of the cocaine detector and president of Universal Sensors Inc., a New Orleans–based company. "It mimics the dog, which is used for cocaine detection."

But biosensors can do far more than even the best-trained dogs. Besides their use for locating drugs, they are being developed to detect and measure levels of pollutants, toxic compounds, and even chemical-warfare agents in the air, soil, and groundwater.

The first of these devices measured blood sugar in diabetics; others now measure several other aspects of body chemistry. Even sensors to determine which chemotherapy agents to use on cancer patients are being developed. The field is in ferment. "I would say there are at least 400 papers a year coming out," says Garry A. Rechnitz, director of the Hawaii Biosensor Laboratory at the University of Hawaii.

Highly Selective

Biosensors offer a variety of advantages over other types of chemical detectors. "The principal advantage is this biochemical selectivity," says Rechnitz. "We can pick out individual kinds of materials or compounds out of a complicated sample."

Chemical-detecting systems called biosensors (above), first developed to track the blood-sugar level of diabetics, are now being refined to zero in on such targets as cocaine, chemical weapons, and pollutants.

Enzymes and antibodies, which biosensors typically employ to distinguish their target material, are key to such specificity. Both are classes of biochemicals that include millions of members, each of which fits other molecules, like lock and key. But another class of biomolecules, called receptors, as well as living microbes and silicon chips, may also be used, says Guilbault, who is also a research professor of chemistry at the University of New Orleans.

Rechnitz is testing the value of the "nose" of the Hawaiian red crab as a sensing device. The crab "has tiny little hairs that are used as a way of finding food, sensing danger, and so on. We detach these little hairs and mount them on an electronic device and make a biosensor." The hairs are sensitive to "such materials as amino acids, hormones, things like that. There is great interest in measuring these kinds of materials from the biomedical and environmental point of view," he says.

Rechnitz is also studying an enzyme extracted from the core of the pineapple that could be used to measure peroxide, "which is implicated in acid rain damaging the trees."

Another advantage of biosensors is that they could be made inexpensively, yet still offer more versatility than other chemical-monitoring systems do. The state of the art in chemical detection "is things like radioactive immune assays or high-pressure liquid chromatography," says Mohyee E. Eldefrawi, a professor of pharmacology and experimental therapeutics at the University of Maryland School of Medicine. "These are the ultimate in sensitivity, but not many laboratories are equipped with them."

Finally, most biosensors give immediate results. This can be crucial, whether one is sniffing out cocaine or testing blood-sugar levels.

There are a variety of ways of transducing the signal generated when the sensor picks up its target. In Guilbault's cocaine detector, when antibodies sitting on a quartz crystal catch the cocaine, the extra weight of the cocaine changes the vibrational frequency of the quartz.

Eldefrawi is developing a cocaine detector that works entirely differently from Guilbault's. He uses cocaine-binding proteins extracted from the liver of the rat. ("It's a curiosity," he says. "We have no idea what it is doing there.") He then attaches the proteins to the end of an optical fiber. The rat-liver substances also bind with a cobra toxin, which Eldefrawi makes fluorescent. The fluorescence gets transmitted along the fiber and measured. When cocaine is present, it knocks the cobra toxin from the liver proteins, reducing the amount of fluorescence through the optical fiber, and the level of reduction indicates the concentration of cocaine.

Highly Specialized

Neither cocaine detector is on the market yet. But Guilbault's company sells a variety of detectors, including one for organophosphates, a

Below: © Jon A. Rembold/Insight; right: Molecular Devices Corporation

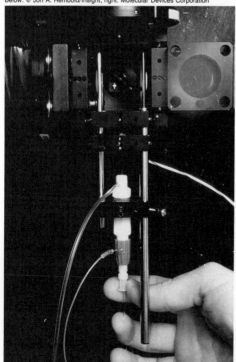

The cocaine biosensor (left) uses a combination of cocaine-binding liver proteins from rats, cobra toxin, and optical fibers to detect the presence of the drug. Biosensor technology has also led to the creation of devices called microphysiometers, which can test the toxicity of various drugs on extracted animal cells (below) rather than on the animal itself.

class of pesticides, and another for nerve gas. "Industry buys the pesticide detectors, and the U.S. Government buys nerve-gas detectors," he says. Another company, YSI Inc. of Yellow Springs, Ohio, is "researching detectors for heavy metals and chlorinated hydrocarbons," says Herbert P. Silverman, vice president for research and development.

The most widely used biosensors are glucose detectors, and these are Guilbault's biggest sellers. "The food industry uses them to fine-tune sweetness of their beverages," he says.

But the most important use of glucose detectors is to monitor the blood sugar of diabetics. Today hospital laboratory instruments can measure blood glucose with an accuracy of 99 percent, says Dr. Leland C. Clark, professor of pediatrics at Children's Hospital in Cincinnati, Ohio, and devices that diabetics can buy over the counter for about $200 are 95 percent accurate.

Some biosensors can be used to detect lactate, a waste that is produced in muscle during intense activity and is eliminated with oxygen. People who are sick enough to be in critical-care units in hospitals sometimes do not get enough oxygen. Today special biosensors that monitor lactate can warn doctors of the problem before it turns into a crisis, says Clark, who invented the first such device that was commercially successful. Lactate monitors also are used in sports medicine, to help athletes tailor their training.

Testing for Toxicity

The most revolutionary concept in biosensors has led to a device that will reduce the need to test the toxicity of drugs and cosmetics on animals, and will also help doctors decide which chemotherapeutic agents to use on cancer patients. For example, some companies still test the safety of cosmetics by using the controversial Draize test, in which researchers flush the eyes of rabbits with the substance of interest. Instead of resorting to this method, researchers could test the toxicity of such compounds on dislodged cells in the new device, which is called a microphysiometer.

The principle behind the instrument is simple. Cells metabolize, just like the animals they come from, says Gary T. Steele, president of Molecular Devices Corp. in Menlo Park, California. Toxic compounds slow down the rate of metabolism, and the microphysiometer can measure metabolic changes as cells are exposed to potentially toxic compounds.

The device is a silicon chip that holds cells. A nutrient medium is pumped over the cells. If the pumping is stopped, the metabolic by-products of the cells acidify the medium. To measure metabolism, the pumping is stopped and the rate of acidification is measured. If a toxin is poisoning the cells, they excrete no waste products to acidify the medium.

"We have tested some compounds that Procter & Gamble sent us, that they normally would use animals to test," says Steele. "We have discovered there is a high correlation between their animal results and what they get on this microphysiometer."

At Stanford University, scientists are testing the microphysiometer to determine its usefulness in prescribing chemotherapy drugs. The problem oncologists face is that the same types of tumors often have very different sensitivities to a given drug. Currently, oncologists try to grow the cancer cells in culture to test their sensitivity, but Kevin Ross, a postdoctoral fellow in the division of oncology at Stanford's medical school, says that many cancers will not grow in culture, and that if they do, their reactions to drugs often change as they grow.

Testing cancer cells' sensitivities in the microphysiometer would be simple and quick. The cells do not have to be grown, but merely placed in the device and tested. "We have cell lines that are resistant to chemotherapeutic agents and ones that are sensitive," says Ross. "We can tell the difference in an hour, whereas any other cell test takes at least three days."

A variety of technical problems must be solved before the device can be used for human tests, Ross says. But he hopes the microphysiometer will improve cancer therapy for many patients. Much sooner, the device will probably begin to replace some animal tests. The company says it will be ready for marketing in about 18 months.

Meanwhile, biosensors are becoming so sensitive that some wonder whether their full capabilities will ever be realized. When it comes to detecting environmental pollutants, for instance, biosensors should have no trouble finding concentrations in the range of parts per billion.

At some point in the development of the devices, says Roger O. McClellan, president of the Chemical Industry Institute of Toxicology, "our measurement technology is probably going to outstrip our ability to understand what these measurements mean."

The 1989 Nobel Prize

PHYSIOLOGY OR MEDICINE

by Glenn Alan Cheney

Deep insights into the genetic nature of cancer brought the 1989 Nobel Prize in Physiology or Medicine to two Americans, Dr. J. Michael Bishop and Dr. Harold E. Varmus. The researchers, both of the University of California at San Francisco, discovered that cancer is generated genetically from within the chromosomes of normal living cells.

The discovery, published in 1976 after years of painstaking experiments, revolutionized the way oncologists look at cancer and follow its development. Until the discovery, it was hypothesized that some normal cells carried not only their own genes, but also viral genes that, when activated by carcinogens, turned the cell cancerous. Dr. Bishop and Dr. Varmus determined that these genes actually originate as part of the normal genetic makeup of healthy cells.

Since the discovery, scientists have identified over 40 oncogenes that seem to lead to tumors in humans. Now that physicians know where to look for the biochemical pathway of cancer development, they can better prevent and treat the disease, as well as identify people more susceptible to it. Subsequent research by the two Nobel winners is probing the genetic code of HIV, the virus that spreads acquired immune deficiency syndrome, or AIDS.

Oncogenes Within Us

Dr. J. Michael Bishop and Dr. Harold E. Varmus, together with colleagues Peter F. Vogt, Deborah H. Spencer, and Dominique Stehelin, started looking at cancer from a biomolecular point of view in the early 1970s. When Bishop and Varmus met at the University of California, they were both fascinated by viruses that were known to cause cancer in animals by inducing normal living cells to change shape and behavior to become tumors.

Such viruses are known as retroviruses because of their backward, or "retro," reproductive process. The retrovirus genetic material is RNA, not DNA. To reproduce, it must make a DNA copy of its RNA structure and insert it into a host cell. That cell then carries out the replication of the virus.

When Bishop and Varmus began their research, retroviruses were known

Harold E. Varmus (left) and J. Michael Bishop won the Nobel Prize for determining that cancer is generated genetically from within the chromosomes of normal cells. Their discovery has led to better ways of preventing and treating cancer.

AP/Wide World

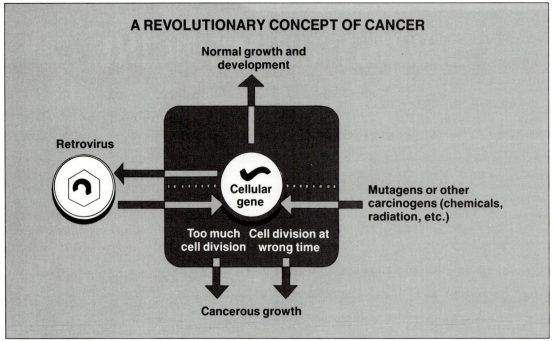

Cancer occurs when a normal gene that controls cell growth triggers either too much cell division or cell division at the wrong time. This happens when the gene's structure is altered, either by radiation, chemicals, or other carcinogens, or by a cancer-causing virus called a retrovirus. When a retrovirus (above left) infects a cell, it picks up that cell's normal gene and carries it to another cell, where it inserts the normal gene into that cell's DNA. With its structure now altered, the gene can no longer properly regulate cell growth; cancerous tumors (below, in white) can result.

to carry viral oncogenes, genes that could infect cells and change their genetic composition. This genetic change was the cause of cancer in animals, but oncologists could find no evidence suggesting that retroviruses led to a similar change in human cells.

The possible connection, however, was too tempting to ignore. When two scientists, Robert Huebner and George Todaro, at the National Cancer Institute suggested that viral oncogenes might lie dormant in all cells, Dr. Bishop and Dr. Varmus decided to investigate.

The hypothesis held that at some point deep in our evolutionary past, our cells became infected with a viral DNA known as provirus. Over the millennia, these proviruses duplicated that DNA structure. Ideally, a cell would live, reproduce, and die without the provirus ever becoming active. But when exposed to a carcinogen—a chemical or radiation—the provirus would "wake up" and trigger cancerous development.

To test this hypothesis, Bishop and Varmus tracked the development of cancer spread by the Rous sarcoma virus in chickens. This was the first reported oncogenetic retrovirus, identified in 1911.

Rous sarcoma has a unique characteristic. As it advances and reproduces, it quickly mutates and loses its power to infect cells and make them cancerous. Dr. Bishop and Dr. Varmus compared the genetic makeup of the Rous sarcoma at its cancer-causing state and at its harmless mutated state, and thus pinpointed the oncogene, the gene that was causing cancer. They dubbed it the *src* (for *sarcoma*) oncogene.

Bishop, Varmus, and their colleagues then prepared a nucleic-acid probe for the *src* oncogene and inserted it into cells from a chicken. They observed that the probe hybridized with the DNA of the cells. The gene that the *src* joined with was a normal and necessary gene found in the genetic material, or genome, of healthy chickens and other vertebrates.

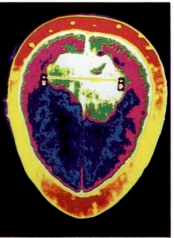

The culprit gene was one that normally controls cellular growth. When damaged, it forces cellular growth to go awry, and a cancerous tumor develops.

The unexpected behavior of the *src* probe revealed that oncogenes lie dormant within the genomes of normal cells. Retroviral oncogenes are versions of cellular genes that mutated or attached themselves to the viral genome. Scientists now call these dormant genes proto-oncogenes. When retroviruses pick up a proto-oncogene, they carry it to other cells as part of their normal reproductive cycle. When the retrovirus RNA produces a DNA duplicate and inserts it into a cell, the cell then replicates it and attaches the proto-oncogene to its own genome.

Dr. Bishop and Dr. Varmus took this information and determined two ways that oncogenes cause cancer. In one, they mutate within the cell's genetic structure and cause too much cellular division or division at the wrong time. In the other, a retrovirus picks up a normal oncogene, carries it to another cell, and there attaches it to the wrong place in that cell's genome, causing abnormal cellular growth.

The day after the announcement of the Nobel award, the U.S. Food and Drug Administration (FDA) approved a diagnostic test based on the analysis of proto-oncogenes. Biotechnologists predict the imminent invention of many such products that use the concept that Dr. Bishop and Dr. Varmus established.

Researchers also foresee the day when oncogenes are understood well enough that their derangement can be detected and then blocked before they do much damage. That, more simply put, would be a cure for cancer.

The awarding of the Nobel Prize to Dr. Bishop and Dr. Varmus was disputed by their former colleague, Dr. Dominique Stehelin, now research director at the Pasteur Institute in Lille, France. He claimed that when he was working with Bishop and Varmus, he had the original idea and performed the experiments.

Other people involved in the experiments confirmed that Dr. Stehelin did participate in some difficult experiments, but only under the supervision of Dr. Bishop and Dr. Varmus, who interpreted the experiments. Jan Lindsten, secretary of the Nobel Physiology or Medicine Committee, stated that, "[Dr. Bishop and Dr. Varmus] are the key persons in the discovery."

Both scientists began their careers studying to become medical doctors.

Dr. J. Michael Bishop, born on February 2, 1936, was the son of a preacher in York, Pennsylvania. At an early age, he knew he wanted to do some good for mankind. His main interest was in history, but he saw more potential as a doctor. At Harvard, he became interested in molecular biology and began studying animal viruses. Though graduating in medicine, he never practiced that profession. He joined the University of California at San Francisco in 1968, and soon began the research that would earn him the Nobel Prize.

Dr. Harold Elliot Varmus, born on December 18, 1939, in Oceanside, New York, was the son of a general practitioner, but studied English at Amherst and 17th-century literature at Harvard before deciding he'd rather study medicine at Columbia University. From there he went to the National Cancer Institute in Bethesda, Maryland, where he researched the production of protein in bacteria. In 1970 he came to the University of California at San Francisco and met Dr. Bishop, who saw him as a freethinker who might someday recognize something new in their research.

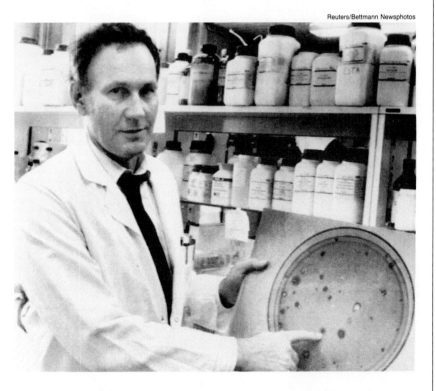

Reuters/Bettmann Newsphotos

Dr. Dominique Stehelin, director of research at the Pasteur Institute in France, disputed the Nobel committee's decision, claiming that he had developed the original idea for the prizewinning research and had performed the experiments. Others involved in the experiments insisted that Dr. Stehelin had worked only under the supervision of Drs. Bishop and Varmus. The Nobel committee declined to change its decision.

COMPUTERS AND MATHEMATICS

Review of the Year

COMPUTERS AND MATHEMATICS

The ever-growing capacity of computers was etched into the national consciousness by the battle between man and machine, played out on the chessboard. Controversy continued to surround the computer in the classroom.

by Philip A. Storey

Marilynn K. Yee/NYT Pictures

Chess champion Gary Kasparov has defeated "Deep Thought" several times, but his reign could be cut short by more strategically advanced computers.

Chess and Computers

Prevailing opinion in chess circles holds that the next world champion may well be a computer. This conclusion persists, despite the defeat of Deep Thought, the premier chess-playing computer, in both games of a two-game match by reigning human chess champion Gary Kasparov in New York City last October.

The stage was set earlier in the year, when Deep Thought emerged as champion at the Sixth World Chess Championship in Edmonton, Alberta. Deep Thought, designed by five computer scientists at Pittsburgh's Carnegie-Mellon University, ranked above 2,500 points in the International Chess Federation rating system, placing it among the top 100 players in the world (human or otherwise). Kasparov ranked above 2,800 points, the highest ranking in history.

Chess-playing computers and computer programs have been around since the 1940s. What's changed is their power: Deep Thought, for instance, can examine 750,000 possible moves per second. Such brute tactical strength lets it perceive the best play for its next five moves. Using programmed logic that eliminates obviously poor moves, Deep Thought can see clearly 10 plays ahead.

Still, tactical prowess does not necessarily make for strategic brilliance. Human players like Kasparov, although lacking the computer's capacity to envision every possible move far into game play, stay ahead of the machines by concentrating instead on general board strength. The next generation of chess-playing computers will likely incorporate this strategy. Feng-hsiung Hsu and Murray Campbell, two of Deep Thought's designers, have already begun developing a computer that will be able to analyze a billion or so moves per second. Such a capacity will let the computer see ahead some 25 moves. At this level of play, computers will likely become competent strategists.

Where will all this leave mere humans like the 26-year-old Kasparov? Experts in both chess and computing predict that the machines will reign supreme at the game by 1995. But others feel that Kasparov could hold on to his title until the year 2000. As for the current champion, Kasparov views the competition as a mission on mankind's behalf. "I had to challenge Deep Thought to protect the human race," he said.

In Pursuit of Pi

With calculating speeds that defy human comprehension, computers have far surpassed us in many endeavors. Take the calculation of pi, for instance. Pi, the ratio of the circumference of a circle to its diameter, represents the value of 3 followed by a nonrepeating decimal. A calculation of pi to eight decimal places—3.14159265—is more than sufficient for most purposes.

The pursuit of pi has a long and even colorful history. By the year 1600, pi was known to 20 decimal places; by 1700 it had been carried to 100 places. Then, in 1949, the ENIAC computer figured pi to 2,037 places. Mathematician Yasumasa Kanada and his supercomputer outdid everybody in 1988 by calculating pi to a record 201 million decimal places.

The record didn't last long. Last year, U.S. mathematician brothers David and Gregory Chudnovsky, using a souped-up calculus equation and whatever supercomputer time they could borrow, figured pi to well over 1 billion decimal places. The equation they used had the added value of making it easier for whoever wishes to carry the calculations still further to do so, a task the Chudnovsky brothers will eschew, as they have grown "somewhat tired" of the exercise.

The passion to calculate pi to seemingly infinite decimal places certainly raises one very legitimate question: Why bother? Mathematicians have their reasons: besides putting supercomputers through their paces, calculating pi to great lengths may help resolve intriguing problems in number theory.

PC Products

In the somewhat less esoteric world of personal computing, great strides were achieved in 1989. Laptop computers, those full-powered PCs that fit into a briefcase, shrunk to even smaller sizes. Two companies introduced models that weigh in at under 6.5 pounds (2.9 kilograms).

What these computers lack in size they make up for in specialization. The tiny machines, some small enough to fit in a coat pocket, are frequently limited to a single task or a single group of tasks. Some function as dictionaries and spelling checkers, while others translate spoken phrases into foreign languages, keep appointment schedules and telephone numbers, or even act as crossword-puzzle consultants.

Perhaps the most revolutionary of the pack is a notebook-size device called the GRiDPad, manufactured by the GRiD company. Instead of a keyboard, the GRiDPad has a flat screen on which the user writes letters and numbers with a metal stylus; the images then appear in a compact-type form on the upper portion of the screen.

Meanwhile, manufacturers of microprocessors (diminutive microchips that control a personal computer's operation) have introduced even more powerful products. One such breakthrough, the 80486 microprocessor from the Intel Corporation in California, holds the distinction of being the first personal-computer chip to contain more than 1 million transistors.

Intel concurrently introduced the 80860 chip, a microprocessor based on Reduced Instruction Set Computing (RISC) architecture. Although less widely hailed than the 80486, the 80860 chip will perhaps prove more important in the long run. Traditional Complex Instruction Set Computing (CISC) chips (Intel's 80486 among them) process a great variety of complicated mathematical instructions. RISC chips work by breaking complex instructions down into smaller, more readily executed chunks. Many computer designers feel that RISC chips will eventually form the foundation from which desktop computers will achieve as much capacity as today's supercomputers.

Grid Systems Corporation

Computers for everyone: Police can store citation forms on an electronic device called the GRiDPad (above). Tourists can use pocket-sized computers to translate foreign languages (below).

Franklin Computer Company

Crisis in the Classroom

The nation's largest association of mathematics teachers has challenged schools over the means by which they develop their students' innate intellectual ability.

In March 1989, the National Council of Teachers of Mathematics proposed a set of 54 standards intended to improve mathematics education in the United States. The council asserted that the technological transformation of society demands new methods of mathematical instruction.

The proposals met with great controversy. According to the council, teachers should emphasize the development of mathematical "thinking skills" and deemphasize traditional rote learning. Students should also develop group problem-solving skills, the council says, that they can apply later on in the workplace. Perhaps the most controversial proposal recommends that students of all grades have free access to calculators and computers.

Some parents and teachers feel that the council's proposals overlook the value of extensive drilling in basic arithmetic skills. Furthermore, they say, students will grow accustomed to relying on calculators and computers to perform basic arithmetic operations for them, even before the children have mastered the skills themselves.

The council responded to these arguments by comparing the use of calculators and computers in mathematics classes to the use of word processors in writing classes. The council referred to these technological devices as "tools that simplify, but do not accomplish the work at hand."

A proposal that students of all ages should have free access to computers has raised concern that children will grow to rely on electronic devices.

Hubert Worley, Jr./NYT Pictures

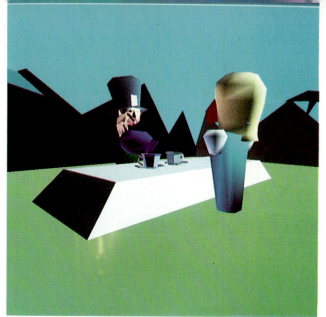

INSIDE ARTIFICIAL REALITY

by Steve Ditlea

You can fly to the moon by pointing your finger. With a flick of your wrist, see the world through the eyes of a child. Reach out and grasp furniture, windows, or walls that exist only within the silicon memory of a personal computer (PC). Wave your hand to create virtual paper on an empty desktop, a simulated skid in a nonexistent car, or X rays of the human body.

Artificial reality was the stuff computer researchers' dreams were made of—until recently. This year it's become . . . well, reality, thanks to a confluence of developments in new technologies emerging from the labs, including lightweight 3-D stereoscopic displays, magnetic positioning systems, advanced graphics chips, continuing expansion of computer memory, and novel interfaces between human and computer.

What's most remarkable isn't that computerized artificial reality exists, but that it is emerging so quickly from the research labs into

Artificial-reality systems allow computer users to experience another world: the world on the other side of a computer screen. By wearing a head-mounted display called an EyePhone and a black Lycra DataGlove (far left), one enters into a three-dimensional "virtual reality" that can be touched and moved around with a wave of the hand. Some advanced artificial-reality systems allow two people to share virtual reality. For example, one participant can play Alice and the other the Mad Hatter in a computerized reenactment of the tea party from Alice in Wonderland *(sequence above). In the last scene (far right) a participant lifts a saucer by means of a DataGlove.*

public reach. Already, prototype systems have migrated from customized graphics workstations to off-the-shelf PCs like the Compaq Deskpro 386/33. Within a year, PC users will be able to design their own artificial worlds for eye-popping presentations and realistic engineering and architectural modeling.

The creative fervor in artificial reality is reminiscent of the atmosphere that prevailed a decade ago, when PCs were beginning to make their mark. A handful of researchers on both coasts are combining home-brewed and professional-grade components to process complex artificial environments. Such virtual worlds are meant to be perceived not only by a person's sight and hearing, but also by touch, shattering the barrier between the computer screen and worlds beyond.

Exploring Places That Aren't There

Within 50 miles (80 kilometers) of one another along the San Francisco Bay are three innovative outposts of research on the frontiers of computerized reality: the National Aeronautics and Space Administration (NASA) Ames Research Center at Moffett Field, VPL Research of Redwood City, and Autodesk in Sausalito.

The NASA Ames Research Center anchors the southern tip of the bay and nearby Silicon Valley to the edge of the 21st century. Here, in a building near one of the world's largest experimental wind tunnels, is the modest lab of the Human Interface Research Branch of NASA's Aerospace Human Factors Research Division. From this room has emerged much of the technology associated with artificial reality.

To experience the NASA Ames personal simulator, you don special headgear: a round metal frame bearing a wraparound visor and a rectangular aluminum enclosure the size of a tissue box. Look into this head-mounted display, and you see a stereoscopic 3-D image of the lab in black and white, complete with walls, checkerboard floor, ceiling, furniture, desktop computers, and equipment racks. When you move your head to the side or up and down, the computer shifts the display to realistically match your point of view.

Next, slip a black Lycra glove attached to strands of black cable onto your right hand. Move your hand to calibrate this DataGlove, then point your index and middle fingers while bending your third finger and pinkie. You see a disembodied image of the glove do the same; the display moves in the direction you indicate. Point straight up, and the room appears to fall away. With your feet still on the ground, you're flying through the air.

Sequence: © George Steinmetz

The Moon in Your Hand

Research scientist Dr. Michael McGreevy, who initiated and guided NASA's journey into artificial reality, is about to make a lifelong dream come true, using this setup to simulate visits to the solar system. The balding, mustachioed scientist has secured funding for a project called Visualization for Planetary Exploration, which could result in virtual environments of the moon and the planets.

"We'll use visual data recorded by space probes and satellites to create computer models of each planet," McGreevy explains. "When we're through, you'll be able to hold the moon or any planet in your hand and point to where you want to go on its surface. The computer will scale the environment back to life size, and you can be virtually present at the indicated location. You'd feel like you were there."

Using the same artificial-reality application, known as virtual travel, you could take simulated trips to exotic locales, faraway resorts, or business meetings without moving from your sofa or desk.

The democratization of space travel is one example of how artificial reality may provide greater access to every individual. "We've shown that a personal simulator is possible," McGreevy says, "but we don't want to be the high priests of artificial reality. We need to democratize the technology because that will unleash thousands of talented people who will collectively develop its rich promise."

McGreevy can be credited with designing and implementing the first practical artificial-reality system on an off-the-shelf computer system (albeit a specialized graphics computer), which included an inexpensive 3-D, head-mounted display.

Considered an essential component of artificial-reality technology by many developers, the head-mounted display was not invented at NASA Ames, but researchers there refined it to a practical size and cost. In 1965 computer pioneer Ivan Sutherland started developing a helmet sporting a pair of CRTs with left- and right-eye views that were adjusted by computer according to the user's head movements. By 1982 Tom Furness of the Wright-Patterson Air Force Base in Ohio had built an elaborate aircraft-training simulator into what became known as the Darth Vader helmet. In 1984 McGreevy asked Furness to sell him a helmet for his experiments, but was told it would cost a cool $1 million.

A Lot Cheaper Than Darth's

By cannibalizing two $79.95 Radio Shack pocket TVs with black-and-white LCDs and placing them in a $60 motorcycle helmet, McGreevy and hardware contractor Jim Humphries were able to assemble the first NASA Ames head-mounted display for a lot less—under $2,000.

NASA's government-funded lab was later able to build on early artificial-reality work at Atari Research, where an affordable head-mounted display, a glove input device, and educational and game software were developed but never marketed. After the first great video-game boom went bust, sending Atari into decline, several of its researchers joined forces with NASA: hardware "imagineer" Scott Fisher, software whiz Warren Robinett, and—under contract with VPL Research—glove inventor Thomas Zimmerman. Fisher is recognized as having brought the glove—now also considered of vital importance for interacting with artificial reality—to the NASA Ames personal-simulator system.

Safer Than Being There

According to Robinett, the NASA bureaucracy was confused at first about the usefulness of the Virtual Interface Environment Workstation (VIEW), the official name for its artificial-reality system. Originally it was seen as just a way to display operational data, essentially a virtual instrument panel. Notes Robinett, now a project manager for artificial-reality research at the University of North Carolina: "The real application for the VIEW system had to be telepresence—artificial presence in hazardous environments like space. With the head-mounted display hooked up to remote cameras, and the glove controlling robot arms, you've got what we call 'telerobotics.' For a lot of uses, it's safer than being there. You can assemble a space station without risking astronauts' lives."

With video-camera input or computer-generated graphics, telerobotics could also be useful at the bottom of the ocean or in handling nuclear materials. Microtelepresence, a variation using optical or computer-aided magnification, would allow the manipulation of materials at a microscopic or even molecular level. (This technology should not be confused with computer-generated simulations of molecules, an area that was pioneered at the University of North Carolina and is now being tested for the modeling and synthesis of drugs.)

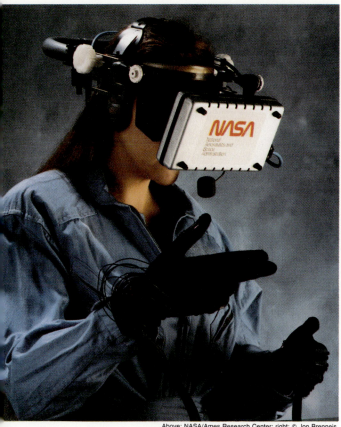
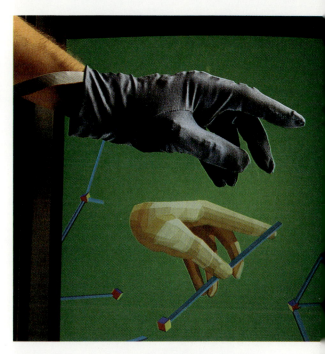

NASA's Virtual Interactive Environment Workstation (left), or VIEW, allows users to experience such remote and potentially hazardous environments as the space station or the Moon's surface without actually being there. The glove is calibrated so that a movement by the gloved hand is replicated instantaneously by a disembodied image of the glove on the screen (above).

Above: NASA/Ames Research Center; right: © Jon Brenneis

At NASA Ames, Fisher and colleagues McGreevy, Humphries, and Dr. Beth Wenzel are now perfecting the next-generation VIEW system. When completed, the system will show stereoscopic images in color or high-resolution black and white that are generated by computer, supplied by video cameras, or replayed from videodiscs. The updated VIEW system will also incorporate 3-D sound positioning and speech recognition and synthesis.

Virtual Beauty Contests

To see the hottest advances in artificial reality, you once had to be privy to government, university, and private research labs, but on June 7, 1989, computerized environments went public. Two firms commercializing the technology held demonstrations in San Francisco and Anaheim. VPL Research proclaimed the occasion a holiday, Virtual Reality Day. Declared its press release: "Like Columbus Day, VR Day celebrates the opening of a new world. VR Day will be celebrated every year with a parade and virtual beauty contest inside Virtual Reality."

Despite such whimsy, big business—as represented by Pacific Bell—is taking artificial reality seriously. In San Francisco's Brooks Hall, at the regional-telephone-company-sponsored Texpo '89 exhibit of telecommunications products, a place of honor is reserved for VPL Research's demonstration amid the applications that will be possible once broadband fiber-optic phone lines connect offices and homes. On display is the world's first shared virtual reality. Side-by-side color monitors suspended above eye level show two views of a colorful computer-generated child-day-care center, as seen from the points of view of two participants seated below the screens, each wearing 3-D goggles and a DataGlove.

Transmitting image data over fiber-optic cable, users in separate cities could manipulate the same computer-mediated environment—a welcome development for an architect and client, say, with offices in different parts of the country. For Pacific Bell, shared virtual reality carries the allure of profits from the use of its extensive fiber-optic-cable networks.

Jaron Lanier, the founder of VPL Research in Redwood City, California, and a pioneer of artificial reality, envisions the creation of virtual worlds where "the entire universe is your body and physics is your language."

To enter this world of virtual reality, you put on a VPL DataGlove and a VPL-manufactured head-mounted color display called an Eye-Phone. You can see the day-care center's interior, in three dimensions, complete with doors, windows, furnishings, and a mannequin representing your colleague. Look down to see your own effigy and a disembodied hand floating nearby. To move yourself around, flex your index and middle fingers and then simply "let your fingers do the walking."

If you want to change the placement of the water fountain, reach out to it and make a fist; grasp and move it, then flatten your hand to release it. This kind of tactile interaction is one way artificial reality differs from computer simulations of the past. Experience the day-care center from a six-year-old's eye level; to get short, point down with your little finger. To get big again, point up.

Reality Built for Two

Presiding over this demonstration is Jaron Lanier, founder and CEO of VPL Research, and originator of VR Day. Clad in a black tunic and topped by blond dreadlocks, he is introducing the first commercial shared virtual reality, RB2, which stands for Reality Built for Two. According to Lanier, "the essence of virtual reality is that it's shared."

Even with sophisticated graphics workstations powering RB2, its synthetic environments are hardly detailed enough to be confused with the real world. And Lanier's "first new level of objectively shared reality available to humanity since the physical world" has other drawbacks: by the end of the first day's showings, one of the system demonstrators is experiencing motion sickness and has to take Dramamine before the next day's session. Such "simulator sickness" is common in flight training, where the body reacts to conflicting sensory cues from simulation and reality.

Virtual Juggling

But the 29-year-old Lanier remains an effective proselytizer for artificial reality. Having learned to juggle by manipulating virtual objects in slow motion with a pair of DataGloves, he argues persuasively for tennis tournaments played across continents in virtual reality. Having worked closely with Greenleaf Medical Systems of Palo Alto to develop medical diagnostics of hand impairment using the DataGlove, and with Dr. Joseph Rosen, who is researching surgical simulation, Lanier convincingly describes the possibilities for virtual-reality-aided surgery with CAT-scan data superimposed on a patient's body to locate internal organs.

Another veteran of game design at Atari, Lanier may have the most vivid imagination among the current practitioners of artificial reality. While putting the final touches on the day-care-center demo, he was also creating a more fantastic virtuality: a giant birth canal through which you travel and are born, only to find that you are the virtual woman giving birth to yourself and are reborn again and again (some of the women at VPL find this excessive and feel that once is quite enough).

Lanier consistently reaches beyond the obvious for virtual worlds. "Everybody wants to see Toontown, like in *Who Killed Roger Rabbit?*," he remarks, "but how about a virtual mirror in which you can see yourself change species? In virtual reality you can visit the world of the dinosaur, then become a *Tyrannosaurus*. Not only can you see DNA, you can experience what it's like to be a molecule. In virtual reality the entire universe is your body, and physics is your language.''

On VR Day in San Francisco, reporter Nancy Steidtmann asks Lanier if he's afraid someone may steal his technology and misuse it. ''Sure,'' he answers, ''but I can't let that stop me. I can't worry about not doing things.'' He adds, ''We're creating an entire new reality. It's too big for any one company to own.'' Famous last words?

Cyberspace on a Desktop

The current activity in perfecting artificial reality, much of it in California, is not unlike the development of the movies in Paris at the turn of the century. Two apparently opposite tendencies emerged then: a realistic, documentary bent, characterized by the Lumière brothers, who filmed scenes of everyday life; and an imaginative, fictional inclination, exemplified by Georges Méliès and his fancifully staged *Voyage to the Moon*. More recently, filmmaker Jean-Luc Godard argues that the two types of movies were much closer in execution than historians would have us believe. If Jaron Lanier, with his hallucinatory environments, is the Méliès of artificial reality, then the developers at Sausalito-based Autodesk, working within the bounds of realistic computer-aided design software, are virtual reality's counterparts to the Lumière brothers. Yet by drawing inspiration from the work of science fiction author William Gibson, they, too, are crossing the line between reality and fantasy.

On Virtual Reality Day in Anaheim, just a gnome's throw from Disneyland, the members of the Autodesk Research Lab are demonstrating their version of artificial reality for invited guests in a hotel suite. One by one, visitors navigate within either of two colorful virtual worlds: a cityscape with a few buildings at an intersection surrounded by fields, a lake, and clouds; or a room interior complete with table, chair, and books that can be ''grasped'' and moved around. While these are strictly solo experiences rather than shared realities, they, too, rely on the VPL EyePhone and the DataGlove as the connection between human and computer.

Like all new technologies, artificial reality will waste no time finding consumer applications. As the software becomes more refined, artificial reality systems will likely see great success as a component in toys and games.

Autodesk, Inc.

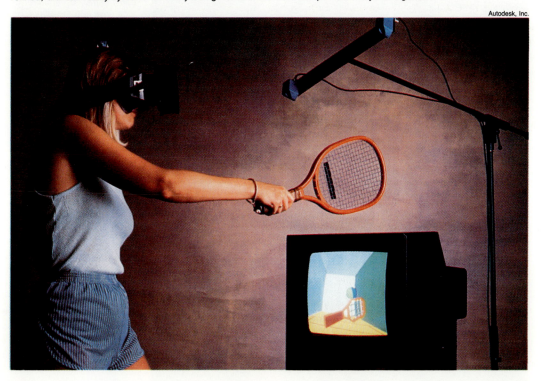

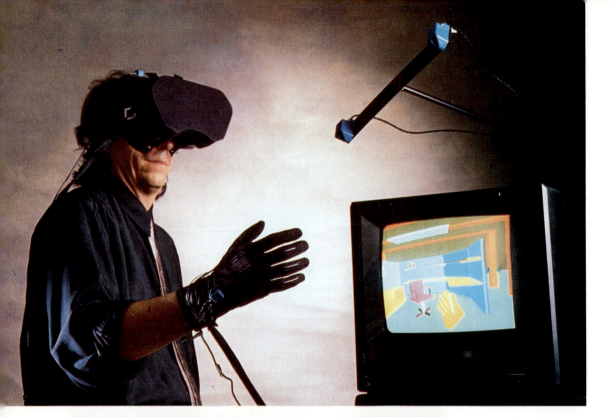

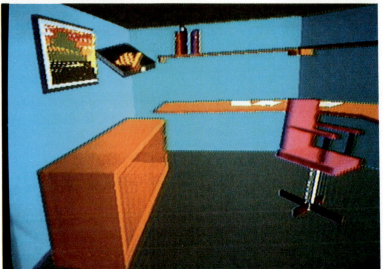

Builders and architects can use artificial-reality systems to give a potential client (top) detailed "tours" of proposed projects. Once inside a "room" (center), the client can rearrange furniture or otherwise alter the layout. He or she then simply opens the door and "enters" the next room (bottom).

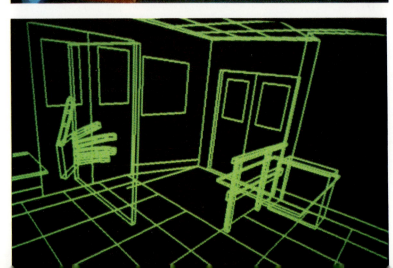

Architectural Walk-Through

Since Autodesk's mainstay design program is the world's most widely used CAD software, engineers and architects will soon be able to take advantage of artificial reality. "You'll be able to create a building design and move through it in any direction," says William Bricken, lab director at Autodesk. "To reposition a design element like a window, you'll be able to reach out and shift it."

Achieving architectural walk-through has long been a goal of artificial-reality researchers. The most advanced work is being done by Fred Brooks' team at the University of North Carolina, where a deluxe system shows high-resolution depictions of floor plans and 3-D, full-color interiors, complete with real-time adjustment of the position of the sun and the ratio of direct/ambient light. To navigate through such a virtual structure, you can use velocity-modulating joysticks (in what is called a helicopter metaphor), move a head-mounted display (the eyeball metaphor), or, in the closest thing to being there, walk on a treadmill you steer using handlebars (the shopping-cart metaphor). Because of hardware limitations, Autodesk's virtual-reality program will be less detailed and somewhat slower in real time, but it nonetheless represents a major advance in PC-based software.

In the absence of a preexisting need or market, Autodesk's attempt at making a product from artificial reality is a gamble. Who will spend $3,000 to $4,000 on a software package requiring a top-of-the-line PC and $20,000 in extras? Current AutoCAD users in architecture, design, and manufacturing may want to use artificial reality in demos, presentations, and sales pitches. Also expected are a new breed of virtuality hackers, using the Autodesk program to create their own alternative realities.

Autodesk's product, due on the market by mid-1990, may be called Cyberspace, in homage to the term coined by William Gibson in his cyberpunk novel *Neuromancer*. Autodesk has filed for trademark protection—as a name for software—of the word, which Gibson uses to describe the nightmarish universal computer data network of tomorrow.

Given to such punning coinages as Cyberia and cyburbia, Autodeskers may have gone a little overboard. Company founder John Walker, who initiated the firm's foray into virtual reality with his internal paper titled "Through the Looking Glass," admits the unsuitability of *cyberspace* and other terms for artificial worlds beyond the screen. Gibson himself is amazed. *Cyberspace* "is the last term I expected to catch on," he admits. "When I came up with it, I remember thinking that it was just too obvious. It didn't seem clever enough."

But What Will It Do for Me?

At work on two film scripts and a novel set in a Victorian England in which Charles Babbage succeeds in building the first computer, Gibson confesses that he's not curious about current developments in virtual reality. "When you've imagined it in an evolved state," he says, "the art isn't interesting. Besides opening things up for paraplegics, I don't see what artificial reality will do for anyone. It won't help save the rain forests; in that respect, I would prefer seeing cold fusion happen. Artificial reality will just be a gadget for rich countries, for military applications. At the low end, it will just end up being used for better Nintendo games."

With an input device like the VPL Data-Glove priced at $8,800, artificial reality should remain a rarefied pursuit for some time. As of this writing, only a thousand or so people have experienced artificial reality directly. Thousands more have experienced Myron Krueger's Videoplace in museums and exhibits, but they had little idea what they were seeing.

At the same time, the ability to plunge into virtual worlds beyond the computer screen is so compelling that the market can hardly wait for real-world technology to catch up, and Gibson's vision may already be turning to fact. Artificial-reality components may soon find their way into as many as 1.5 million homes, thanks to a DataGlove-like plastic glove and a competing flat-panel sensor device, selling for $80 to $90 apiece as Nintendo computer-game-system peripherals. The orders are in from Toys "R" Us and other retailers: Mattel's Power Glove and Broderbund's U-Force are expected to be big hits. At first they will be rather expensive Nintendo joystick replacements, with the accompanying virtual-reality software expected sometime this year.

With 20 million home Nintendo systems, projections show an aftermarket for 7 million artificial-reality add-ons. Even with deep discounts, a half-billion-dollar U.S. artificial-reality industry seems inevitable within a few years. And that's just for lowest-common-denominator video games. PC hardware and software entrepreneurs, take note.

AMERICA'S BICENTENNIAL COUNTDOWN

by Glenn Alan Cheney

Are you an Eskimo? How much did your family earn last year? Is your home connected to a sewer? Do you have a home? Are you alive?

Uncle Sam wants to know. It's that time of the decade again, time for the decennial census. On April 1, 1990, for the 21st time, the federal government started counting every living individual in the United States and its territories.

It's a tremendous task, logistically as complex as moving an army, statistically as tricky as counting what cannot be found. The Bureau of the Census has printed, distributed, and is collecting and evaluating some 250 million questionnaires. It has sent over 106 million questionnaire packages to homes from Mars Hill, Maine, to American Samoa.

Some 1.6 million people have been tested for 635,000 temporary jobs involved in collecting demographic data. Some 70 million questionnaires flooded the Census Bureau within two weeks after the April 1st Census Day. The total cost of planning, executing, and providing results for the census will probably reach $2.6 billion, more than the gross national product of Iceland.

The Census Bureau admits the impossibility of a perfect count. But statistical techniques and local-government involvement should help make this year's results the most accurate in U.S. history.

The degree of overall accuracy, however, does little to assuage those included in groups less accurately tabulated. While it is relatively easy to locate and count white middle-class Americans, it is much harder to find everyone living below the poverty line. The uncounted tend to be minorities, illegal immigrants, and the poorest-of-the-poor, the homeless.

The 1990 census marks 200 years of census collection. The census is a rare instance of the federal government methodically probing the lives and living conditions of the general American populace. The government is required to do so by Article 1, Section 2 of the United States Constitution.

Why Count?

The primary purpose of the census is to ensure that citizens have equal representation in the House of Representatives and that no state has a disproportionate number of representatives.

To keep representation as close to equal as possible, state governments use the results of each census to redefine federal congressional districts. By changing the boundaries of the districts to include more or fewer people, they bring to constituencies reasonable equality.

The intent is simple, but the calculation is not. It would be easy if we could just divide the number of people in the United States by the number in each state and apportion districts accordingly. The problem is in the inevitable fractions left over in such a simple calculation.

Various methods of figuring apportionment have been devised and deployed. The most recent, called Equal Proportions, has been in use since 1941. It assigns each state a priority value calculated by dividing the population of the state by a divisor. The formula is $P/\sqrt{n(n-1)}$ where P is the state population and n is the number of seats the state would have if it gained a seat in the reapportionment.

Census information also determines how the federal government doles out funds to states and municipalities. Census results identify what areas need child-assistance programs, more low-income housing, more federal grants, and so on. These figures determine the spending of an estimated $100 billion.

State and local governments use census information to redistrict for state and local governments, to assess transportation systems, modify school districts and educational programs, and assess the need for state and municipal bonds.

Private companies also use statistical data based on census information. Marketing companies can look at income and housing information to find markets for certain products. Real-estate

Every ten years, the federal government counts all the men, women, and children in the U.S. and its territories. With its extra emphasis on minorities, immigrants, and the homeless, the Census Bureau expects the 1990 results to be the most accurate in 200 years of census-taking.

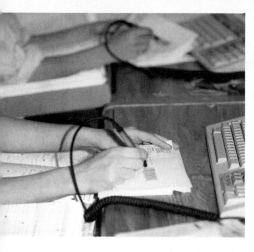

The Census Bureau strives for the fastest, most accurate results possible. When a completed form reaches a census office, workers scan the answers using grocery-store-style "wands" (above). Data processors store the information in the bureau's computer system (right).

Left: Mark Mangold/Census Bureau; above: © Brig Cabe/Insight

developers can analyze migration and population trends to deduce where people will want to be living in the future.

Historical Perspective

The technology needed to count Americans and process information about them has had to advance as fast as the size of the population. In 1790 U.S. marshals oversaw the census in each state, receiving anywhere from $100 for small states such as Rhode Island and Delaware, to $500 for Virginia, the most populous state. The actual census takers, who held the title of assistant U.S. marshal, earned a dollar for every 150 people counted in rural areas or 300 in urban areas. They went from house to house to count five categories of people: free white males over the age of 16, free white males under 16, free white females of any age, other free persons, and slaves.

The assistant marshals noted the information on pages, called schedules, with horizontal lines representing each household, and vertical columns for each category of information. Then, with pen and paper, they tallied up the results and sent the totals to the U.S. marshal for their state, who consolidated the data and sent it over to the census office in Washington, D.C.

The total population that year, including slaves and other persons, was 3,929,214—a bit short of the 4 million Thomas Jefferson predicted. He suspected that many people had lied in order to avoid taxes.

Every census asked more people for more information. In 1800 it wanted the ages of white females. In 1820 it even wanted the ages of non-white females.

By 1840 the population reached 17 million. Census forms in 1850 had a horizontal line for each individual rather than each household. By 1870 census takers, now called enumerators, were asking about the value of personal estate, school attendance, literacy, and handicaps. They even determined that there were precisely 473 Americans employed as mule packers.

The piles of data grew so high that the results of the 1880 census were not released until 1888, just two years before the next census. Clearly the Census Office needed to find a better way.

Punched Cards

A major advance in data processing came from a recent graduate of Columbia University's School of Mines, a young man named Herman Hollerith. At the age of 20, he joined the Census Office to collect statistics on steam- and waterpower. But upon seeing the overwhelming job of tabulating the 1880 census results, he ended up more interested in solving the data-processing problem.

Hollerith invented a machine that could read and tabulate data directly from a sheet of paper. That sheet was actually a long roll of paper with the familiar horizontal lines of information. The big innovation was that the infor-

To simplify the seemingly overwhelming task of counting hundreds of millions of people, the Census Bureau has divided the nation into 7.5 million residential blocks. Computers can produce an electronic map of any given block (right). The vast amounts of data maintained in the bureau's tape library (above) allow government statisticians to compare and correlate past census figures with the 1990 results.

Top: Mark Mangold/Census Bureau; above: © Jon A. Rembold/Insight

mation was not written, but rather punched in particular spots, leaving the paper riddled with holes. Each hole was a bit of information. This paper could be passed over a metal drum and under a set of wire brushes connected to an electric power source. When a brush passed over a hole in the paper, it touched the drum, completing a circuit and advancing a counting device. The bulk of the machine was a bank of counting dials, one for each category of information; that is, each vertical column on the paper.

Though the principle would prove invaluable, Hollerith's first system was overly awkward. To find a bit of information or separate a category of information, the machine would have to run through the entire sheet of paper.

The principle, however, inspired a more efficient machine. It used punched cards—a principle still used in data processing today. In retrospect, it is not surprising to find that young Hollerith later founded a company called International Business Machines (IBM).

The punched-card system offered a significant advantage. It allowed the machine to separate a batch of cards—that is, a batch of individual people—according to a given category of information. Given the batch that included, say, the residents of Baltimore, the machine could count and separate female residents in various age groups, or residents in various income brackets, or females of a certain age in various income brackets. In minutes a single machine

COMPUTERS AND MATHEMATICS 113

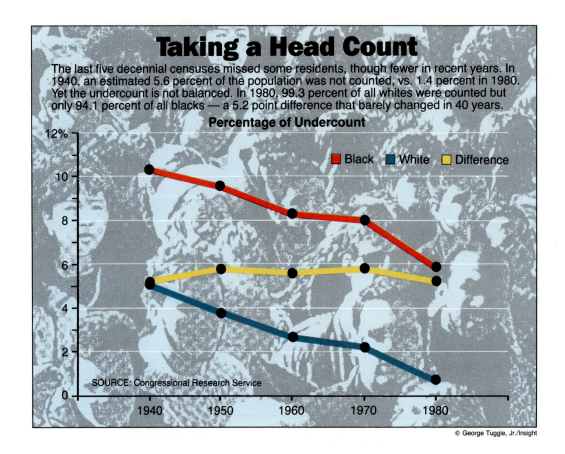

© George Tuggle, Jr./Insight

operator could consolidate information that would otherwise take several people several days.

Computerized Counting

The 1940 census was the last processed without electronic computers. In 1950 the new and, by today's standards, huge Univac I processed about 20 percent of the data on punched cards. In 1960 a Film Optical Sensing Device for Input to Computers (FOSDIC) read census schedules from microfilm and transferred the data to magnetic tape.

In 1990 the Census Bureau is trying to collect most data by mail, with enumerators gathering information in person only at homes that fail to mail in their questionnaires. Field offices will use some 570 minicomputers, each with far more computing power than Hollerith's closet-sized machine, yet still using similar binary bits of data.

To make more sense of the census, the Commerce Department, which oversees the Census Bureau, has developed a vast computer file known as TIGER—"Topologically Integrated Geographic Encoding and Referring" system. This file is an automated digital description of every road, river, railroad, and political boundary in every state and territory of the nation. In essence, it is an electronic map of the places where the census is taken.

The TIGER File produces maps that outline the areas that each of the bureau's 300,000 field enumerators cover. Each area is considered a "block." A block may be a few city blocks, a ZIP-code area, or an area bordered by certain rivers, roads, or other clear-cut boundaries.

The census has been using the block concept for many years. Before the flexibility of an electronic map, however, any change in boundaries, or any combination of different blocks, would lead to a tremendously time-consuming recalculation. With all this data stored in digital form, the Census Bureau and other users can quickly and immediately gather the specific selection of demographic data they need.

In 1988 the Department of Commerce released a prototype of the TIGER File for Boone County, Missouri, where the bureau had performed a dress rehearsal of the 1990 census. The rehearsal gave future users of the TIGER File an opportunity to develop software that could access and manipulate the file. The Census Bureau is providing no application software, thus allowing the private sector to develop software products that meet the needs of the market.

Private users of the TIGER data, such as marketing companies and real estate developers, can buy TIGER Files on 9-track, 1,600-bpi or 6,250-bpi computer tape, ASCII or EBCDIC. They can request files for individual counties, whole states, or even the whole country. A single county file costs $190, with additional counties in the same state costing $15. The data set for the entire United States, including territories, is $57,785.

Finding the Unfindable

Despite the technological advances, the bicentennial census, as previous censuses, will not be able to account for each and every resident of the United States. Census enumerators have a hard time finding people who don't live in traditional family situations in traditional households. The poor, the homeless, and immigrants living here illegally are especially hard to count. The poor living temporarily with friends or un-

The percentage of uncounted individuals has decreased with each new census since 1940 (facing page). This year, with special enumerators seeking out the homeless (above) and other difficult-to-locate persons, the reapportionment of congressional seats (below), carried out in accordance with the 1990 census results, should accurately reflect the nation's population distribution.

Above: © Eric Kroll/Insight; below: Copyright © 1989 by The New York Times Company. Reprinted by permission.

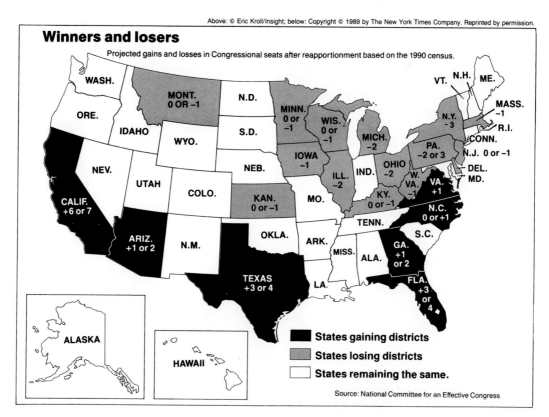

der bridges or in cars, abandoned buildings, campsites, or public shelters have no address for the census to call on. Sections of some ghettos where crime is rampant and English not spoken remain virtually impenetrable to the census.

The bureau estimates that its 1980 population total of 226,545,805 was probably low by about 1.4 percent, assuming a certain amount of double counting counterbalanced the noncounting. This was a substantial improvement over the 2.9 percent inaccuracy of the 1970 census. Nonetheless, 1.4 percent of America equals 3.2 million people.

The bureau was besieged by 37 lawsuits demanding upward adjustments of the 1980 count. State and local governments pointed out that each individual in their area was worth about $150 in federal funds. If tens of thousands remained uncounted, these governments would be shortchanged by millions of dollars.

Ironically, a good deal of that federal funding would be earmarked to help those uncounted people who need it most. While the 1980 census may have missed fewer than 2 percent of white males in the 40-to-49 age group, it may have undercounted black males of that age by almost 20 percent. The figure for homeless people and illegal immigrants was probably even higher.

One result of these widespread complaints has been the fine-tuning of data-collection and statistical-estimation techniques. The bureau is able to estimate its inaccuracy by comparing its count with the records of past censuses, Medicare enrollment, Social Security, immigration and emigration, births and deaths, and other sources. At best, however, such comparisons are but an estimate of the actual numbers.

The bureau also applies a technique known as "dual-system estimation," which is a variation of a method used to count wildlife. Biologists count animal populations by capturing and tagging a species inside a given area, then recapturing and counting tagged and untagged animals. Under ideal conditions, as, say, that of fish trapped in a lake, one could tag a sampling of 100 fish, and then, later, catch a sampling of 50 fish to see what percent were tagged. If only 10 of the 50—one-fifth—had tags, then it could be calculated that the original catch of 100 was one-fifth of the total population.

This system works well when the sampling is truly random and the target population stays in one place. But people are not fish. Those missed in the first count because they lived in storm drains or abandoned cars would probably not be counted in the second sampling either. At best the second count would detect those who were moving to a new residence, on vacation, or otherwise temporarily uncountable.

After considerable debate over whether to adjust the 1980 census to account for the uncounted, it was decided that the bureau should adjust neither the 1980 nor 1990 census. It was felt that it would be better to aim for a perfect count and fall short than to rely on statistical guesswork to support a count less accurate than it should be.

To get an accurate count of the people most often overlooked, enumerators will have to be scrupulous, creative, and courageous. In black ghettoes, white enumerators might be mistaken for undercover police. In some neighborhoods, drug addicts might see them as targets for robbery. In some houses, it may not be clear who is actually living there.

The foreign-born may not be willing or even able to answer questions accurately. It is estimated that in the 1980 census, as many as 40 percent of recent immigrants falsely claimed to be naturalized citizens. An estimated 2 million illegal immigrants claiming citizenship probably led to California and New York improperly gaining congressional seats. Whether those who are counted are legal residents or illegal immigrants, critics question whether noncitizens are entitled to the representation that redistricting apportions them.

To count the homeless, enumerators will go to public shelters. They will ask special questions that will help formulate a more accurate profile of how such people live. People found sleeping on the street will be counted, too, though they will not be awakened. Enumerators will simply make estimates of age, sex, and race. Sleeping persons completely covered will be simply counted as an individual; a computer will assign them characteristics according to demographic probability.

The 1990 census will give the federal government a better idea of how the U.S. population will look in the next century. It is expected that the population will stop growing at close to 300 million in 2040, and then will decline slowly. The percent of white people and young people will decline, while the number of blacks and other minorities, and people over 65, will continue to increase. With accurate census data, the government can begin to prepare to project shifts in population centers and changes in the national economy.

SMART TV

by Brenton R. Schlender

Technology that combines TV, VCRs, optical disks, and personal computers into a single system may soon revolutionize the electronics industry. Such a system might allow home viewers to add their own twists to a movie's plot (above).

If you think television sets and computers dominate our lives already, wait till you see Andy Hertzfeld's new toy. The impish computer hacker is best known as a key software designer for Apple's innovative Macintosh personal computer. Now he works for a Silicon Valley start-up company called Frox. He is one of a group of engineers who have built a prototype of what he calls "an information center for your house that also happens to be the world's greatest TV and stereo."

Photo: © Steve Cooper/Discover Publications; arm: The Stock Market

Frox's home-entertainment system is clever enough to read the TV listings and pick out and record programs Hertzfeld might want to watch later—and it can edit out the commercials to boot. It also catalogs and plays his compact disks (CDs) on command, and shows on the TV screen the cover art and liner notes for each disk. The TV is the focal point of the system, but what makes it all work is a built-in computer as powerful as an engineering workstation.

Soon the machine will simultaneously monitor electronic databases for news or other information of particular interest, answer the telephone, watch for incoming electronic mail,

Andy Hertzfeld's Frox home-entertainment system reads TV listings and selects and records programs for later viewing—with all commercials edited out. The system can also play compact disks on command and, if requested, display the CD's cover art and liner notes on the screen.

© Ed Kashi

and control additional home appliances even as it runs the TV or stereo. In essence, the Frox machine is an ambitious effort to give the boob tube some real smarts.

Don't rush out to buy one for Christmas. The Frox system isn't supposed to make it to market for two years. If and when it does, at first it will probably cost—gasp—$10,000. (As with most electronic marvels, the price is sure to head downward if it catches on.) It's a harbinger of a whole new genre of electronic devices arriving in the next few years that will blend the realistic, compelling, moving imagery of TV with the brains of computers. Already, titans of both the computer and consumer-electronics industries—as well as hungry smaller companies like Frox—are plotting what shapes these new machines will take. The big lure is the prospect that the new hybrid technology will bring powerful computers into the home at long last.

Like Frox, other consumer-electronics companies, most of them Japanese, are working on ways to make TVs more intelligent and versatile so the viewer can take better advantage of the plethora of programming available through the airwaves, over cable, and on tape and laser disk. Both IBM and Apple, the PC kings, are already touting "multimedia"—the combination of text, sound, and graphics—as the wave of the future in personal computers. They are working on the TV-PC matchup from the opposite direction by putting full-motion digital video into their already brainy machines in order to make them more engaging and entertaining. The goal: desktop video computers that users interact with, not merely another box for the couch potatoes to sit and stare at.

These video computers would usher in video encyclopedias and other interactive educational and training tools. They could read and display patterns of stock-price quotes, and would make possible hundreds of new and elaborate computer games. Ultimately, just about anybody will be able to create electronic productions that mix snippets of moving video and sound with conventional text and computer graphics.

Grand Vision

For example, you could write your mother a letter including video highlights of your daughter's birthday party or your trip to Europe, with commentary dubbed in. You would mail it to her on a single computer disk—or, better yet, transmit it to her computer almost instantly over tele-

The marriage of audio and video in multimedia technology may soon lead to entirely new methods of composing music.

phone lines. One day, video computers may even act as the futuristic "videophones" that telecommunications companies have promised for decades, but never really delivered.

"It's time to change the face of information in the home," declares Nicholas Negroponte, director of the Media Laboratory at the Massachusetts Institute of Technology (MIT). The Media Lab, the world leader in research on the technology of mixed media, is organizing an American, European, and Japanese consortium of big-name computer, consumer-electronics, and telecommunications concerns to set standards for the television of tomorrow. Says Negroponte: "Our ultimate goal is to make personal computers and televisions as one."

Plenty of technological and marketing hurdles must be jumped for Negroponte to realize his grand vision, or for IBM, Apple, and the TV makers to carry out their more modest plans. For one thing, the nature of video imagery requires vast amounts of data storage space, computing power, and memory. That means video computers or smart TVs will be frighteningly expensive at first, and hence will probably be aimed initially at business and education rather than home users.

Moreover, many experts aren't so sure that ordinary people can master the exacting techniques necessary to put together a comprehensible video program, even if it's just an edited home movie. Indeed, some contend that most people would really need or want only a "multimedia player"—a TV or computer that allows them more control over prerecorded, professionally produced interactive video programming.

Despite the technical obstacles and marketing debates, manufacturers are pressing ahead. IBM and Apple are already sniping at each other over what form desktop video will take. IBM contends that people want players mainly for preprogrammed material, while Apple believes that computer users will want to "roll their own." IBM, which entered the PC business only after Apple pioneered it, badly wants to be first out of the chute with PCs that offer digital video. Analysts expect the company to unveil a production model within the year.

IBM isn't discussing its plans for a multimedia computer in any detail yet, but concedes it is working closely with Intel Corporation, maker of the microprocessor chips that are the brains of IBM's previous PCs, and Microsoft Corporation, designer of the operating-system software that controls them. The initial goal, according to Intel and Microsoft officials, is to build a computer costing less than $3,000 that functions as a conventional PC, but incorporates color graphics, stereo sound, and loads of memory. It would include a special disk drive that can read CD-ROMs—compact disks permanently etched with sound, still images, or ordinary words and numbers.

COMPUTERS AND MATHEMATICS

Intel is also developing a technology called DVI, for Digital Video Interactive, that will allow IBM's computers to convert video images from cameras, tapes, or laser disks into digital data that can be speedily edited, enhanced, and manipulated.

Visual Cartwheels

This means that at first the IBM machine will be primarily a player of interactive media, allowing the viewer to choose among programming from many sources, but not to produce his own. It will display only still images, but the user will be able to make them expand, contract, move across the screen, and generally do visual cartwheels. "People will find that they couldn't care less about full-motion video, and that still video images will give them a lot to work with," insists William Spaller, director of advanced planning for IBM's entry systems division, and the man in charge of developing the new machines. Consequently, he says, at the outset, full-motion video will be optional.

The dawn of this new industry shows IBM and Apple awakening to oddly unfamiliar roles. IBM has the home as much as the office in mind as the primary market for its new machines. That's something of a switch, given the company's original name and historical orientation. Conversely, Apple, which started out building PCs for hobbyists, sees desktop video as strictly for business and schools, at least until late in the 1990s, mainly because of cost.

Even then, Apple officials don't seem quite so optimistic as IBM's that video computers will displace conventional TVs. "The desktop video computers will require a completely new media form that is very different from passive TV," says Apple chairman John Sculley. He adds, however, that the marriage of computers and TV could give television "a second chance to live up to its potential."

Apple remains skeptical that digital video technology in its present form is particularly useful. "We don't want to make the equivalent of a Swiss army knife with nifty, underpowered features that can't do any real work," says Jean-Louis Gassee, president of Apple's product division. "We'll give people an affordable way to send Mom a video letter, but not until the technology is good enough."

Video Compression

The trick to making digital video work in a personal computer is a technique called video compression. The process relies on complex mathematical formulas and special high-powered processor chips that reduce the amount of raw data needed to draw video images. Compression systems look at video pictures frame by frame to identify consistent shapes and patches of uniform color that don't move much, and then ignore the redundant data in each frame.

Without compression, a standard compact disk, which holds 75 minutes of music or 30,000 pages of text, can store only seven minutes of

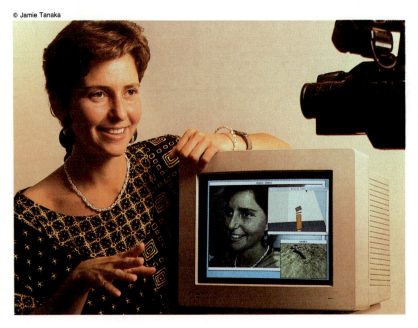

© Jamie Tanaka

Combinations of personal computers and video technology have emerged as powerful educational tools. At left, the speaker shares the screen with the appropriate computer-generated graphics and/or animation.

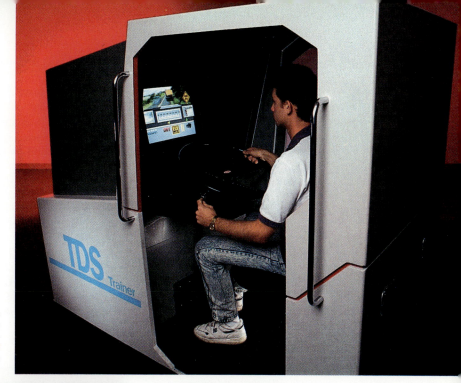

Du Pont uses interactive video systems to train its truck drivers (right). The system simulates various road conditions to test the reflexes of the trainees (below).

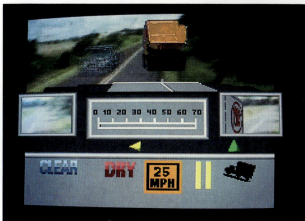

Both photos: © John McGrail

full-motion video. A floppy disk holds only a fraction of that. Without compression, too, most PC memories would choke while processing only a few seconds of video. "Whoever said a picture is worth a thousand words didn't know how right he was," says Nolan Bushnell, the roller-coaster high-tech entrepreneur who founded Atari Corporation. Bushnell now heads privately held Aapps Corporation, which makes $400 add-on boards enabling Macintosh computers to display live television images.

Intel's DVI technology uses one form of compression; Apple is working with Sony and other companies to develop another. To get what Intel calls "presentation-quality" video out of a DVI-equipped PC, the user must send his original videotapes to Intel to be compressed into digital form compact enough for the desktop video computer to edit. The process takes three or four days. (In a year or so, the DVI boards for PCs will do it all themselves, Intel officials promise.) Apple hopes that its approach, still under development, will yield faster, tighter compression.

One of the biggest skeptics about digital video computing is Steven P. Jobs, the Apple cofounder who now is chairman of Next Inc., a workstation maker. Says he: "In order to compress and manipulate video images, you have to throw away a bunch of data, and with it much of the picture quality." That leaves the images fuzzy and jerky. "As regular TV viewers," Jobs argues, "we all have high standards for what this stuff should look like. So unless you solve the compression issue in a serious way, it's all just fluff."

Interactive Viewing

While they wait for compression technology to evolve, companies like Apple and Commodore, maker of the Amiga personal computer, are devising ways to connect their computers to laser-disk players. That makes possible interactive viewing of prerecorded video through a conventional TV monitor without first converting the data into digital form. These systems don't allow editing on the computer, but they make good multimedia players for use in various types of instruction.

At Apple's multimedia laboratory in San Francisco, California, researchers have developed an elementary-school teaching aid called the Visual Almanac, an interactive encyclopedia illustrated with both still pictures and moving video clips. Students can browse through the almanac at their own pace, jumping instantly from subject to subject. The system isn't cheap. It requires a computer, a separate television monitor, a laser-disk player, stereo speakers, and a large hard disk—for a total cost of more than $6,000. (Commodore sells something similar for around $4,000.)

In Japan, Fujitsu is a leading experimenter with multimedia. In 1989, the company started selling the FM Towns computer. Like the forthcoming IBM multimedia PC, it has a built-in CD-ROM drive for displaying still images. The machine is available only in Japan.

So far the consumer-electronics industry's move into smart TV isn't as well focused as the PC industry's. The various manufacturers seem content to add one feature at a time to their existing TV sets. Now, for example, VCR manufacturers are building in circuits that will allow PCs to control them. Magnavox, a brand of N.V. Philips, the Dutch electronics conglomerate, has a TV that allows viewers to display a small black-and-white version of a broadcast- or cable-TV program in one corner of the screen while playing a tape or laser-disk recording in full color on the rest of the screen. Sony is experimenting with a computerized home-entertainment system called Avina that controls a roomful of audiovisual equipment.

Because the consumer-electronics companies' efforts are so diffuse, video-technology experts think the PC makers have the inside track on making truly intelligent and interactive video devices. "The challenge isn't making smart televisions, but how to create an environment that allows people to manipulate images as easily as words," says Kristina Hooper, a former MIT Media Lab researcher who now heads Apple's multimedia laboratory. "It's one of the great business issues for both the PC and consumer-electronics industries. I'm betting the PC companies will figure it out first, mainly because we know so much more about what makes computers easier for people to use. After all, how easy is it to program a VCR?"

Quantum Leaps

Because its roots are in the computer business, Frox thinks it can provide the know-how the consumer-electronics industry lacks. Frox was founded last year by Hartmut Esslinger, a West German whose Frogdesign firm helped devise

Nicholas Negroponte, the director of MIT's Media Lab, looks forward to the day when television and personal computers are "as one." The "smart" yellow pages at left can call up additional data on a selected service.

Without leaving the classroom, students can create documentaries by combining a personal computer and a laser disk.

the striking ergonomic look of most of Apple's personal computers and the Next machines. He hatched the plans for Frox a couple of years ago; the privately held company is backed by Frogdesign and several other European investors. Steve Jobs helped Esslinger refine the idea last year, but had to back out to devote full time to Next.

Esslinger continued on his own. To build the prototype, he enlisted the help of Andreas Bechtolsheim, one of Sun Microsystems' founders; Peter Costello, another top Sun engineer; and Frox's Hertzfeld. Despite all that impressive engineering talent, so far Esslinger has been unable to win financial backing from an American computer or electronics company. Now he is attempting to market the technology himself, but may have to try to interest foreign manufacturers as partners. "It has been very frustrating," he says. "We wanted to help America reinvent the consumer-electronics business, and they treat us like donkeys."

MIT's Negroponte sees technological nationalism and the fractiousness of the consumer-electronics, telecommunications, and computer industries as the biggest obstacles to realizing the promise of smarter televisions and video computers. "These industries and technologies have been moving toward each other for more than two decades, and the closer they get, the bigger the stakes become," he observes. With the backing of his main sponsor, the government's Defense Advanced Research Projects Agency (DARPA), Negroponte is trying to impose some order on the blending of the technologies, partly by encouraging agreement on standards. He's not finding it easy.

While the big names and big industries duke it out, some little guys are discovering ways to bring video and computers together on the cheap. A good example: Bushnell, the Atari founder. "This new market is a mishmash of lots of old markets, and carries with it all that extra baggage, so introducing the new technology seems kind of like herding ducks," Bushnell says. Even so, he adds, "this is one of those special moments when several pieces of technology have gotten good enough all at the same time to make a quantum leap at affordable prices."

Just as when Atari introduced Pong, the first video game—and look at the lasting impact mere video games have had. Smart TVs and video computers, by contrast, will be useful tools as well as playthings. They may take a few years to get established in the marketplace, but once they do, they aren't likely to pass quickly from the scene. They'll probably be among the most important innovations of the 1990s, simply because they are the offspring of two of the most pervasive and powerful electronic technologies man has yet devised.

© J. Chiasson/Gamma-Liaison

COMPUTING AT THE SPEED OF LIGHT

by Laurence Hooper and Jacob Schlesinger

When the skeptics resume their chorus, as they surely will, Alan Huang will draw consolation from a simple image: seven fresh eggs, all balanced on end.

Mr. Huang, a researcher at AT&T Bell Laboratories in Holmdel, New Jersey, is trying to develop an optical computer. Instead of electrons doing the calculating, it would use photons of light. An optical computer could run thousands of times faster than its quickest electronic forebear.

To some, the idea is impractical to the point of foolishness. "You get shot on sight for talking about optical computing in this building," snarls Hyatt Gibbs, director of the Optical Sciences Center at the University of Arizona in Tucson, a program backed by the National Science Foundation. He says the discipline is "a dead issue, at least for this generation."

International Business Machines Corporation (IBM) and several prominent scientists say optics may be good for some things, like carrying phone messages, but doing complex digital computations simply isn't among them.

In January 1990, though, Bell Laboratories announced a notable step toward optical computing. It unveiled a digital optical processor, apparently the world's first. The gizmo is a collection of lasers, lenses, prisms, and microscopic "photonic" devices that takes up about 4 square feet on a lab bench and looks suspiciously like an undergraduate physics experiment. But when it is running, infrared laser beams turn its switches on and off a million times a second and process 32 channels of information at once.

Hooking enough of these unwieldy contraptions together could—in theory, at least—

Optical computing—processing information using light instead of the electrons used in ordinary electronic computers—may soon be a reality: in 1990, Alan Huang unveiled the world's first digital optical processor, the device that would form the guts of any optical computer.

COMPUTERS AND MATHEMATICS

produce an optical computer without further breakthroughs. So it can now be said that Mr. Huang has proved that it is possible to make an optical computer.

Japan Again

He worries, however, that America may not be the first to make one. For one thing, Japan's Mitsubishi Electric Company and NEC Corporation have already unveiled optical switching devices, though they haven't put them together into processors as Bell Labs now has. Mitsubishi has also announced a technology that could make it possible to etch a million optical switches on a 1-centimeter-square wafer by the year 2000.

Thirteen large Japanese electronics companies have teamed with Japan's Ministry of International Trade and Industry (MITI) in a research effort slated to last ten years. Funding is small so far, but 200 researchers meet monthly at Tokyo University to share progress.

Skeptics doubt that anyone, Japanese or otherwise, will ever build a practical optical computer. But as Mr. Huang might say, they never saw the eggs.

Over Easy

It was in February 1989, the Chinese New Year. Mr. Huang, a lanky 41-year-old who gestures excitedly when he gets going on optical computing, had gathered his research group, ostensibly to review recent progress. But instead of talking logic devices and laser diodes, a deadpan Mr. Huang presented each person with an egg and a seemingly impossible mission: to balance it on end. Chinese folklore said the new year was the perfect time to do it.

Much to his own surprise, Mr. Huang managed the feat in two minutes. Colleagues suspected cheating, but then another group member did it. When a woman known for her integrity also succeeded, the mood turned to excitement. "When we left that room," Mr. Huang remembers, "no one could believe we had ever thought that balancing an egg was impossible." Eventually, all five managed to balance an egg.

Mr. Huang hopes his research will play a similar role in convincing people that optical computing isn't so difficult that it should be ignored. If it could be perfected, believers point out, it would have powerful advantages over electricity.

The first is speed. Light moves faster than electrons or anything else. A computer that could tap into the speed of a photon—a packet of light energy—would operate at the limit of physics.

Also, because photons don't interact with one another, light beams can cross without interference, permitting almost unlimited wireless connections between computing elements. The computer technique known as parallel processing—splitting a problem up and sending parts to different processors—would be much easier, notes Demetri Psaltis, professor of electrical engineering at California Institute of Technology (Caltech).

Finally, light can carry a lot more information than can electricity. Bell Labs calculates that one small lens could convey the phone conversations of the world's entire population, even if everyone were speaking simultaneously.

Yet one of light's chief strengths is also its big drawback. A digital computer, the only kind most people consider useful, is basically a collection of switches. Since light beams don't interact, it is hard to make them turn one another on and off.

Mr. Huang claims that this problem was solved in 1986, when Bell Labs announced its optical switching device, called a SEED (it stands for self electro-optic effect device). In the processor demonstrated last January, pairs of SEED devices act like the transistors in ordinary computers. One laser beam focused on a SEED determines whether another beam will be reflected by the neighboring SEED or absorbed by it. Since photons can't switch each other without help, a small amount of electricity is used to change the refractive nature of the materials; but the process is controlled optically.

When each device is switched, its output serves as input to another device, allowing basic logic to be performed. On a nearby television screen, the effect is seen as an array of blinking lights.

Asked about the size of the processor, its main architect, a Scot named Michael Prise, explains somewhat defensively that things are about to get much smaller. Right now, he says, he could build the same machine in a quarter of the space. The next-generation model, due later this year, will take up only 2 cubic inches—and will have many more switching devices and is likely to run much faster.

But this machine is significant for what it does, not how big it is. Although it runs relatively slowly and is only about as smart as a dishwasher control unit, that is a big leap for-

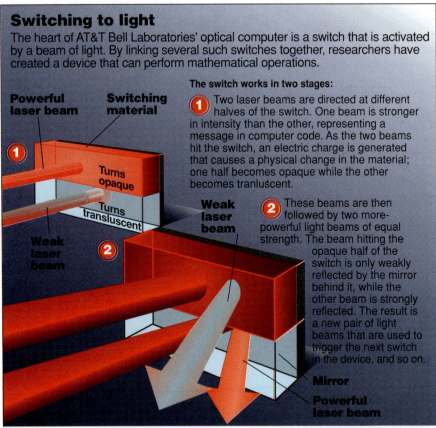

Jeff Glick/U.S. News and World Report

ward for optical computing. Previously, no one had ever assembled a digital processor that could do more than pass light around in a circle.

Getting Off the Ground
William Ninke, Mr. Huang's boss at Bell Labs, calls the achievement a milestone. "The Wright brothers only flew so far and they only flew so fast. No one expected them to build the Concorde their first time around," he says. "But because they had built on a solid engineering and technological base, as we have, the opportunities for doing more were immense."

Baloney, say the skeptics. Though many now are willing to concede that an optical computer is possible—a concession Mr. Huang sees as a triumph in itself—only a few believe it can ever be a practical alternative to today's electronic technology.

"It's easy to get seduced by the idea of computing with light," says Jim McGroddy, IBM's director of research. "The work [at Bell Labs] is interesting, the physics is interesting, and their devices may be useful for certain purposes, but one shouldn't kid oneself about this area's applicability or practicality."

IBM did extensive research on photonics, as the field is sometimes called, during the 1960s. Its scientists concluded that computing with light uses more energy than computing with electricity, and therefore that photonics would never compete with electronics. Light is good for transmission and connections, Mr. McGroddy says, but not much else.

Even Joseph Goodman, Mr. Huang's former teacher and a Stanford electrical-engineering professor who is considered the guru of optical computing, says that "the first commercial general-purpose digital optical computer will appear sometime between the year 2000 and infinity—and I'm not sure it won't be infinity."

Like some others, Professor Goodman suspects AT&T is less interested in general-purpose optical computers than in highly specialized ones that would switch telephone calls and other information through its burgeoning optic-fiber network. Messrs. Huang and Ninke concede such a use of their photonic technology is likely, but Mr. Huang says his main goal, endorsed repeatedly by superiors, remains development of a general-purpose machine.

Use in Supercomputers?

And, despite the detractors, this approach has many well-regarded believers. "The field has changed, and half of my fellow researchers don't even know it," says John Caulfield, director of the Center for Applied Optics at the University of Alabama in Huntsville. He calls Bell Labs the leader, and its SEEDs "unquestionably the best optical devices in the world." Though Professor Caulfield won't endorse Mr. Huang's idea of an optical computer, he says the opportunities for parallel processing with light can't be ignored.

"I think we will see the first applications well before we're halfway through this decade," he predicts. "It will start in areas like supercomputers—anywhere the problems need solutions that are big, fast, and dirty. I could even see optical boards inside [electronic] personal computers by 2000. The only obstacles are engineering and dollars."

In any case, photonics research is burgeoning. Spending in the U.S. has been estimated at about $100 million annually. Over 50 institutions working in the field are funded by the U.S. military alone. The Pentagon is interested because optical computers would be harder to break into and wouldn't be disabled by the electromagnetic burst of a nuclear blast. A market-research firm, Frost & Sullivan Inc., has gathered data suggesting the market for all types of optical computing devices could reach $1 billion.

OptiComp Corporation, a small firm in Zephyr Cove, Nevada, aims, like Mr. Huang, at a general-purpose optical computer. Its founder, Peter Guilfoyle, demonstrated a special-purpose prototype in 1984. But many researchers say they are leery of Mr. Guilfoyle's work, which uses a different principle to make light beams switch each other.

Most research so far has been focused on individual elements that would go into a computer, especially those that could be used in tandem with electronics. Hewlett-Packard Corporation has about 50 photonics researchers, mostly looking into interconnection rather than optical computing. "Our work is aimed at projects that can be of practical importance in a reasonable time frame, say three to five years," says an official, Ted Laliotis.

Silicon Won't Do

Aaron Falk, lead scientist at Boeing Company's Boeing Aerospace & Electronics, says he will believe in optical computers only when he sees optical "transistors" that can be made of something as cheap and versatile as silicon. Bell's SEEDs are made of the much-more-costly gallium arsenide and aluminum gallium arsenide. "There's no reason to believe that won't happen sometime soon," Mr. Falk adds, "though I sometimes joke that we've spent so much time looking for this dream material that it should be called 'unobtainium.' "

Mr. Huang hates such pronouncements. "The Japanese don't say such things," he says. "The Japanese don't ask, 'Will it be viable in five years?' They know it'll be viable sometime, so they go ahead. Doesn't anyone in this country realize that ten years isn't a long time?"

Mitsuhito Sakaguchi, head of NEC's optoelectronics research laboratories, predicts that optical computing will develop, with a 20-year lag, the same as electronic computing did. Taizo Nishikawa, deputy director of MITI's Industrial Electronics Division, says, "Optical computing is one of the most important technologies of the near future."

Japanese electronics conglomerates have an edge because optics involves a merging of previously unrelated technologies. The same Japanese companies make both semiconductors and computers, and the biggest computer companies are also the biggest makers of telecommunications equipment—a field already revolutionized by optics.

Nevertheless, most Japanese experts say the U.S. is still ahead, citing the pattern wherein they do better perfecting a technology rather than inventing it. Says Yoshiki Ichioka, a physicist at Osaka University: "Japanese companies develop many systems, but the fundamental motive is commercial. In the U.S. the important thing is the frontier spirit."

In the U.S., however, Mr. Huang sees scant evidence of such spirit. After his group balanced the five eggs, he relates, he sent excited electronic messages to about 80 people, touting the achievement and inviting all comers to verify and duplicate it later the same day.

When he entered the lab at the appointed time, he found that the five eggs were still balancing—but that only two more people had shown up. "Only two people," he says with obvious disgust. Then he brightens: "But both of them balanced eggs, too."

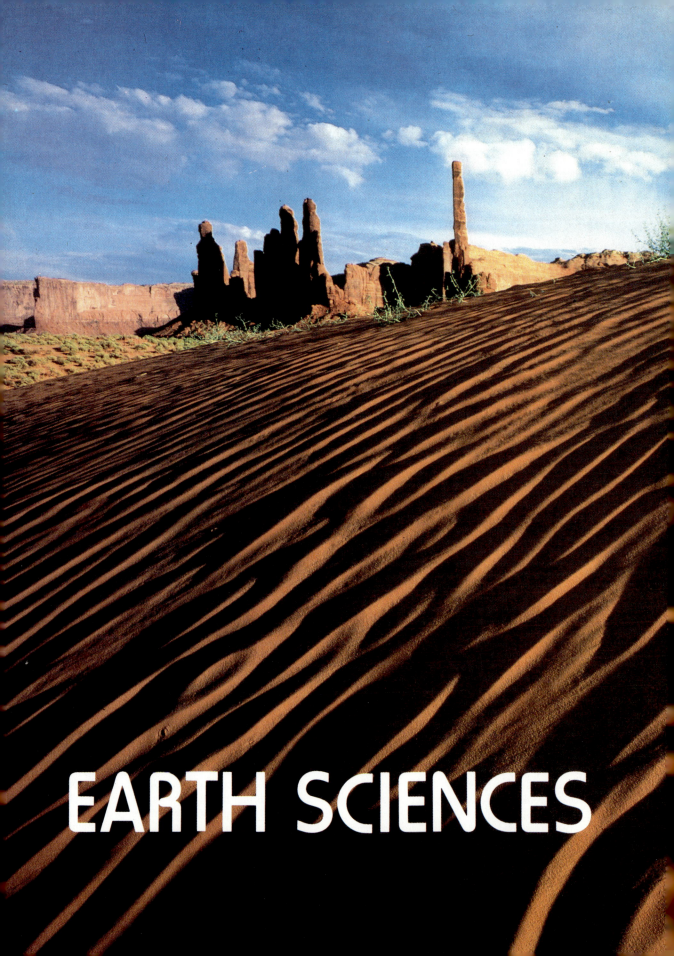
EARTH SCIENCES

Review of the Year

EARTH SCIENCES

Major earthquakes in California and Australia, volcanic eruptions in Alaska, significant fossil discoveries, devastating hurricanes, and new information about the earth's structure topped the earth science news in 1989.

by *William H. Matthews III*

A moderate quake that jolted a California earthquake drill in August 1989 (above) prepared rescue workers for the severe quake that hit two months later.

Earthquakes and Seismology

A major earthquake measuring 7.1 on the Richter scale struck the San Francisco area on October 17, killing 42 people and causing an estimated $7 billion in damage. (See article beginning on p. 132.) A 5.5-magnitude quake rocked southeastern Australia on December 28. The temblor, Australia's largest in 12 years, killed nine people; estimates place property damage at more than $1 billion. In the Soviet Union, two quakes hit within two months. The more serious quake struck on January 23, 1989, and measured 7 on the 12-point Soviet scale. The catastrophe buried several central Asian villages in Tadzhikistan and killed at least 1,000 people. On January 19, 1989, a quake of magnitude 5.6 struck Tokyo and much of northern Japan.

A magnitude-6.8 earthquake hit Mexico City and Acapulco on April 27, 1989. There were no casualties, but people panicked, remembering the killer quake of 1985. An April 27 earthquake that shook the New Madrid, Missouri, area was felt in Arkansas, Kentucky, and Tennessee. Measuring 4.7 on the Richter scale, it originated in the New Madrid fault zone, a seismic area that was the center of several major earthquakes in the winter of 1811-12. A 5.7-magnitude quake killed at least 29 people and flattened about 8,000 homes in rural north China on October 19. And on November 2, a 7.1-magnitude temblor rocked the coast of northeastern Japan. Its epicenter was on the ocean floor; no casualties or damage were reported.

Meanwhile, geologists at the National Aeronautics and Space Administration's (NASA's) Jet Propulsion Laboratory (JPL) discovered several previously unknown faults in the central and eastern Mojave Desert in California. These strike-slip faults, some of which may be active, were located by analyzing images taken by a Thematic Mapper instrument on an earth-orbiting satellite. The images were used as a "map" that guided scientists in the right direction to locate and confirm the faults in the field.

Paleontology and Fossils

A Scottish quarry has yielded the remains of the oldest known reptile. The 338-million-year-old fossil, a lizardlike reptile about 8 inches (20.3 centimeters) long, is some 40 million years older than any previously known ancient reptile. Paleontologists from Brigham Young University discovered the oldest dinosaur egg yet found in the Northern Hemisphere. The egg, about 150 million years old, was collected between the boundary of the Lower Jurassic and upper Lower Cretaceous systems in east-central Utah.

Denversaurus, a new species of dinosaur, was described after its remains were located in collections at the Denver Museum of Natural History. A plant-eating nodosaur, the dinosaur was about 20 feet (6 meters) long, covered with bony plates, and had spikes protruding from each shoulder. The specimen was previously assigned to another species closely related to the nodosaurs. The beak of a huge fossil bird was recovered from late Eocene rocks on Seymour Island on the Antarctic Peninsula. Similar to the fossils of giant meat-eating birds found in Florida and South America, the beak is about 40 million years old. This and other fossils suggest that a much warmer climate existed in Antarctica in late Eocene time.

Epanterias amplexus, a ferocious, giant version of the allosaurs, flourished 30 million years before the infamous Tyrannosaurus rex even existed.

© 1990 by The New York Times Company. Reprinted by permission.

Volcanoes

Alaska's Redoubt Volcano erupted December 14, 1989, hurling volcanic ash 7 miles (11.3 kilometers) into the air. Lying some 115 miles (185 kilometers) southwest of Anchorage, the volcano last erupted in 1968. Although air traffic had to be rerouted because of dense clouds of ash, no major damage was reported. On June 24, an eruption of Kilauea Volcano in Hawaii triggered a 6.1-magnitude earthquake that rattled Hawaii Island and generated a small tsunami, or seismic sea wave. The earthquake destroyed two houses and knocked down power lines on the island.

In December 1989, Mount Redoubt in Alaska erupted in a series of almost continuous explosions that caused no injuries but managed to disrupt air traffic.

AP/Wide World Photos

Earth Structure and Plate Tectonics

Using the geophysical equivalent of a computerized axial tomography (CAT) scan, new data on the formation of the ocean floor has been obtained by marine geologists. Their study was made on the East Pacific Rise, an ocean ridge off the coast of Mexico. The new information suggests that a region of hot rock approximately 3.7 miles (6 kilometers) wide and almost 1 mile (1.6 kilometers) deep underlies the ocean ridge. Submarine vents extrude lava from vents in the axis of the ridge; this action produces new ocean floor and spreads the existing seafloor apart. Canadian geophysicists discovered the remains of a previously unknown buried mountain range stretching more than 621 miles (1,000 kilometers) from Great Bear Lake to the Beaufort Sea. Consisting of a great underground area of faults and folds similar to those that form the roots of the Appalachian and Rocky mountains, the subterranean range probably formed from a collision between North America and an unknown landmass about 1 billion years ago.

U.S. and Australian geologists reported the discovery of the earth's oldest known rocks. The two 3.96-billion-year-old specimens were collected from the Slave Province, a remote region of Canada in the Northwest Territories.

Meteorology and Climatology

On September 21, Charleston, South Carolina, was ripped by the tenth-strongest hurricane ever to hit the United States. (See related article beginning on p. 215.) Hurricane Hugo also cut a wide swath through the eastern Caribbean; its 140-mile (225-kilometer)-per-hour winds caused widespread destruction on the islands of St. Croix, St. Thomas, Montserrat, Puerto Rico, and Guadeloupe. Hugo's winds and rain raked both North and South Carolina, doing great damage in Charleston. All told, the killer hurricane caused some 37 deaths, injured hundreds, and left tens of thousands homeless.

Spurred by increasing concern of global warming because of the greenhouse effect, seven U.S. agencies are joining to set research goals to monitor the earth's rising temperatures. The cooperative federal study plans to monitor world climates, learn how atmospheric chemical and physical systems work, and build computer models that will predict climatic changes caused naturally and by human acts.

Meanwhile, having refined the way computers simulate clouds, a team of British scientists predict that climates of the future may not produce temperatures as warm as predicted. This assumption has support from a team of U.S. Government scientists who studied climate data extending back almost 100 years. They concluded that no significant change in rainfall or average temperatures has occurred in the United States over that period of time.

Alan S. Weiner/NYT Pictures

The South Carolina coast bore the brunt of Hugo, the tenth-strongest hurricane ever to hit the U.S. mainland. The region has yet to fully recover.

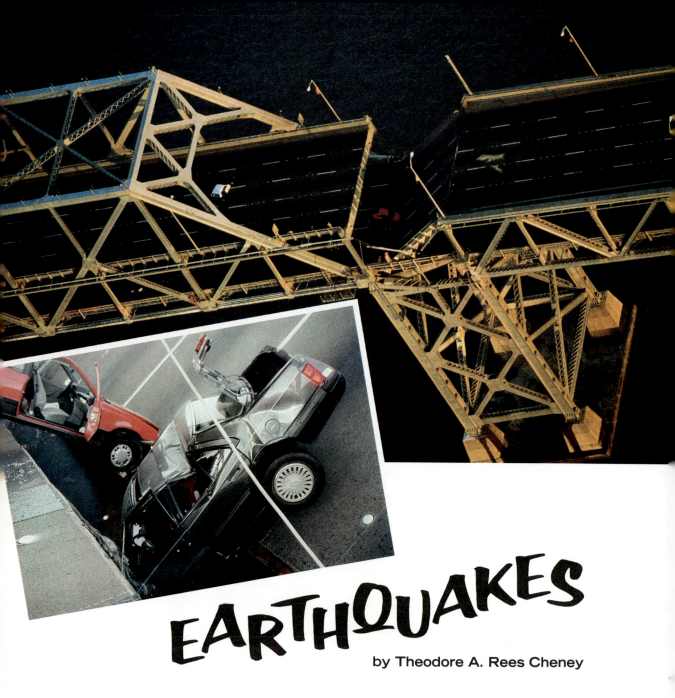

EARTHQUAKES

by Theodore A. Rees Cheney

All systems are "go" for the third game of baseball's 1989 World Series—the Oakland Athletics vs. the San Francisco Giants. The weather's perfect for a ball game. A happy, expectant crowd of 62,000 fills the stands at Candlestick Park, San Francisco, California.

Latecomers rush along the Oakland Bay Bridge and the Admiral Nimitz Freeway, hoping to make it over in time to cheer the first pitch of the third game scheduled for 5:25 P.M. This is a "Bay Bridge Series," the West Coast equivalent of New York City's "Subway Series," one of those rare occasions when both home teams are battling for the World Series title. Spirits are running particularly high, and excitement runs rampant through the stands. Ferryboats load, ready as usual to depart across the Bay in a few minutes. The sky paints the Bay and the boats a delicate pink. To live around the Bay Area is to have found heaven on Earth this day. In the locker rooms below the stands, ballplayers ner-

Top: © Mark Richards/Sipa Press; inset: © Jose Bueno Garcia/Saba

The earthquake that shook California's Bay Area in October 1989 generated shock waves that shoved the Oakland Bay Bridge laterally 3 inches, causing a 50-foot section of the upper roadway to collapse (facing page). Flames ignited by a broken gas line raged in a section of the heavily damaged Marina District in San Francisco (above). Sixty-seven people died in the quake; property damage ran into the billions.

vously ready themselves for the game before they jog out onto the field. Their wives and children sit in special family boxes, ready to jump up and down to cheer their private heroes. What more could a young ballplayer want this day?

Thousands stand up in anticipation of the teams' arrival on the field. Then, through their shoes and sneakers, they feel a rumble like a subway train. Sports journalists in their high lookouts see the light poles sway an abnormal amount. Journalists in from the East Coast won-

der what's happening. It can't be the wind; the pennants barely ripple. The local California writers and announcers know just what it is. The time, which no one there will ever forget—5:04 in the afternoon.

After the quiet, calming announcement that there has been an earthquake and that Game Three is postponed, the fans file slowly out of the park. This may not have been "The Big One" they've all been preparing themselves for, but neither has it been merely another sneaker-

Forty-two perished when the upper tier of the double-decker Nimitz Freeway collapsed, crushing the cars beneath it.

tickling tremor. Those few fans with portable television sets see their sets go black. Ferryboat men cast off all lines early when they see the old ferry building swaying above them, its flagpole whipping around like a reed in the wind. It slowly comes to a final salute at 45 degrees.

The Marina District

The fans walk out of Candlestick watching intently the towering cloud of black smoke in the Marina District across the Bay in San Francisco. Had that slight rumbling they'd felt through their feet been merely a small outlier of a much larger quake that has hit San Francisco hard? Video cameras shoot long shots of the black plume, then close-up shots of silhouetted fire fighters pouring long, high streams of water on the blackened debris. This becomes a familiar, repetitive scene on television that night for viewers glued to their sets across the nation.

The underlying soil of the Marina District has reacted more strongly than soils in other parts of the city, because back in 1906, a landfill was created with soil and trash to build the Pan American Pacific Exposition. Unconsolidated soils like these shake worse than bedrock, shaking to the ground many structures built upon them, especially low buildings of masonry construction.

The Marina District fire started when a gas main, ruptured by the shaking ground, ignited. Those familiar with historic details about the devastating 1906 earthquake knew that, back then, most of San Francisco had been destroyed by runaway fires caused by broken gas mains, not by earth shaking.

Was that happening again, one stubborn fire spreading out to destroy everything in the district, the city? To television observers, the fire seemed to go on unabated. Fire fighters kept hosing down nearby buildings. The fireboat *Phoenix* came right up into the marina to supply water through 6-inch (12-centimeter) hoses to the district where water mains had burst.

Oakland Bay Bridge

The country sees first on television the destruction within the Marina District, which looks bad enough, but then they begin to see video images terrifyingly reminiscent of Hollywood films about earthquakes. A cameraman crouches like a Vietnam door gunner in the open door of a helicopter hovering close by the damaged Oakland Bay Bridge.

A hole in the upper roadway gapes where the edge of a 50-foot (15-meter), 250-ton (225-metric-ton) section of concrete has fallen down onto the deck below, leaving the other edge attached to the upper deck, as though hinged by the steel reinforcing bars within the concrete upper roadway. One car has gone over the edge and pancaked onto the lower deck. Another car teeters precariously on the edge, ready to dive into the yawning chasm. Several people walk rapidly and randomly near the car, as though wondering whether to climb in.

The driver whose car plunged down to the lower deck has been killed. A bus carrying 37 passengers has stopped just short of the gaping hole. About 150 other people see what's happening, abandon their cars, and run in panic toward the nearest land. They are more fortunate than those driving along Interstate 880, the Admiral Nimitz Freeway approach to the bridge on the Oakland side.

The Nimitz Freeway Collapses

In the early hours after the quake, the media all report that the quake has injured some 1,400 people and killed 270, of whom 250 were reported killed when trapped in their cars on the Nimitz Freeway. A 1¼-mile (2-kilometer) section of the upper deck has smashed down onto the lower deck. Drivers have seen the upper deck move in clearly visible undulations, like rolling waves in the Bay.

The two levels are now so flat against each other that rescuers cannot see whether there are cars in between. Firemen, policemen, and other rescue workers can barely squeeze between decks on their stomachs to search for survivors and victims. They come out from this middle world of horrors to report that some cars have been compressed to a height of only 12 to 18 inches (30 to 45 centimeters), some with men, women, and children entombed within. One boy survivor has lain beneath his dead mother for hours before they find him pinned by a huge slab of concrete. They make the heartrending decision to amputate one leg to free him.

After several days of crawling through precariously balanced wreckage, rescuers discover that many fewer cars have been trapped than initially estimated. Apparently, fewer than usual cars were on the road, and many of those trapped have managed to crawl out of their cars and make their way to safety on hands and knees. The final death toll was set at 42 for the Nimitz Freeway tragedy.

All this death and destruction of property has occurred within 15 seconds after 5:04 this day. Unfortunately, the devastating effects are not limited to the Bay Area.

Epicenter Near Santa Cruz

About 70 miles (112 kilometers) south of Candlestick Park lie the Santa Cruz Mountains, smiling benignly down on the tourists buying their way through the attractive Pacific Garden Mall in Santa Cruz. Engineers, mathematicians,

The quake's epicenter lay near Santa Cruz, a seaside resort about 70 miles south of San Francisco. The quake destroyed 50 homes and devastated the town's fashionable shopping district (right).

© Bill Lovejoy, Santa Cruz Sentinel/Sipa

and computer-chip specialists drive home from nearby Silicon Valley labs. Some have left work early to catch the Series on TV, and listen on car radios to the pregame commentaries.

Within those same 15 seconds after 5:04, the Pacific Garden Mall comes to its knees. Bricks litter the streets and sidewalks; plate-glass windows shatter and fall on the tourists strolling by. Two people are killed and 50 homes demolished. Electricity fails all over the city. It seems doubtful that the economy of Santa Cruz, a lovely tourist town of 42,000, will ever fully recover.

The epicenter of the quake, that point directly above the quake's focus, turned out to lie not far from Summit, California, near Santa Cruz. By Thanksgiving and Christmas, only several months later, people were hiking a 6-mile (9.5-kilometer) trail through the "Forest of Nisene Marks State Park" to see for themselves the epicenter, the place where it all began.

Earthquake Roots

Not far from Santa Cruz and 6 to 10 miles (9.5 to 16 kilometers) below ground, the top of one giant block of rock, the North Pacific plate (about 58,000 feet—17.7 kilometers—thick), has rubbed along the edge of another huge plate, the North American plate. These two giants have been rubbing each other the wrong way for millions of years. Although geologists, seismologists, and geophysicists remain uncertain as to the initiating mechanism behind this continual rubbing and periodic rupturing, most agree now that the Earth has a shifting crust. At the present time, most geoscientists accept the theory of plate tectonics.

Plate-tectonics theory says that Earth is broken at great depth into 12 giant slabs, or plates, which float above the molten mass of Earth's outer core. Scientists are not in total agreement about what force moves the plates, but move they do. Evidence for the plates' existence and movement is indirect, because the plates float at such great depths that man's deepest drill holes have not come anywhere near them.

When these huge plates collide, they bend or even shatter overlying rocks, sometimes all the way to the surface. Where rocks are faulty, they shatter or slide along a fairly well-defined line, the fault line. Rock on either side of the fracture may slide up, down, left, right, or diagonally. Fault lines may be short or extremely long, such as the famous San Andreas fault in California. We can see it running along the surface for 650 miles (1,045 kilometers). When it then dips below the surface, it continues for still another 200 miles (320 kilometers).

Predicting the Unpredictable

Earthquake prediction is extremely difficult but potentially valuable to society, if it can ever be done with accuracy as to where a quake of a certain magnitude will occur—and just when.

Geoscientists feel fairly certain of their ability to predict in the long range; e.g., this particular 330-mile (530-kilometer) segment of the San Andreas fault will experience a quake of an 8.0 magnitude within the next 100 years, and it will slip toward the east. But that level of prediction is of interest mostly to basic scientists. Engineers, policemen, fire fighters, politicians, and others charged with minimizing death and destruction need prediction in the short range; i.e., they would like to feel confident in telling people to evacuate these areas on this day. Predictions must be accurate, for these people know that if they issue too many false alarms, citizens will lose confidence and not evacuate or take other precautions when they should in the future.

In an attempt to demonstrate their capabilities if given sufficient instrumentation, scientists from various organizations have set up an impressive array of instruments along the San Andreas fault line in the town of Parkfield. They've chosen this small town of 34 people about halfway between San Francisco and Los Angeles because it happens to have earthquake tremors on a shorter cycle than anyplace else—a "good" one of about 5.5 on the Richter scale every 22 years. Other places have such quakes on perhaps a 100- or even a 300-year cycle. Parkfield last experienced a 5.5 shake in 1966 (24 years ago), so scientists attached to the Parkfield Experiment have been expecting one any day now since 1988.

The Parkfield scientists have set up their instruments along the 18-mile (29-kilometer) segment where the fault has slipped equal amounts in the same direction the past two times, and the past three slips have hit right around 5.5 on the Richter scale. Geoscientists appreciate this rare opportunity to work with such consistent variables. Geology in general, and earthquakes in particular, are not noted for consistency. The scientists have installed several million dollars' worth of seismometers, creep meters, tiltmeters, strain gauges, lasers, etc., many of which are deep in the Earth along

Earthquakes occur when the plates that make up the Earth's crust shift. Such shifts most often take place where two plates meet, as along California's San Andreas fault (right). Laser observatories (above) near the fault can detect minute ground movements 6 miles away.

Above left: © James Balog/Black Star; above: © George Hall/Woodfin Camp

the faulted bedrock. Others are in the soils above, and on nearby hills, some for extremely accurate distance measurement.

The hope is that the instruments will reveal certain behaviors along the fault that will occur ahead of the quake itself. With good fortune, these "precursors" of the quake will be so clear (after checking with later years of quake information) that accurate predictions could be made in the future.

One new possible precursor made itself heard right before the 1989 quake in the Bay Area. Research people at Stanford University recorded totally unexpected low-frequency radio waves. Instruments to record such waves will be added to the extensive array now on alert in Parkfield. Scientists are paying attention to anything that has a reasonable chance of serving as a precursor.

Quake Precursors

Historically, there have been many claims as to early signs (precursors) of imminent quakes. Many people have claimed that animals, especially horses, act strangely just hours or minutes before a quake. It seems reasonable that animals, with their various hypersensitivities, could sense the early vibrations or other indications of a quake's beginnings, but no scientific studies have yet confirmed this. Scientists find it difficult to organize animals and quakes to be at the same place at the same time. It's equally difficult to ensure that instruments or scientists would be there at the same time to observe any bizarre animal behaviors.

Animals are not the only precursors people have suggested over the centuries. The following have all been suggested at one time or another as possible precursors:
- an increase or decrease in microquakes;
- flashes of lights in the sky;
- a slight tilting of land near the fault line;
- up or down changes of well-water levels;
- radioactive radon released from wells;
- less resistance to electric flow in rocks;
- slow creep of materials along the fault.

Accurate prediction remains unbelievably difficult, and Parkfield remains the best hope for a scientific breakthrough. On April 5, 1985, scientists associated with the National Earthquake Prediction Council issued the very first "official prediction" that a quake of 5.5 to 6.0 on the Richter scale would occur near Parkfield before the year 1993.

The Richter Scale

When a prediction claims that a quake of a certain magnitude has a 65 percent chance of hitting a certain area within the next, say, 30 years, that's important information for engineers and architects. With faith in the prediction, they can build in extra strengths and design features that will minimize destruction during a quake of that magnitude. They could design every structure to withstand the worst possible magnitude quake, but that would be prohibitively expensive. Prediction of not only time and location, but of magnitude becomes essential.

Charles Richter and Beno Gutenberg decided in 1935 to devise a magnitude scale that would be accurately descriptive, but easy to deal with. It soon became known as the Richter scale, still the scale most used. Scientists find it very useful, as do nontechnical people who like to have an easy way to discuss the magnitude of this quake versus that quake. "I've experienced a 6.2, how about you?"

The scale seems at first an easy one to comprehend, but not unless one is used to thinking logarithmically. The logarithmic Richter scale runs approximately from 0 to 10, each higher whole number representing a 10-fold increase in ground motion over the next lower number. This also means a 30-fold increase in the energy released. The great 1906 quake in San Francisco, 8.0 according to the Richter scale, was 30 times "worse" than the 7.0 one in 1989. Fortunately, the world experiences one of the 1906 magnitude only once every 50 to 100 years, while experiencing one the magnitude of the 1989 quake about 14 times every year, somewhere in the world.

Preparing for the Next One

Californians, in a tragic sense, had a golden opportunity with the 1989 quake to see how their preparations were working, and what needed improving before the predicted "big one" hits. The 1989 San Francisco quake and the one a year earlier in Armenia each had a magnitude of 6.9 on the Richter scale, but differed greatly in their levels of death and destruction: Armenia lost 25,000 people; California, about 65. In Armenia, almost every building collapsed; in California, damage was limited to a few buildings in Santa Cruz, a dozen or so in the Marina District, one section of the Bay Bridge, and a stretch of the Nimitz Freeway.

The immediate cause of most property destruction is the shaking of buildings—unreinforced concrete-masonry buildings being the most vulnerable; wood-framed, single-story structures less so. Each building or structure has its own, unique, "fundamental" vibrational frequency. If a building's frequency happens to coincide with the quake's frequency, the lateral movement may be magnified as much as five or six times by "sympathetic" vibrations. Ancient-lake-bed clays under downtown Mexico City magnified that city's last quake by up to six times, causing maximum destruction in the central part of the city, killing 20,000 to 30,000 people.

A hundred miles from the epicenter, less-pronounced tremors rattled dishes and shook up swimming pools (below).

© Peter Menzel

Schools throughout California conduct regular drills to teach their students how to react during an earthquake.

The Nimitz Freeway collapse was probably caused by a number of factors: the two-tiered design itself, which will likely never be used in California again; the collapse occurred at a curved section, always the weakest in a quake, but made even worse in this case because each end of the section sat on different types of soil that vibrated differentially. The Bay Bridge came through relatively unscathed because the pier and tower structures could be considered tall buildings, and engineers know how to build tall buildings to stand up to quake shakings.

Most of the newer buildings in California have been built according to "building codes," designed to help buildings withstand certain magnitude quakes. Many engineers and architects feel that the building codes, stiff as they are, will not work in a quake the magnitude of the 1906 one. They design their buildings 1½ to 2 times stronger than the code requires. Some engineers claim that if they build in earthquake resistance during the early stages, the total costs for the building itself increase no more than 20 percent.

Buildings constructed "below code" can be upgraded to meet the code even years after construction, i.e., retrofitting. Buildings built before the 1960s need retrofitting, especially those anywhere near the San Andreas or other active faults. Setting aside the cost of retrofitting thousands of structures, how quickly should it be done? If prediction is so questionable, should we move rapidly to improve all these buildings before the next quake hits (estimated to occur sometime between today and 50 years from now)? If the quake waits for 40 or 50 years, many of the buildings will have been taken down for ordinary reasons, and the costly retrofit will have been wasted effort. But if it happens next year, and everyone in the building you own or designed lies maimed or killed under rubble, would retrofitting not have been worth almost any cost? This is the essential dilemma of the Golden State today.

Be Prepared

The motto of the Boy Scouts of America should become the motto for all cities standing near active faults. San Francisco, prepared for the 1989 quake, minimized death and destruction. Part of the preparatory activities involved engineering and design, but the citizens had also prepared themselves. Only two months earlier, San Francisco had had a simulation drill—eerily similar, some said, to what actually happened. When planners saw how difficult matters were, with relatively little damage and only 67 dead, they realized how bad it would have been had the scenario's hypothesized death toll of 3,000 to 8,000 been true. Mexico City's most recent quake killed over 20,000; Armenia's, 25,000; and scientists estimate that the next one could hit either San Francisco or Los Angeles, killing perhaps 21,000 and injuring up to 300,000.

Planners realize now that for such a future scenario, they must also prepare for the reality that rescue workers, doctors, nurses, fire fighters, and policemen will succumb to fatigue and stress after many hours of hard work. Normal communication systems will likely fail. In the recent quake, San Francisco administrators found that they could communicate with Los Angeles, but not with their own rescue workers in the Marina District. California also found that it lacked an effective network among cities that could have provided aid of various kinds quickly to the stricken area. Computers, upon which so much of modern life depends, may fail or memories get erased. California's coastal population centers must allow in their planning for the near certainty that water conduits coming down from the mountains to the east will rupture where they cross the San Andreas fault.

U.S. Cities Most at Risk

Has history shown that earthquakes have occurred elsewhere in the U.S., and were they of sufficient magnitude to compare even remotely with the California quakes? Yes, earthquakes have shaken many areas whose populations today probably never think of their areas as "earthquake-prone." The United States Geological Survey (USGS) lists the following as "most at risk": San Francisco, Los Angeles, Salt Lake City, Puget Sound, Hawaii, St. Louis/Memphis, Anchorage, Fairbanks, Boston, Buffalo, and Charleston, South Carolina. Only the first two have made serious preparations.

The largest magnitude earthquakes in the continental United States occurred in the Mississippi River Valley near New Madrid, Missouri, during 1811 and 1812—an estimated 8.5 on the Richter scale. Fortunately, at the time the area was largely unpopulated. The quakes destroyed every home in the 1,000-person town of New Madrid. Had the area been heavily developed like San Francisco, death and property destruction would have been far worse than San Francisco's 1906 quake. New Madrid and that area of the valley near the juncture of Kentucky, Tennessee, and Missouri experienced other quakes in 1843, 1895, and 1968, proving once again that the giant fault that created the Mississippi River Valley in the first place remains active at depth.

Since the 1800s the Mississippi River Valley has become America's Rhine, with giant chemical plants built along much of its earthquake-prone length. They produce thousands of chemicals, many of them among the most deadly made by man. Should an earthquake of similarly great magnitude occur again—which it will at some unpredictable time in the future—the scene will far exceed any other disaster anywhere at any time.

Not only will the chemicals stored in tanks or in transit aboard barges, trucks, and trains possibly be released, containers of toxic wastes buried underground throughout the region may be ruptured. Local streams will carry the deadly materials into the Mississippi, which will then carry it through cities and towns all the way to the Gulf of Mexico, where such a large percentage of the nation's seafood lives and is harvested.

The 1811–12 earthquakes brought up to the surface of the Mississippi huge tree trunks long buried beneath heavy muds. A quake today would also bring to the surface industrial wastes and tons of toxic materials accumulated in the Mississippi mud ever since chemical insecticides began to flow from the farm fields along the nation's longest waterway. The river drains over a million square miles (five times the area of France).

We need think only of Union Carbide's chemical plant in Bhopal, India, in 1984, when 3,000 were killed, some 30,000 blinded or otherwise injured, and 200,000 evacuated—all because of one leak of one deadly chemical in one plant. A chemical similar to the one at Bhopal is manufactured within the fault zone of the Mississippi Valley.

Consider 50 or 100 such leaks in a region whose hospitals may well have been destroyed; while, in surviving hospitals, doctors and nurses will be overwhelmed by stress; communication poor to nonexistent; freeway ramps fallen . . . bridges out . . . airport runways shattered . . . conceivably all happening in freezing weather. The national significance of such a disaster is almost beyond human comprehension, the numbers of people killed or maimed almost too many to contemplate.

Probably nothing man can do will stop, or even lessen, the released energy of an earthquake, the most powerful of all natural phenomena. All we can do is stay away from those faults known to be active, establish and prepare people to rescue and medically serve those caught in its power, and be ready as a people to help the survivors start life anew.

© Mark Wexler/Woodfin Camp and Associates

Stopping Beach Erosion

by Cory Dean

Using computerized measuring techniques, scientists are tracking with new precision the erosion that is eating away 90 percent of the nation's East and Gulf coasts.

Their findings are bad news, especially for the 295 barrier islands that lie like beads on a necklace along the coast from New England to Texas. The islands—long, sandy strips of land immediately off the coast—are home to many of the nation's shore resorts and coastal towns.

The sea has been rising relative to the land for at least 100 years, geologists say, and most agree that erosion is going to accelerate as global warming melts the polar ice packs, sending sea levels even higher.

The barrier islands are adapting to the rising sea by slowly migrating to higher elevations inland; as their seaward dunes erode, new dunes rise just behind them. But in the process, structures newly exposed to the sea are often undermined. When heavy storms hit, dozens of houses can be destroyed.

Over the past 100 years, the Atlantic Coast has eroded on average 2 to 3 feet (0.6 to 1 meter) per year, says Stephen P. Leatherman, director of the Laboratory for Coastal Research at the University of Maryland. The average yearly erosion on the Gulf Coast is 4 to 5 feet (1.2 to 1.5 meters), he says, and in some places, it is much worse. The south shore of Martha's Vineyard is losing about 10 feet (3 meters) a year, while Cape Hatteras, North Carolina, is losing 12 to 15 feet (3.6 to 4.6 meters), and Louisiana is losing 30 to 50 feet (9 to 15 meters) a year.

"It's no secret anymore what's happening on the coast," Dr. Leatherman says. "Ten years ago we didn't have good data on erosion rates. Now we do, and we know about 90 percent of the coast is eroding."

Beach erosion has accelerated at an alarming rate along both U.S. coasts. On the south shore of Long Island, waves lap beneath oceanfront houses that originally stood far back from the water (above).

EARTH SCIENCES

Beach replenishment provides at least a temporary remedy to excessive beach erosion. At the severely eroded Wrightsville Beach in North Carolina (left), the Army Corps of Engineers replaced enough sand to create a 200-foot beach (above). Replenished beaches must receive periodic infusions of new sand to maintain themselves.

Photos: U.S. Army Corps of Engineers, Waterways Experimental Station

The threat has driven many communities to seek refuge behind seawalls and jetties, only to find that these costly engineering projects often make the erosion worse, either for them or for neighbors down the coast. Other resorts are spending vast sums to pump new sand onto their beaches, only to see the sand disappear again within a few years.

Many engineers maintain that seawalls, jetties, and replenished beaches are necessary to protect the valued developments already on the barrier islands. Too many people live or vacation at Atlantic City, Miami Beach, and other barrier-island towns to let their roads and buildings simply fall into the sea, these experts say.

But many environmentalists and geologists contend that roads and buildings just do not belong on the unstable, migrating islands. At a minimum, some critics say, the government should discourage further development on barrier islands by refusing to subsidize such projects as roads and waste-treatment plants, or property insurance for owners.

Saving Buildings, Losing the Beach

Although geologists say that nothing can stop the sea's inexorable march inland, few barrier-beach communities are prepared to let themselves fall into the ocean. Often they try to halt erosion with bulkheads or rocky seawalls that protect the structures behind them.

But many scientists now say that such structures accelerate the erosion of sand on the beaches in front of them.

"Seawalls destroy beaches," says Orrin H. Pilkey, Jr., a Duke University geologist. In a recent study of scores of sites on the Atlantic Coast, Dr. Pilkey and his colleagues found that beach width on stabilized beaches was dramatically lower than on beaches with no seawalls. Many of the stabilized "beaches" had no beach at all.

Although the mechanics of wave action on beaches are imperfectly understood, geologists can suggest ways in which a seawall might hasten erosion. On a gently sloping beach, a wave's force is reduced as the surge of water, or swash, rushes onshore. Much of the water's energy is gone by the time it returns down the seaward slope as backwash, so it carries off relatively little sand.

A seawall cuts this process off. Water hits it sharply and is reflected back strongly, carrying sand with it.

"The wave hits the seawall, and the energy doesn't dissipate," Dr. Leatherman says. "A seawall is a last, draconian step to save property. You just kiss off the beach."

Sea Bright, New Jersey, for example, once had 300 feet (90 meters) of beach in front of its massive seawall, says Gilbert Nersesian, an engineer for the New York District of the Army Corps of Engineers. "But the beach that existed in front of that seawall is completely gone."

Maine and North Carolina have passed laws banning new hard stabilization along the shoreline. But property owners pay a heavy price for such laws. In March 1989, a series of storms off North Carolina eroded the beaches at Nags Head and Kill Devil Hills, pitching several houses into the sea, and damaging others so severely they were condemned.

"We let a lot of houses fall in," Dr. Pilkey says. "I'm proud of that." But houses threatened here and there are not the truest test of whether a state's politicians can withstand pressure for stabilization, he says.

Pointing to a five-story condominium on the dune line on Bogue Bank, a heavily developed barrier island in North Carolina, he asks: "Are we going to let a building like this fall into the sea?"

Feeding One Beach, Starving Another

For centuries, people have built jetties of stone jutting out from the shore to keep sand from silting up channels. And for years they have tried to stem beach erosion by building smaller groins out from the shore to trap the sand carried alongshore by currents.

Now there is wide agreement that groins can hold sand on one beach only by starving another downstream.

"Most people now believe you have to be very careful with groins," says Dr. James R. Houston, director of the Army Corps of Engineers Coastal Engineering Research Center in Vicksburg, Mississippi. "They usually cause problems."

The 1938 hurricane that devastated Long Island and southern New England cut an inlet from the ocean into Shinnecock Bay, on the South Fork. The inlet and jetties built to keep it open blocked the longshore transport of sand, causing erosion at Westhampton Beach, Fire Island, and even New York City beaches, Mr. Nersesian says.

Groins at Ocean City, Maryland, have had a devastating effect on Assateague Island. When the groins were built, the rate of erosion on Assateague went to 38 feet (11 meters) a year, from 2 feet (0.6 meter), Dr. Leatherman says.

"The island is going to disappear," he says. "It could happen in 20 to 30 years."

Simple Approach, But a Costly One

Many communities faced with the loss of their beaches are adopting a simple but expensive approach: replace the sand.

Done right, officials at the Army Corps of Engineers say, beach replenishment can be a cost-efficient way to protect a community and provide needed recreation facilities, especially in resort towns dependent on income from tourism. Miami Beach was replenished 10 years ago, in a $65-million project. "It's in magnificent condition," Dr. Houston says.

The Corps of Engineers is about to begin a replenishment project in and around Sea Bright, Mr. Nersesian says. "It's quite expensive, probably over $100 million for 12 miles [19 kilometers] of shoreline," he says. The money will come from the federal government and the state.

But scientists and environmentalists dispute the value of replenishment. For example, Beth Millemann, director of the Coastal Alliance, an environmental group, describes beach

Sandbags used to stem erosion now lie exposed along the beaches of Cape Hatteras, North Carolina. Even large-scale antierosion measures would suffer serious setbacks if a powerful hurricane struck the coast.

replenishment rather pessimistically as "throwing dollar bills in the water."

In a recent study, researchers led by Dr. Pilkey studied 90 replenished beaches on the East Coast. North of Florida, he says, none lasted more than five years. On the Gulf of Mexico, only 10 percent lasted more than five years.

But Dr. Houston and Dr. Leatherman say they question Dr. Pilkey's findings. "Orrin and I disagree a lot on this," Dr. Houston says. "You can name a lot of beach fills that have lasted a long time."

A lot depends on how the replenishment is carried out, he says. "If you replenish a beach with the same material in the native beach, there's nothing in the laws of physics" that would alter erosion rates, Dr. Houston says. "If the sand you put in is finer, it will go faster. If you put coarser sand, it lasts longer, but coarser sand is usually more expensive."

One problem with beach-replenishment programs, Dr. Leatherman says, is that "beaches are like icebergs—only 10 percent of the active part of the beach is above water.

"The Corps of Engineers is pumping all the sand on the high part of the beach," Dr. Leatherman says. "Some of that sand has to go underwater to maintain the slope of the beach, the gentle bottom profile. So it appears that the beach is eroding like crazy."

Once a beach has been replenished, it must receive regular infusions of sand if it is to survive the assaults that diminished it in the first place. Dr. Houston says communities should look on this as a normal part of maintaining their infrastructure.

"It's like everything else," he says. "If you build a road, you have to maintain it."

Insurance Pays Moving Costs
In many places, property owners are being encouraged to bow to the inevitable and simply move their houses inland, a practice that has occurred for decades on Long Island, the Outer Banks, and other high-erosion areas.

In areas of high erosion hazard, the National Flood Insurance Program will pay the owners of threatened buildings 40 percent of either their value or the amount of their coverage to move them inland, says James L. Taylor, assistant administrator for the program. The money helps cover the costs of moving the house and reestablishing it on a new lot. If a property owner agrees to demolish the structure, the payment rises to 100 percent, plus 10 per-

BARRIER ISLANDS IN MOTION

Barrier islands are at once the most vulnerable and the most resilient parts of the coastline. Formed and nourished by sand carried on currents that run along the shore, they are constantly in motion: rising, falling, and migrating. A storm can divide them in an afternoon, forming inlets in low-lying areas.

These barriers buffer the mainland from the ocean. As waves break on sandy beaches, their water is absorbed, and their force is reduced as they climb up the sloping sand. In storms, waves carve the dunes into a scarp or bluff, carrying sand underwater offshore and altering the slope of the beach. But every beach has an equilibrium, determined by the size of waves that strike it and the sand grains that form it.

Like a person shifting to find a comfortable position in a chair, the beach shifts to restore its equilibrium slope. Sand is carried back up to the beach in calmer weather. Later, onshore winds pick up some of this sand to feed the dunes.

As the sea level rises, geologists say, the beach's slope may stay the same, but the beach itself will move to a higher elevation inland. As storm overwash chews the dunes, a new dune line eventually forms on what was once the back of the barrier island, or even the lagoon or bay behind it.

In the past century, according to the National Research Council, the sea has risen about 1 foot (30 centimeters) relative to the Atlantic Coast, and 6 inches (15 centimeters) along the Gulf of Mexico. Much of the increase so far is the result of subsidence, the sinking of the land, caused in part by removal of groundwater.

Stephen P. Leatherman, director of the Laboratory for Coastal Research at the University of Maryland, says that by the year 2050, "we may have seen a 3-foot [1-meter] rise in sea level. There's sort of a rule of thumb that for every foot [30 centimeters] of sea-level rise, you see a shore retreat of close to 200 feet [60 meters]," Dr. Leatherman says.

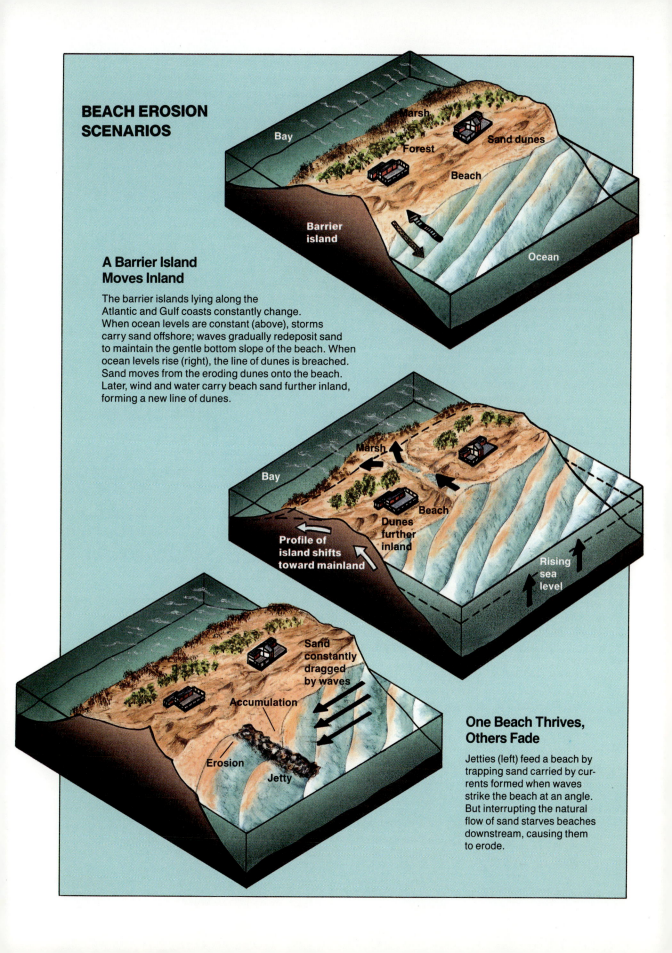

cent for cleanup costs. But anything later built on the property is ineligible for insurance.

In Texas the Open Beaches Act requires owners to move their houses when erosion brings the beach to their foundations. After Hurricane Alicia struck in 1983, about 100 houses in Galveston had to be moved, Dr. Leatherman says.

But retreating from the shore is probably likely only in those places where most structures are single-family houses and where land is relatively abundant.

"Unfortunately," Dr. Leatherman says, "if you look at New Jersey, the checkerboard is filled."

Seeking Methods to Cut Damage

To help people who cannot or will not move their buildings, the Federal Emergency Management Agency (FEMA), working with the Corps of Engineers and the National Oceanic and Atmospheric Administration (NOAA), has commissioned scientists to recommend ways that developed barrier islands might reduce storm damage. Officials from FEMA, accompanied by Dr. Pilkey, recently toured Bogue Bank, a heavily developed barrier island off the coast of Beaufort, North Carolina, to learn more about lessening property damage.

Dr. Pilkey points out some of the island's problems: protective dunes have been leveled for easy development; roads run straight back from the beach, providing overwash channels for floodwaters; marshes that could have absorbed storm water have been filled and built on. Even two low places on the island, where inlets formed in Hurricane Hazel in 1954, have houses on them now.

And there has been no big storm to test what Dr. Pilkey calls "stack-a-shack" condominiums that line the beach in the island towns of Atlantic Beach and Emerald Isle.

The boom in coastal development over the past 20 years has occurred in a time of low storm activity. Max Mayfield, a hurricane specialist at the National Hurricane Center (NHC) in Coral Gables, Florida, says only about half as many major hurricanes have struck the United States mainland since 1970 as in the 30 years previous.

To be covered by federal flood insurance, buildings must conform with codes requiring, among other things, that they be elevated and that all ground-level walls be "breakaway" to allow a storm surge to pass through. But many older houses do not meet such codes, and many new buildings have been modified with solidly constructed living areas on the ground level.

"They're like rotten apples in a barrel," Dr. Pilkey says. When these buildings are destroyed in a storm, surging water can crash the debris they produce into other houses.

But Dr. Pilkey also points out steps property owners have taken to restore some of the protective beach landscape destroyed in development. For example, some homeowners are running sand fences across the beach in front of their homes. Sand collects at the fences, starting the process of dune formation. Others have planted sea grass and other vegetation to hold the dunes, and have built wooden walkways across them so that the plants are not damaged.

Such steps will not save structures from destruction in a major hurricane, Dr. Pilkey says, but they can reduce damage in lesser storms.

Difficult Choices Still Lie Ahead

But what should public officials do when developed areas are threatened?

In principle, says Dr. Leatherman, "I'd say, let 'em take their licks."

"But," he adds, "the reality is that Ocean City is Maryland's major resort. It's a place where the average person can afford a vacation in, and it's a place the masses enjoy. I would say you can't let Ocean City fall in."

For now, he says, the best way to protect resorts like Ocean City is to nourish the beaches with sand. But as sea levels rise, such programs get more expensive. "We may have to abandon more of the beach, and wall up the rest" of the island, he says.

Dr. Houston also says that while he believes undeveloped barrier islands ought to be left that way, where development exists, "Why not protect it?"

"In California there are earthquakes," he says. "In geologic time, which is the kind of time Orrin Pilkey likes to talk about, it's probably going to be destroyed. But we don't say people can't build in California. We don't say people can't build in Kansas because of tornadoes."

But for people like Dr. Pilkey, the choice is simple. "I feel that beaches should be viewed as sacred," he concludes. "I can't accept the idea that any development should have any consequence as opposed to a beach."

© Edi Ann Otto

TORNADO!

by Gode Davis

As dusk approached on May 8, 1986, Randy Hogan stood in his front yard in Edmond, Oklahoma. He peered anxiously toward the southwest and noticed what meteorologists call a wall cloud, a dramatic lowering of a thunderstorm's cloud base. "It was black, ominous, and getting closer," Hogan recalls. Seconds later a sinuous funnel dropped from the cloud to the ground—a tornado. "It was heading for our street," he says. He ran indoors and herded his family into the bathroom. "I laid the kids facedown in the tub and covered them with pillows," says Hogan.

The Hogans were terrified by a low-pitched rumbling sound that seemed to surround them. A tremendous crash followed as the Hogans' roof was ripped off, and a 2.5-ton (2.25-metric-ton) pickup truck was hurled into the children's bedroom. "We found the cab parked on my daughter's bed," Hogan says. A moment later the tornado had passed. "Here we were," he remembers, "sitting outdoors in the pouring rain in the midst of what a few moments before had been our home."

Lucky to be alive, the Hogans are survivors of a strong F3 tornado (see the box on p. 152). Edmond, a suburb of Oklahoma City, lies in a corridor from Texas to Wisconsin highly prone to tornadoes. The town is near Norman, Okla-

During its brief existence, a tornado can generate winds up to 300 miles per hour—a force powerful enough to destroy virtually everything in its path. Fortunately, the human toll has been much reduced thanks to refined prediction techniques and advanced warning systems.

EARTH SCIENCES 147

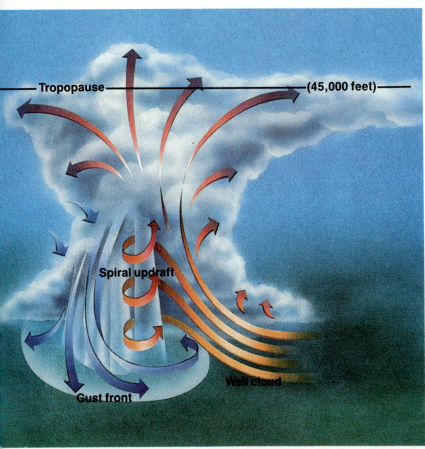

BIRTH OF A TORNADO

A mesocyclone (or supercell) is a highly developed and dangerous thunderstorm. As its central updraft spirals skyward at 60 miles per hour, reaching heights of more than 45,000 feet, a bulging wall cloud starts protruding from the storm's base. The wall cloud's core may begin to rotate rapidly, leading it to lower a spinning funnel, which usually evolves into a tornado.

Above: Melody Warford/Insight; facing page photos: top: © Howard Bluestein/Photo Researchers, Inc; center and bottom: © E. R. Degginger/Bruce Coleman Inc.

homa, where researchers from the National Severe Storms Laboratory (NSSL), the University of Oklahoma, and the National Weather Service are working to improve the forecasts and warnings of severe storms.

In recent years, researchers using pursuit vans that carry portable radar equipment and sensor-equipped balloons have been able to intercept storms in the field, measuring variables they couldn't study before—like the speed of winds rotating inside a tornado. In the lab, researchers studying simulated tornadoes are testing new ways to calculate wind strengths. These data, combined with information gathered by meteorologists monitoring storms from weather stations, should give researchers a detailed model of tornado behavior, which will help them predict the arrival of violent storms—hours in advance, perhaps. To learn how the new tools are being used, I visited the research centers in the heart of tornado country, and even went along on a tornado chase.

Born in the U.S.A.
Tornadoes can strike anywhere in the United States, which has more tornadoes than any other country. (A tornado is any rapidly rotating column of air that touches the ground. In the air, it's called a funnel cloud.) Tornadoes are most apt to occur during the late-afternoon and evening hours in the spring and fall, when the air is least stable.

"First there has to be a favorable area for thunderstorms to form—an unstable air mass in which the wind direction turns with height," says NSSL meteorologist Dr. Edward Brandes. "If the air is then lifted by a passing cold front or rises because of daytime heating, a mesocyclone—a severe thunderstorm with a powerful rotating updraft—may develop."

With few exceptions, tornadoes are born mesocyclones. Once they spot a mesocyclone, forecasters know they're dealing with a fierce storm that produces ¾-inch (1¾-centimeter) hail and winds blowing at a minimum of 60 miles (96.5 kilometers) per hour. Says NSSL meteorologist Doug Forsyth: "It also means there's a fifty-fifty chance of a tornado."

Meteorologists aren't sure why some mesocyclones produce tornadoes and others don't. And though they've learned from video recordings and direct observations that tornadoes almost always form in the right rear quadrant of a severe storm, they don't know why. But they've learned a great deal since the early 1970s, when they first began chasing tornadoes across the plains with scientific instruments.

For one thing, meteorologists have learned that tornado wind speeds don't exceed 300 miles (480 kilometers) per hour. And they've discovered that although a tornado's atmospheric pressure may be as much as 2.9 inches of mercury (100 millibars) lower than the surrounding air pressure, that isn't enough to make a building explode. So it isn't necessary to open your windows when a tornado approaches.

But there's still much that meteorologists don't understand about tornadoes. What triggers them? Is lightning a factor? Why are most tornadoes relatively weak while a few are monsters? Why do some have longer damage paths? Do wind speeds change during a tornado's lifetime? Does a mile-wide tornado have stronger winds than a smaller tornado?

Dishing Up the Data

To answer these questions, tornado researchers are using Doppler-radar devices, their most important tools. There are several sizes of Doppler radars, including large stationary dishes and smaller portable ones. All emit microwaves that are reflected back to a receiver by moving particles, such as water droplets or hail.

Conventional radars can detect thunderstorms, but they can't tell meteorologists whether a storm's winds are rotating. Doppler devices *can* identify a mesocyclone. Dopplers can also spot wind shear—sudden gusts near the ground. And vertical wind shears may indicate a likelihood of tornado development two to four hours before it occurs.

A 30-foot (9.15-meter)-diameter dish antenna on the NSSL grounds in Norman sends out 4-inch (10-centimeter)-wavelength microwaves. It's housed in a hollow radome. A twin radome sits at the NSSL's Cimarron site, 25 miles (40 kilometers) to the northwest. Finding the radomes is easy. They're 70 feet (21 meters) high, 60 feet (18 meters) wide, and look like enormous golf balls.

A whirling funnel that seems to hang from the base of a thunderstorm cloud (top) warns of imminent tornado development. As the funnel descends (center), it makes a thunderous roar like a jet squadron. It is not officially designated a tornado until it touches ground (bottom).

EARTH SCIENCES 149

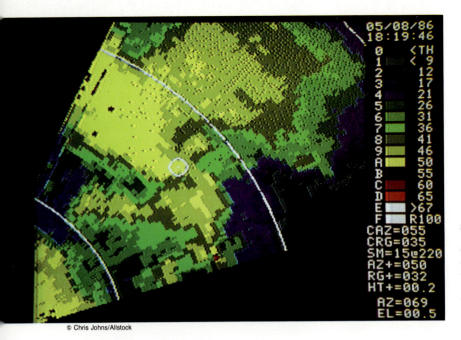

Tornado prediction and detection took a major step forward with the development of Doppler radar. The radar image at left shows the hook-like shape that the rainbands assume in a storm likely to produce a tornado. The circled area pinpoints the location of a tornado in progress.

Inside the radar building with Forsyth, I'm watching bright, color-enhanced images of a mesocyclone on three display monitors. "The first screen tells us where the rain is falling and at what rate," Forsyth explains. I notice a red area called a hook echo. Says Forsyth: "It depicts heavy rain and hail that's being carried out into a thunderstorm's drier air, which happens because the storm is rotating." A second screen shows velocity. "It's telling us whether particles are coming toward or going away from the radar," he continues. The last screen gives an indication of turbulence.

Before forecasters began using Dopplers, they relied on reports radioed or phoned in by cooperating observers. Because some spotters were poorly trained, up to 80 percent of the tornado reports received were false alarms. All too often, lead times simply didn't exist. "By the time hook echoes might appear on ordinary radar, the tornado already existed," Forsyth says. "Using Dopplers, we now have up to 20 minutes' lead time."

But, for most forecasters, Dopplers aren't yet available. "There are a limited number of Dopplers operating in the entire country, and most of them are smaller wavelength radars," says meteorologist Brandes.

The situation should improve with NEXRAD, or Next Generation Radar, a federal program that is supposed to replace the nation's network of ordinary radars with 100 Dopplers during the 1990s. A second experimental network of radars, called Profilers, should also be in place by 1995. Profilers are radars specifically calibrated to measure winds in the upper atmosphere, 1 to 20 miles (1.6 to 32 kilometers) above some 30 sites. "In addition to having weather-balloon soundings every 12 hours, we'll have wind-profile data available every six minutes to help forecast potential severe-weather episodes," says Forsyth.

But even after NEXRAD and Profiler are operative, gaps will exist. "Many tornadoes form in isolated areas and strike tiny towns," says Dr. Robert Davies-Jones, an NSSL meteorologist and director of the agency's storm-intercept program. What's needed are hundreds of automated weather modules placed in the remote areas of western states. These stations would measure variables like temperature, wind direction, wind speed, and atmospheric pressure; and they'd be linked to a central computer. The remote stations could detect localized wind shears, possibly indicating tornadic activity in advance.

A smaller version of these automated weather stations has already been used by storm-intercept teams. Called TOTO, an abbreviation for totable tornado observatory, the 400-pound (180-kilogram) unit can be placed directly in a tornado's path. Dr. Howard Bluestein, a University of Oklahoma meteorology professor

who's an avid tornado chaser, led a chase group on May 22, 1981, the afternoon TOTO was first deployed, and the same day a tornado struck Binger, Oklahoma. "That was a monster tornado—more than one-half mile [0.8 kilometer] wide," recalls Bluestein. "Even from 3 miles [4.8 kilometers] away, we could hear its low-frequency roar." A smaller tornado brushed by TOTO, which had been deposited along a likely pathway. "We got some wind measurements, close to that tornado, for the first time ever," Bluestein says.

However, TOTO probably will be abandoned this year, or outfitted with a more stable triangular base. "It's supposed to be able to measure wind speeds of 150 miles [240 kilometers] per hour or higher, but it would tip over before any winds got that high," says Davies-Jones. "Besides, it's such a long shot. Tornadoes are rare in the first place. To get TOTO in the direct path of one has proved practically impossible."

In the early 1980s, chase teams typically drove to areas of tornado activity in beat-up station wagons, using hand-held video cameras to document sightings. Today five-person intercept teams from the University of Oklahoma and the NSSL pursue twisters in 15-passenger vans equipped with $100,000 worth of computer-controlled electronic sensors. Mounted on top of the mobile laboratories are video cameras operated by remote controls inside the van.

Researchers brave extreme weather (above) to place a Totable Tornado Observatory in the path of an approaching twister. The 400-pound device contains instruments to collect data inside the tornado. Using vortex machines, scientists create mini-tornadoes (below) to study how buildings, hills, and other obstacles affect the storms.

Above: © Chris Johns/Allstock; right: © Dan McCoy/Rainbow

MEASURING TORNADOES

The Fujita, or F-scale (named for Dr. T. Theodore Fujita, a professor of meteorology at the University of Chicago), is the standard measure of a tornado's relative wind strength and damage potential. The lower end of the scale can also be used to assess other types of violent winds, such as microbursts and gust-front phenomena. Most tornadoes are F0 or F1; fewer than 1 percent are F4 or F5.

F0 or F1: Weak Tornado

F0: 40 to 72 mph (64 to 116 km/h). Light damage. Will do some damage to chimneys, break branches off trees, push over shallow-rooted trees, and damage signboards.

F1: 73 to 112 mph (117 to 180 km/h). Moderate damage. Will peel surface off roofs, overturn mobile homes or push them off foundations, and push moving cars off roads.

F2 or F3: Strong Tornado

F2: 113 to 157 mph (181 to 253 km/h). Considerable damage. Will tear roofs off frame houses, demolish mobile homes, push over boxcars, snap or uproot large trees, and turn light objects into missiles.

F3: 158 to 206 mph (254 to 331 km/h). Severe damage. Will tear roofs and some walls off well-constructed homes, overturn trains, uproot most trees in a forest, and lift and throw heavy cars.

F4 or F5: Violent Tornado

F4: 207 to 260 mph (333 to 418 km/h). Devastating damage. Will level well-constructed homes, blow poorly constructed buildings some distance, toss cars about, and turn heavy objects into missiles.

F5: 261 to 318 mph (419 to 512 km/h). Incredible damage. Will lift well-constructed homes off foundations and carry them considerable distances to disintegrate, send truck-size missiles flying through the air, debark trees, and perform other amazing—and deadly—feats.

Only buildings directly in a tornado's path suffer total destruction; neighboring buildings often escape unscathed.

Stalking Unpredictable Prey

The roving research teams don't really chase tornadoes; they try to intercept one without getting trapped in its immediate path. The teams attempt to remain in constant communication with their bases in Norman by radio repeaters, mobile phones, or even roadside pay phones.

I asked tornado chaser Brian Curran what a chase is like, and he agreed to show me. That night we left the NSSL after checking the Doppler. A line of severe thunderstorms was moving through central Oklahoma, and we headed for three large thunderstorm cells. High winds and large hail were guaranteed.

As the lightning flashed, I scanned the sky for wall clouds. "It's just a strong thunderstorm. Let's pop this one's core," Curran suggested. A moment later we were inside. Conditions: light rain; winds at 30 miles (48 kilometers) per hour, getting stronger; vicious thunderclaps.

Heavy rain began to pound the car, and we slowed down. "I can hardly see," Curran told me. The windshield wipers couldn't keep pace with the torrents of water. Curran logged in our position: "We're westbound on Highway 40, about 1½ miles [2.4 kilometers] north of the Cimarron Doppler site," he shouted, as I watched a nearby transformer blow in a shower of sparks. Lightning flashed about a mile away. Pea-size hail clattered on the car's roof.

We passed a stranded car, and we were out of the storm, which was about 3 to 4 miles (4.8 to 6.4 kilometers) deep. "We hit one good thunderstorm," Curran said to me. "But the bigger ones we were trying to intercept look as if they're going to die before they can reach us." A relieved smile crossed my face. "Good practice," Curran concluded.

Chase teams have verified the accuracy of Doppler-radar sightings. They've also documented the longevity of tornadoes—from a few seconds to about an hour. And they've confirmed the existence of satellite tornadoes called "suction vortices," which orbit the core of intense tornadoes and are responsible for the most violent winds.

The 1988 season is the second in which NSSL teams in vans have launched weather balloons. Radiosondes (transmitters and instruments lofted by the balloons) are used to record information about a tornado's core.

Even more useful tools are portable Doppler radars with a range of up to 3 to 5 miles (4.8 to 8 kilometers). Bluestein's university-based storm-intercept team has used a portable Doppler to measure actual tornadic wind speeds. On May 25, 1987, using an experimental 40-pound (18-kilogram) unit, Bluestein's team first obtained wind-speed measurements from inside a short-lived but strong tornado churning over the north Texas plains. The team used equipment designed by Dr. Wesley Unruh at the Los Alamos National Laboratory in New Mexico. The gear consisted of two side-by-side dish antennas—one to transmit a continuous microwave beam, and the other to receive the echo.

Eavesdropping on a Tornado

Bluestein describes the event: "We were south of Gruver, Texas. From 3 miles [4.8 kilometers] away, the radar measured tornadic winds of between 138 and 184 miles [222 to 296 kilometers] per hour." On the portable radar's receiving screen, Bluestein and his team of meteorology graduate students were able to "see" the tornado's wind speeds. They also "heard" them. The professor explains how: "Because wind-speed measurements are monitored as audio, we put on earphones. In one ear you hear motion toward you; in the other, away from you. The higher the pitch, the faster the motion. While we were listening to this tornado, one of the graduate students said, 'Wow, the pitch just went way up!' That's when the tornado touched the ground."

While researchers work with portable units in the field, Dr. John T. Snow is creating simulated tornadoes in his laboratory, located at Purdue University in West Lafayette, Indiana. He hopes to measure the wind speed inside a "tame" tornado sometime this year. He'll use a laser Doppler velocimeter that fires laser beams, not microwaves, at the spinning air to detect how fast the tiny particles in the air are moving.

Eventually, data from laboratory experiments and field research will be linked by computer with data from satellites, weather maps, and weather stations all over the country. This information will be used to construct the first computer models of tornadoes, similar to today's thunderstorm simulations. These models, called algorithms, describe a tornado's behavior. Simple models will take data and convert it to graphic displays that are easy for forecasters to interpret. To build more complicated models, actual weather conditions will be recorded over a long period, perhaps for many years, until a predictable pattern emerges.

ICE CYCLES

by Robert Kunzig

If you wanted to find out what causes ice ages, and you had the whole planet to search for clues, you'd probably never even think of heading for Devils Hole, Nevada. You might go to a place that was once covered by continental ice sheets, or perhaps to Greenland or Antarctica, where the ice sheets survive to this day. You might even decide to look in the oceans, reasoning that the cyclical freezing and thawing of vast amounts of water must inevitably have left an imprint there. In fact, investigators have searched in all those places, with con-

siderable success. But Isaac Winograd and his colleagues at the U.S. Geological Survey went to Devils Hole, and what they found there contradicts previous research in a small but important way. It's a question of timing.

The timing of the ice ages, most researchers thought, had been pretty well nailed down—and with it the theory of what causes them. Studies of deep-sea sediment cores had shown that at least during the past half million years, the glaciers have advanced and retreated every 100,000 years or so. The cycle is not symmetrical; the glaciers take tens of thousands of years to build up through the methodical stacking of snow on cold continents, but they melt away rather rapidly when it gets warm. Nor is the cycle smooth; both buildup and meltdown are punctuated by fits and starts of glaciation and partial retreat.

As recently as 14,000 years ago, ice sheets thicker than the Antarctic icebergs above covered much of North America. During the last half million years, glaciers have advanced and retreated every 100,000 years or so.

The Pacemaker of Ice Ages

In a seminal paper published in 1976, James Hays, John Imbrie, and Nicholas Shackleton announced that these various ice cycles could be matched to three cyclical fluctuations in Earth's orbit: the tilt of the planet's spin axis, the orientation of the axis (the planet wobbles like a top), and the shape of the orbit itself. Together these cycles affect the amount of sunlight that reaches Earth at each latitude and season. In the process, the researchers said, the orbital cycles act as the "pacemaker of the ice ages."

The theory was hardly new. It had been worked out in the early decades of this century by the Serbian mathematician and astronomer Milutin Milankovitch; but at the time no one knew enough about the timing of the ice ages to prove him right or wrong. Milankovitch proposed a simple physical connection between the orbital variations and the ice cycles. The key factor that determines whether the entire Earth is plunged into an ice age, he said, is the amount of sunlight that strikes the high latitudes of the Northern Hemisphere—where there is enough land to support vast ice sheets—during the summer. "It's always cold enough there in the winter to make ice," explains Imbrie, an oceanographer at Brown University. "What counts is how much melts in the summer."

Taking off from Milankovitch, Imbrie and others developed the theory that ice ages occur when the three orbital parameters conspire to keep the Northern Hemisphere cool during the summer; interglacial periods, like the one we're in now, happen when the northern summers get hot enough to melt the glaciers. The theory is appealing and straightforward, and it explains why the last set of glaciers began to melt around 14,000 years ago. What's more, the same summer-sunlight calculations also account for the next-to-last glacial retreat—an event called Termination II—which the deep-sea sediment records place at about 128,000 years ago.

The theory's success in explaining Termination II has been particularly important in convincing people that it is correct. If that date ever turns out to be wrong, Imbrie notes, "then a good part of the Milankovitch theory falls."

Winograd and his colleagues say that date may be wrong.

Window on the Ice Age

Looking back from the comfort of a temperate climate, it is hard to imagine the world our ancestors inhabited. Around 17,000 years ago, when Cro-Magnon men were filling the walls of Lascaux Cave in southern France with polychrome murals of horses and bison, most of northern Europe was buried under a sheet of ice 1 mile (1.6 kilometers) thick. In North America the ice covered all of Canada, reaching down over the Great Lakes region into New England and the Midwest. Along the edge of the glaciers, howling wind raised huge dust storms.

All told, the glaciers covered some 11 million square miles (28.5 million square kilometers) of land that now is ice-free, most of it in the Northern Hemisphere. With all that water removed from the ocean, the sea level was around 350 feet (100 meters) lower than today, and the shape of the land itself was different. There was no North Sea, no Baltic Sea, no Bering Strait.

The ice sheets never reached Nevada's Great Basin, and they never reached Devils Hole, a Mojave Desert oasis just east of Death Valley at the southern end of the basin. But Winograd and his Geological Survey colleagues—Barney Szabo, Tyler Coplen, and Alan Riggs—had good reason to pick Devils Hole for research. It is an unusual place: an open window in a 10-mile (16-kilometer)-long fault, along which the crust is being stretched apart. Just 30 to 40 feet (9 to 12 meters) wide at the surface, the opening narrows to around 10 feet (3 meters) at a depth of 50 feet (15 meters) or so, where it meets the water table. The pool in the bottom of Devils Hole has attracted biologists as the last refuge of *Cyprinodon diabolis,* a 1-inch (2.5-centimeter) pupfish.

And the pool is what attracted Winograd's group to Devils Hole, although they were not interested in the pupfish. For 30 years now, Winograd has been studying groundwater in the Great Basin. Recently his studies have focused on calcite, a mineral that precipitates out of groundwater—it forms stalactites and stalagmites in caves—and which is also the main component of limestone and most seashells. The calcite in seashell-laden ocean sediments provides a good record of the advance and retreat of the ice sheets, because its precise chemical composition depends on climatic conditions.

Winograd thought that record might be bolstered by calcite in Nevada groundwater. He had already found many veins in Death Valley, but they had been dry for more than a million years. In Devils Hole, however, groundwater still existed; calcite deposits there, he and his colleagues reasoned, would record the more recent ebb and flow of the ice sheets.

The presence of both calcite deposits and groundwater makes Devils Hole near Death Valley an ideal place for glaciological study. By analyzing the precise chemical composition of the calcite deposits, scientists can construct a fairly accurate chronology of the advance and retreat of the ice sheets. Devils Hole, unlike most other calcite deposits, still retains its pool of groundwater, and therefore might well provide a record of the more recent ebb and flow of the ice sheets.

© Gary Thompson/Las Vegas Review Journal

Clues from Heavy Oxygen

So in 1982 Riggs took the plunge into Devils Hole. Armed with a hammer and a chisel, he dove about 70 feet (20 meters) below the surface (the pool is more than 300 feet—90 meters—deep), found a vein of calcite on the wall of the fault, and knocked off a 6-inch (15-centimeter)-thick chunk. Winograd and Coplen then analyzed the rock for the climate tracer that other researchers had found in the marine sediments: the abundance of "heavy" oxygen, or oxygen 18. (Normal oxygen has a total of 16 protons and neutrons in its nucleus; oxygen 18, which is much rarer than oxygen 16, has two extra neutrons and therefore is a little heavier.)

The water in the ocean contains more oxygen 18 than groundwater does, for two reasons: first, water molecules that are weighted down by heavy oxygen are, not surprisingly, less apt to evaporate from the sea surface; second, the heavier water molecules that do evaporate are also more likely to fall out of a cloud as rain. Since clouds tend to form over the warm tropical ocean and move from there to higher latitudes, the oxygen-18-laden water tends to fall while the cloud is still over the ocean. So the rain or snow falling over land always contains less heavy oxygen than ocean water does.

Because the waters of Earth are a circulating system, the proportion of heavy to light oxygen in the seas is ordinarily maintained as the "lighter" water runs back into the ocean through rivers. But during an ice age, a lot of that water gets locked up in continental ice sheets. The ocean becomes depleted of oxygen 16 (or "enriched" in oxygen 18), and so does the calcite in the tiny plankton shells that sink to the ocean floor and form sediments.

At the same time, Winograd and Coplen reasoned, Nevada groundwater ends up getting even less oxygen 18 than it normally does. The argument is a little subtle. Rain falls from clouds only when the clouds have gotten cold enough for the water vapor to condense into liquid drops. In the colder climate of the ice ages, the reasoning goes, this happened more readily, and

EARTH SCIENCES 157

The Serbian scientist Milutin Milankovitch proposed that a connection exists between variations in the Earth's orbit around the Sun and cycles of glaciation. According to his theory, the ellipticity of the orbit (1) varies over a perod of about 100,000 years; the tilt of the axis on which the Earth spins (2), now at about 23.5 degrees, varies from 22.1 to 24.5 degrees over 40,000 years; and (3), the axis also precesses, describing a circle on the sky every 26,000 years. Marine sediments contain evidence of all three cycles. The map at right shows the greatest extent of ice during the last glaciation.

so the rain clouds reaching Devils Hole had lost a higher percentage of their water. In the process they had lost even more of their heavy oxygen—which "likes" to go into rain—than they would have lost under warm conditions. As a result, the groundwater in Nevada during the ice ages was abnormally light—and so was the calcite deposited by the groundwater on the walls of Devils Hole.

Operating on this assumption, Winograd and Coplen measured the oxygen 18 content of their calcite chunk, millimeter by millimeter, starting at the surface of the vein and going backward in time. They measured 110 samples in all, then plotted their measurements. "As soon as we saw the curve," Winograd recalls, "we knew we were onto something. We saw that we had the sawtooth pattern."

Comparing Patterns

The sawtooth is the hallmark of marine-sediment records: its distinctive shape traces the gradual buildup of ice in each 100,000-year cycle, followed by a rather abrupt return to warm conditions. The pattern has been observed in marine-sediment cores from the North Atlantic to the Indian Ocean, and in ice cores from Greenland to Antarctica. Now Winograd and Coplen were seeing it in Devils Hole calcite. They could match the major peaks and troughs in their oxygen-18 curve with peaks and troughs in the sediment and ice-core curves, and show that all three curves appeared to be recording the same glacial advances and retreats.

But not at the same time: when Szabo dated the calcite layers—he measured how much radioactive uranium 234 in each layer had decayed to thorium 230—the results were surprising, to say the least. The ice-age chronology from Devils Hole was shifted well back in time compared with the chronology from the deep-sea sediments. According to the sediments, for example, Termination II happened about 128,000 years ago; according to the Devils Hole calcite, it happened at least 147,000 years ago.

What difference does 19,000 years make? To the straightforward version of the Milankovitch theory, it makes a big difference. According to orbital calculations, 128,000 years ago the high latitudes were getting the unusually large amount of sunlight during the summer needed, in Imbrie's model, to melt the glaciers. But 147,000 years ago the situation was different. The amount of sunlight falling on the glaciers then was at best average.

Imbrie, for now, thinks Winograd's date for Termination II will turn out to be wrong. "I'm 95 percent sure they'll end up learning something new about the groundwater system of Nevada," he says. But if Winograd's results hold up, he acknowledges, "the neat match we now have with Milankovitch would be shot."

It hasn't come to that yet. The 128,000-year date for the next-to-last ice melt is sup-

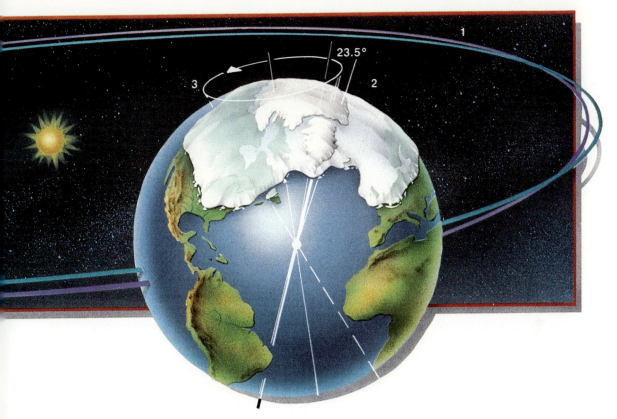

ported by several independent lines of evidence, ranging from coral reefs in Barbados—which show that that sea level was very high around 125,000 years ago—to limestone formations in Antarctic lakes. The Devils Hole measurements are a little lonely by comparison. "Before you can throw the last 20 years out the window, you have to find similar situations," says Chris Hendy of the University of Waikato in New Zealand, who dated the Antarctic limestone. "You need several springs like Devils Hole to be sure you're not looking at unusual circumstances." Nevertheless, the Devils Hole measurements have got people thinking.

Waxing and Waning

Actually, many researchers, including Imbrie, have long recognized that the Milankovitch theory has trouble explaining some of the most important features of the ice ages, starting with the 100,000-year cycle itself. The shape of Earth's orbit—the extent to which it departs from a perfect circle and becomes more elliptic—varies with a period of about 100,000 years. But that variation changes the amount of sunlight reaching Earth by less than half a percent. In contrast, the shorter tilt and wobble cycles change the amount of Northern Hemisphere summer sunlight by as much as 20 percent. The question researchers have asked ever since the Milankovitch theory came back into vogue is, Why should the weakest orbital cause have the strongest geologic effect?

Then there is the question of whether the waxing and waning of ice sheets in the Northern Hemisphere could, as the Milankovitch theory assumes, drive all the other climate changes that have accompanied ice ages—the cooling and glaciation of the Southern Hemisphere, for example. Glaciers in the Andes and Antarctica have advanced at roughly the same times as those in the north; that is, at times when the orbital calculations indicate that the Southern Hemisphere should have been getting a lot of summer sunlight. There is no generally accepted explanation of why the ice ages should be globally synchronized if they are driven by orbital fluctuations.

Finally, and, in the view of some researchers, most tellingly, there is the solid evidence that the ice ages have ended abruptly, with temperatures rising to their warmest levels even more rapidly than the ice sheets melted. During glacial periods the temperature has also fluctuated rapidly and dramatically. In the stately sweep of the orbital cycles, there is no provision for such bursts of climatic change.

All these failings of the Milankovitch theory have led many researchers to the belief that a lot more than just orbital variations are involved in the ice-age cycle—that, at the very least, Earth's response to those variations is complicated and riddled with feedback loops. People are now struggling to understand bits and pieces of that response. One of the most interesting models has been put forward by Wallace Broecker, a geochemist at Columbia University's Lamont-Doherty Geological Observatory.

Circulation Modes

Broecker has an explanation for the rapid changes that have marked the ice ages. He says that the entire climate system—consisting of the ocean, the atmosphere, and the ice sheets—has the ability to jump back and forth between two dramatically different modes of operation, and to do so almost instantaneously on a geologic time scale. The key to this is a kind of conveyor belt in the North Atlantic, of which the Gulf Stream is a part.

During warm periods like the present, this conveyor belt carries warm salty surface water as far north as Iceland and Scandinavia. As the warm water moves north, two things happen. First, the water gives up its heat to the westerly winds that warm Europe. Second, the cooled water, dense with salt, sinks to the bottom of the ocean, pulling warm surface water in behind it. From there it flows south, ultimately winding around the Antarctic and making its way into the Pacific.

But during an ice age, says Broecker, the conveyor belt jams. The sinking of salty water in the North Atlantic stops or is drastically curtailed. As a result, the warm surface water stops flowing north, the land around the Atlantic cools rapidly, and the ice sheets grow.

Computer models have confirmed Broecker's idea that the ocean has two distinct circulation modes. More important, he has recently found a bit of real-world evidence for the ocean's ability to switch between the two modes. It concerns a strange climatic interlude called the Younger Dryas, which began about 11,000 years ago.

At that time the glaciers of the last ice age had been in full retreat for 3,000 years, and temperatures were already almost as warm as today. Suddenly the North Atlantic region relapsed into deep cold, and the ice sheets paused in their retreat. The relapse was most severe in Europe. "Within a few hundred years," says Broecker, "all the trees that had reforested Europe disappeared and were replaced by tundra plants. That lasted between 700 and 1,000 years. Then—bang!—it went the other way again." The warming resumed, and the ice sheets completed their retreat within a few thousand years.

Several years ago Claes Rooth of the University of Miami suggested an explanation for this puzzling perturbation. As the North American ice sheet began its retreat, he said, its meltwater must have flowed down the Mississippi to the Gulf of Mexico. But after the ice sheet retreated beyond the Great Lakes, a different outlet was opened: the St. Lawrence. A prodigious surge of meltwater rushed down the river into the North Atlantic. Because the water was fresh and thus lighter than the salty Atlantic water, it could not sink into the abyss. Instead, it just piled up, jamming the oceanic conveyor belt that had been carrying warmth to the melting ice sheets. The conveyor restarted only when the meltwater flow tapered off.

This sort of abrupt transition is not explained by the Milankovitch theory, but it fits right into Broecker's model. The evidence he found last year is actually a sediment core from the Gulf of Mexico. Its oxygen-18 curve, he says, traces perfectly the abrupt cessation of the flow of meltwater into the Gulf after the St. Lawrence outlet had opened. Broecker believes this lends direct support to his ideas about the ice-age cycle as a whole. "Earth has shown its ability to jump from one way of operating to another," he says. "And these jumps are profound. It's amazing, really, that a planet should be able to change its own temperature by 9 degrees."

The Greenhouse Effect

Broecker's model is one of many, and even if it's true, it's far from the whole story. For instance, any complete theory of the ice ages would have to include the role played by the greenhouse effect. Swiss and French researchers who have analyzed bubbles of gas trapped in the Greenland and Antarctic ice caps have found that the amount of carbon dioxide (CO_2) in the atmosphere was much lower during the past ice age. That makes sense, because carbon dioxide is a greenhouse gas: it traps heat. Less CO_2 means a colder climate, and a sharp drop in the

During the last ice age (1), the conveyor belt that now carries warm, salty water to the North Atlantic had stopped. When it restarted (2) and the glaciers began to retreat, the meltwater flowed down the Mississippi valley to the Gulf of Mexico. But about 11,000 years ago (3), when the ice sheets had retreated north past the Great Lakes, meltwater surged down the St. Lawrence valley, effectively blocking the northward flow of warm ocean water and triggering a cold interlude. Today (4) the conveyor is once again warming the North Atlantic.

atmospheric concentration of greenhouse gases might be enough to set off, or at least amplify, a global ice age.

Such a drop would have to be driven by the ocean, which is the major storehouse of CO_2, and it just might be caused by a switch in the ocean's circulation pattern. It's too early to say. "There is a lot of ferment in the scientific community right now," says Rooth. "There is a lot of new data, but things are still at an early stage when theories are competing."

The simplicity once promised by the Milankovitch theory has proved elusive. And now Winograd and his colleagues have suggested that some of the evidence supporting the theory may have been wrong. "The Milankovitch theory has not been without problems for some time now," says Winograd. "We've simply tossed another monkey wrench into the works."

Few researchers, including Winograd, doubt that orbital variations have some impact on Earth's climate; but the extent of that impact and the mechanism that underlies it are now very much open to question. It's even conceivable that Earth might produce a 100,000-year ice-age cycle on its own, without the help of Milankovitch. The orbital variations may simply be putting wiggles on the cycle, adding high-frequency noise to the natural hum of the terrestrial machine. "We have a system with many variables," says Winograd. "There may not be a single 'pacemaker' of the ice ages. It may be the interaction of all of them."

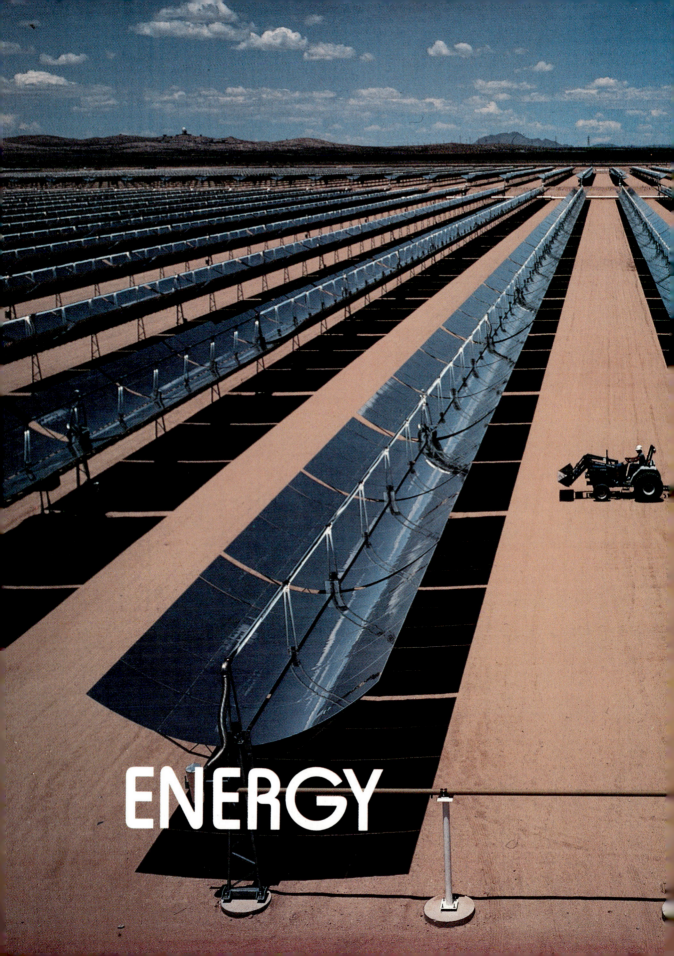

Review of the Year

ENERGY

Oil spills and rising oil prices overshadowed advances that may hold far greater significance in the long run: reports of nuclear fusion at room temperature and a dramatic, though virtually unnoticed, breakthrough in solar energy.

by Anthony J. Castagno

The record cold weather that plagued much of the U.S. last December caused prices for home heating oil to skyrocket. Prices leveled off and eventually dropped during the first half of 1990.

Fossil Fuels

The energy world was rocked by the news on March 24, 1989, that a huge tanker, the *Exxon Valdez,* had gone aground and ruptured in Prince William Sound off the coast of Alaska. The accident spilled 260,000 barrels of oil and touched off a national—and international—protest about the environmental consequences of the spill.

The cost of home heating oil skyrocketed in December as much of the country shivered in record cold. Consumers claimed "gouging," while suppliers claimed the price increases were the result of tight supplies due to unprecedented demand coupled with production problems. The rise was so dramatic that federal and state legislative hearings were convened to investigate, but by year's end, nothing conclusive had been decided. As the weather moderated in January, prices began dropping, but still remained well above those at the beginning of the winter heating season.

There was continued, though lessening, discord among the members of the Organization of Petroleum Exporting Countries (OPEC), as a decade of uncertainty—marked by falling revenues and fluctuating prices—seemed to have come to an end. Prices rose nearly 50 percent during the first few months of 1989, breaking the $20 barrier in March, hovering in this range throughout the year, and increasing only slightly in December, even though prices to consumers that month increased so dramatically. At an OPEC meeting in September, the daily production ceiling was raised from 19.5 million to 20.5 million barrels, still lower than the estimated actual output of some 22.5 million barrels. By December, however, OPEC output had soared, exceeding an estimated 24 million barrels, the highest since 1981.

Oil production in the U.S. continued to decline, dropping by approximately 1 million barrels a day, to 7.5 million barrels. Actual consumption in the U.S., however, continues to rise, and is now at some 18 million barrels a day, up a million barrels since last year, and close to the 1978 record of 18.8 million barrels. With imports up some 20 percent over a year ago, 1989 had the dubious distinction of being the first time since 1977 that imported oil supplied more than half the U.S. petroleum needs. The U.S., producer of only about 3 percent of the world's oil, now consumes nearly 30 percent.

Estimates of the world's fossil-fuel reserves—oil, coal, and natural gas—continued to rise, contrary to the conventional wisdom of the 1970s that these supplies were dwindling. During the past three years, estimates of recoverable oil reserves had increased 30 percent, to 890 billion barrels, or enough for more than 45 years (compared to 1979's projection of 28.2 years). Likewise, estimated coal reserves have increased some 80 percent in the past three years, to nearly 1.1 trillion tons.

Recoverable reserves of natural gas have been doubling every 10 years, with latest estimates of 4,000 trillion cubic feet, enough for nearly 60 years.

Once considered a worthless by-product of oil production, natural gas continued to take on increased prominence in the U.S. energy picture. Since 1986, gas consumption has increased 11 percent, from 16.6 trillion cubic feet to 18.5 trillion. The American Gas Association estimates usage in 1990 will exceed the 1972 record of 22.7 trillion cubic feet. Prices, too, continued to increase, attracting more interest in expanding exploration and recovery operations in the U.S. and in the Arctic.

The increased reserves, and the increased consumption, continued to concern environmentalists, who had hoped that higher cost and scarcer supply would curb consumption of fossil fuels, the major source of carbon dioxide and other gases believed to be responsible for the "greenhouse effect" and the global warming trend. The World Energy Conference, at its September convention in Montreal, warned that carbon dioxide emissions could rise by 70 percent by 2020 unless there is some reduction in the increasing production of this gas.

Nuclear Power

Controversy swirled around reports of successful room-temperature nuclear-fusion reactions (see p. 173). In the meantime, nuclear power derived from the more conventional fission reactions continued producing a substantial share of the country's electricity requirements in 1989, approximately 20 percent, second only to coal. By year's end there were 111 plants licensed in the United States and more than 410 worldwide, providing more than 17 percent of the world's electricity.

One of the country's most controversial nuclear plants, Seabrook in New Hampshire, received its low-power operating license on May 18, 1989, and began its first atomic chain reaction June 13. The plant ended its testing in September, and, on March 1, 1990, was issued a full-power license. Plant operators planned to start producing electricity by summer.

In Massachusetts the Pilgrim plant, idled since 1986 because of maintenance and management problems, restarted and reached full power in the fall. The troubled Rancho Seco plant in Sacramento, California, became the first nuclear plant ordered by voters in a referendum to shut down, and it has been permanently removed from service.

In New York the fate of the controversial Shoreham plant on Long Island was sealed as state and utility officials reached an agreement to decommission the plant. Shortly after this decision, the Nuclear Regulatory Commission (NRC) granted the plant a full-power operating license, but the die had already been cast. U.S. Department of Energy officials, concerned about the insufficient supply of electricity on Long Island, have requested that the plant not be dismantled immediately as had been planned; the matter remains in negotiation.

AP/Wide World Photos

The still controversial Seabrook nuclear power plant in New Hampshire is in the process of gearing up to produce electricity by the summer of 1990. The plant received its full-power license last March.

Alternative Energy

In 1989 scientists at Boeing announced an efficiency breakthrough in photovoltaic technology, with a cell that operates at 37 percent efficiency, substantially higher than last year's record of 31 percent produced at Sandia National Laboratories. This compares to 34 percent efficiency for fossil-fuel-burning plants, and opens the door to future widespread use of the sun for the generation of electricity.

Just as solar and other alternative technologies seemed to be ready to become viable, however, the U.S. pulled back even further from its support and research. At the same time, Japan and other foreign countries were stepping up their activity in this field and buying the rights to much of the U.S. technology as well as investing in or buying outright companies involved in research. International Fuels Cells of Connecticut, for example, whose chairman, William Podolny, is often called the godfather of fuel-cell technology, lost federal research funding in 1985 under the Reagan administration. When no U.S. firms would make up this difference, Podolny turned to Japan's Toshiba and Tokyo Electric, both of which were enthusiastic in their response and investment. "Renewable-energy technology is on the brink of exploding," said one Washington, D.C., analyst, "but a lot of U.S. companies won't be around when that happens."

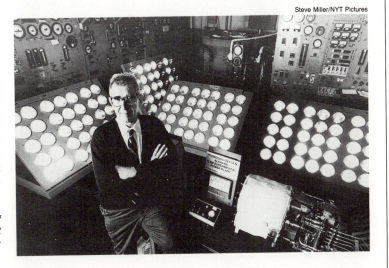

Steve Miller/NYT Pictures

William Podolny (right, with a bank of headlights powered by a 12-kilowatt fuel cell) decries the lack of U.S. investment in renewable-energy technology.

Automotive Fuels of the FUTURE

by Andrew Kupfer

Move over, gasoline. Autos powered by clean-burning alternative fuels are the best means to reduce lung-burning ozone in the many cities where it remains stubbornly above the federal limit. Motorists won't rebel at the change, but quite the opposite: "The

The government has urged automakers to turn out more cars that run on compressed natural gas (above) and various other gasoline substitutes. Such fuels have the dual advantage of decreasing automotive pollution emissions while increasing America's energy self-sufficiency.

Above: © Bob Daemmrich/The Image Works; facing page: John B. Carnett/Reprinted from *Popular Science* with permission © 1990 Times Mirror Magazines, Inc.

new fuels will take the market by storm." Cars will pollute less *and* perform with pizzazz.

Is this Ralph Nader speaking? No, it's our own president, George Bush. The most daring part of the clean-air plan that he submitted to Congress directs automakers to start selling cars that will run on gasoline substitutes—500,000 units in 1995, rising to 1 million a year in 1997 and thereafter.

Bush's former sidekicks in the oil patch, who would have to invest billions in new plants to produce alternative fuels, are irate. Oilmen hope to run the new-fuels bandwagon into a ditch by devising a cleaner-burning gasoline that could be produced in existing refineries. Under heavy lobbying, the administration added reformulated gasoline to its list of potential alternative fuels just before releasing details of its clean-air bill.

The automakers, who aim to prosper no matter what fuel the government dictates, have been working to solve problems posed by alternative fuels, notably methanol made from natural gas. Anticipating Bush's proposal, Detroit for some time has been developing flexible-fuel cars that can run on gasoline or methanol or blends of the two.

Why Methanol?

Cost: $300 or so added to the car's price. Methanol, or methyl alcohol, is the most promising substitute. Indianapolis 500 racers have used it for decades. It's safer: when cars collide, they are far less likely to explode than when using gasoline. It's cleaner and more powerful. Thanks to the fuel's superior combustion and high octane rating, an engine running on it produces up to 20 percent more horsepower—and up to 90 percent less ozone. Tail-pipe emissions of volatile hydrocarbons, which combine with nitrogen oxides in sunlight to produce ozone, are far lower than with gasoline. Methanol also is much less prone to evaporate, and thus produces less smog from fuel escaping to the atmosphere during refueling and when the engine is hot.

So what's the catch? Well, the stuff has a tendency to eat through metal and turn engine oil to mayonnaise. Though cars become more virile on methanol, they gulp lots of it: because of its molecular structure, the fuel has only 60 percent as much energy per gallon as gasoline. Anyone worried about dwindling fossil fuels can relax: a given amount of those resources can be turned into enough methanol to do roughly the same amount of work as gasoline, even though the methanol takes up more room. But for motorists, methanol will mean more trips to the filling station.

A Smorgasbord of Alternatives

Washington's ozone fighters are looking at a whole smorgasbord of clean fuels. The most pristine energy, from the user's standpoint, is

A car engine running on methanol fuel produces 90 percent less ozone pollution than does one powered by gasoline.

GM's prototype electric car can travel 124 miles in city traffic between charges, a range sufficient for most commuters.

electricity. As anyone familiar with battery-driven golf carts knows, vehicle emissions are zero. But generating the juice in a coal-fueled plant to charge the batteries can produce plenty of pollution—plus carbon dioxide, which causes global warming. When that's figured in, electric cars are not so pure after all. A battery that delivers as much energy as just 1 gallon (3.7 liters) of gasoline would weigh several hundred pounds. Electric vehicles today have a range of less than 100 miles (160 kilometers) and take hours to recharge. They wouldn't be much fun to drive, either, requiring an eternity by car-buff standards—nearly 20 seconds—to accelerate from zero to 50 miles (80 kilometers) per hour.

Ethanol—which can be thought of as 200-proof vodka—is an efficient, low-polluting fuel. It is less corrosive than methanol, and it packs one-third more energy per gallon. But ethanol, which can be made only by fermenting corn or other crops rich in starch or sugar, costs twice as much per mile as gasoline. And it would take the entire U.S. corn crop to make enough fuel for a quarter of America's cars. Heavily subsidized since energy-crisis days, ethanol is used regionally in small amounts to boost octane in a gasoline blend called gasohol. Corn Belt lobbyists are working through Congress to broaden the subsidy to cover a new gasoline additive derived from ethanol.

A fuel that deserves a future, especially for urban buses and delivery vans, is compressed natural gas. Carried under pressure in thick-walled cylinders, the stuff already powers hundreds of thousands of vehicles in Italy, the Netherlands, New Zealand, and Canada. It burns cleanly, starts easily, and has an extremely high octane rating (130, versus about 105 for methanol and the low 90s for premium gasoline). Maintenance expenses would be lower than with gasoline because natural gas is kind to engine oil.

Natural gas is ideal for vehicles that return to a home base every day. Large-fleet owners could save enough on maintenance and fuel costs to pay for converting their vehicles and buying a refueling compressor. Price: $250,000 for one that can refill gas cylinders in a few minutes. United Parcel Service (UPS) is testing 10 trucks altered by Brooklyn Union Gas to make deliveries in the Canarsie section of Brooklyn, New York.

A nasty little PR problem haunts natural gas: fear of explosions. UPS picked Canarsie partly because its trucks wouldn't have to go through New York City's tunnels, where compressed gases are prohibited. But Brooklyn

Union has conducted safety tests, including one in which a Mercury Marquis was dropped from 90 feet (27 meters) without damaging the canister. In another, a stick of dynamite was taped to the tank and set off, the equivalent of a car hitting a brick wall at 89 miles (143 kilometers) per hour. Gas leaked out, but did not ignite.

The real drawback for drivers is that natural gas is inconvenient. Storing enough of it on board to travel more than 100 miles (160 kilometers) at a shot is a problem, even when the gas is squeezed to 200 times atmospheric pressure. It might be years before all filling stations offered the fuel. To keep the option of using gasoline where no natural gas is available, the car would need separate systems for carrying the two fuels and conveying them to the engine. Along with a gasoline tank, it would also need heavy natural-gas cylinders sitting like little zeppelins in the trunk.

Modifying an existing auto to run on natural gas as well as gasoline would cost about $2,000, but far less if cars were mass-produced with dual-fuel capability or designed from scratch for natural gas alone. Since most urban homes are served by natural gas, motorists could install fueling compressors right in their garages. But it might cost $2,000 for one that would take three or four hours to fill 'er up.

So the best substitute fuel is methanol, also known as wood alcohol—the kind that's bad to drink. A liquid fuel requiring no pressurization, it could be dispensed from pumps alongside those selling gasoline. C. Boyden Gray, the president's counsel, sold his boss on clean fuels when the presidential campaign was getting under way. For five years he has driven a Chevy Citation that runs on alcohol. Tanking up with alternative fuels, Gray says, is a far less expensive way to clean the air than cracking down on small ozone-creating operations like dry cleaners and paint shops. Battling air pollution from any source is an obsession for the lanky, professorial North Carolinian. He picks up a plastic jar with a half-inch or so of nasty black grit on the bottom. "Here, see this," he says. "This is five minutes from a diesel bus. Five minutes!"

Chemical companies make 1.5 billion gallons (5.7 billion liters) of methanol a year in the U.S. by combining methane—the main component of natural gas—with steam at very high temperature and pressure. That output, the energy equivalent of three days' U.S. gasoline consumption, is used almost entirely by the chemical industry to make such items as the glue holding plywood together. But methanol could also shave as much as a second off the time it takes production-line cars to go from zero to 60 miles (96 kilometers) per hour. That is possible, says Roberta Nichols, Ford's principal research engineer for fuel systems, because the fuel burns faster and at a lower temperature than gasoline.

Flexible-Fuel Vehicles

At the heart of the auto industry's plans for methanol is the flexible-fuel vehicle, or FFV. A sensor in the fuel line measures the proportion of methanol and gasoline in the mixture. Its find-

Compressed natural gas is a cost-effective alternative for vehicles that return to a home base every day. The United Parcel Service has already begun evaluating the benefits of natural gas in ten trucks that make deliveries in the Canarsie section of Brooklyn, New York.

© Tobey Sanford

Although it resembles water, methanol (above), also known as wood alcohol, causes blindness and even death if swallowed. Automakers, keeping in mind that people most typically ingest fuel during gas-siphoning attempts (below), have designed strainers for the mouths of car fuel tanks to prevent motorists from poisoning themselves.

ings are transmitted to a computer chip similar to those on most new cars, which continuously adjusts the air-fuel ratio and timing for different driving conditions. If more methanol is coming through the fuel line, the computer lowers the air-fuel ratio and advances the spark to ignite the fuel sooner. Result: more oomph.

FFVs will be costlier because methanol would rot the guts of standard cars. It eats through aluminum, used in gasoline-tank linings, and dissolves the rubber in fuel hoses. So automakers will outfit FFVs with stainless-steel tanks; hoses may be made of Teflon wrapped in stainless-steel mesh. Thus the added price of $300 or so. General Motors and Ford are planning to send large fleets of FFVs to California—2,200 and 2,500, respectively—by 1992 to see how they bear up under regular use, especially the fuel sensor. Chrysler, late getting into the game, will be sending out a fleet of 10.

How will Detroit market these cars? All three automakers will emphasize the chance for drivers to do their bit for the planet, as well as touting the cars' potent engines. The performance gains will vary by make. Because of methanol's high octane, the engine can be souped up by maximizing the compression ratio—the degree to which the fuel is squeezed before it is ignited—without causing it to knock. Since this change worsens performance when the car is running on gasoline, GM and Ford will leave the compression ratio unchanged. When fueled with methanol, the cars will still get a power boost of up to 7 percent. Chrysler, playing catch-up, will build FFVs with higher compression ratios, aiming for a 20 percent jump in power, and a ride you can feel in the seat of your pants.

The flexible-fuel approach makes sense because it allows methanol-using vehicles to proliferate before the fuel is available everywhere. The president's bill calls for alternative-fuel vehicles to be sold first in the country's nine most polluted metropolitan regions. Within these areas, owners of FFVs will be expected to burn methanol exclusively. Some states, such as California, may require service stations to sell clean fuels. Outside the designated areas, FFV owners will be able to use gasoline.

Methanol Drawbacks

Will FFV owners avoid methanol if gasoline is cheaper? Oil executives grouse that it will cost 20 cents to 30 cents more for an amount that will do the work of a gallon of gasoline; they fear

Alternate fuel drawback: a tankful of compressed natural gas hardly provides enough energy to travel 100 miles.

they may have to raise the price of gasoline to subsidize the price of methanol. Chemical companies that make methanol—and have surplus capacity—say the price should be about the same as premium gasoline.

A limited switch to methanol, involving fewer than a tenth of the new cars sold annually, need not make the U.S. more dependent on imported fuel. Some in the administration talk of using methanol made from natural gas in Organization of Petroleum Exporting Countries (OPEC) nations that now burn it off as waste, such as Saudi Arabia. But the U.S., Canada, and Mexico could supply all the natural gas needed for the president's clean-air plan.

Methanol has its critics, not all in the oil business. Cars fueled with 100 percent methanol are hard to start at temperatures below 40° F

President George Bush has emerged as a strong supporter of alternate fuel research and development. His clean air policy calls for American automakers to manufacture cars that run on gasoline substitutes at a rate of one million per year by the end of 1997.

ENERGY 171

GASOLINE VERSUS ALTERNATIVE FUELS

FUEL	VEHICLE EMISSIONS (Grams Per Mile)				FUEL COST TO TRAVEL 100 MILES	MAXIMUM TRIP ON FULL TANK OR CHARGE	OCTANE RATING
	Hydro-carbons	Nitrogen oxides	Carbon monoxide	Carbon dioxide			
Gasoline*	1.4	0.8	7.2	350	$4.00	400 miles	87–93
Methanol	0.15–0.25	0.8	3–7	320	$3.50–$4.25	260 miles	105
Ethanol**	0.5–0.8	0.8	6–8	140–350	$7.50–$9.00	300 miles	99
Natural gas	0.15–0.25	0.8	1–4	300	$3.00–$5.00	100 miles	130
Electricity	0.3	1.2	.05	200–300	$2.40–$4.70	60–100 miles	N.A.

*Under rules proposed in President Bush's clean-air plan.
**Based on a blend of 85 percent ethanol and 15 percent gasoline.

(4.5° C), and require frequent oil changes. Though conventional pollutants would decrease, more of another would issue from the tail pipe—formaldehyde, an eye-stinging smog-former that was used to pickle the frog you dissected in high school bio lab, and which is now thought to cause cancer. Methanol itself is highly toxic. A jigger or so can kill or blind you. And if you have an accident in bright sunlight, you might get burned by failing to notice methanol's nearly invisible flame.

Automakers are working hard to overcome the problems. New lubricating oils, they hope, will stand up well. To add volume to the fuel tank, Detroit is experimenting with plastic polymers that can be molded to take advantage of every spare nook and cranny under the chassis. Firing up the car in cold weather is easy with a blend of 85 percent methanol and 15 percent gasoline, known as M85. It allows quick starts at temperatures down to −20° F (−28° C), and burns with a bright yellow flame. Using M85 instead of pure methanol, however, has a price: the reduction in smog falls from 90 percent to 50 percent.

California has set strict standards for formaldehyde emissions, which automakers will control with redesigned catalytic converters. Converters installed on today's conventional cars can meet the standard when new, but fall short by the time the car has reached 50,000 miles (80,000 kilometers).

Detroit is also taking steps to make sure drivers don't take a fatal swig of wood alcohol. Fuel most commonly enters the body when a motorist siphons it from a tank; an estimated 14,000 Americans a year swallow gasoline while sucking on a tube to get the siphoning process started. Automakers will install strainers in the mouth of the tank to prevent drivers from having the drink to end all drinks.

The Bottom Line

Big oil, its livelihood at stake, hopes to head off alternative fuels with revamped gasoline. In August 1989, Arco announced a new lead-free gasoline for pre-1975 cars without catalytic converters. It cuts tail-pipe emissions by 15 percent. For newer cars, oil companies might try to get rid of gasoline's more volatile components and add oxygen-rich chemicals made from methanol and ether that make fuel burn more fully and boost octane.

Top engineering officials from the oil and auto industries are holding a series of meetings to discuss new formulations and changes necessary in engines to help the redesigned fuels work better. Says Jim White, in charge of Arco's alternative-fuel project: "We think that methanol is a fuel of the future. We just don't think it's *the* fuel of the future. We'd rather have the chance to fool around with our gasoline."

The Environmental Protection Agency (EPA) doubts that cleaner gasoline could cut pollutants from newer cars by more than 15 percent. Even if the oil companies can make gasoline cleaner, at least one authority in Detroit thinks it's worth going the extra mile with methanol. "We certainly are supportive of the oil companies' efforts to make clean gasoline," says Ford's Roberta Nichols. "But we still see methanol as the replacement in the long term. And in the short term, methanol would still have the advantage in air quality and higher power." The payoff in more breathable air alone seems well worth the expense.

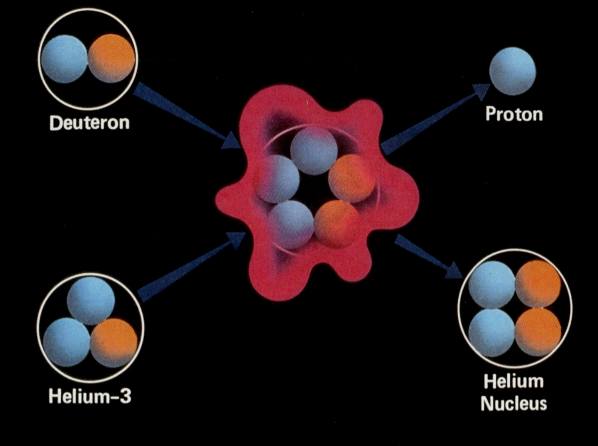

University of Wisconsin

THE FUTURE OF FUSION

by Gene Bylinsky

You thought cold-fusion-in-a-flask was dead? Well, yes and no. The scientific establishment has generally pooh-poohed the claims of B. Stanley Pons and Martin Fleischmann, the University of Utah team that first announced it had fused hydrogen ions in a jar of water at room temperature. But scientists are not so dismissive of Steven E. Jones, a Brigham Young University (BYU) physicist. Indeed, he may have won a small prize in the cold-fusion lottery. No, not those $25 boxes of pennies with him in the photo on page 174. The pennies are there to shield his instruments from any gamma rays that might confuse their readings. Just possibly, his brand of cold fusion may turn out to be useful in producing energy on a commercial scale. Even if it doesn't, he says, he has planted a seed that science can nourish into a beautiful flower. The cold-fusion drama is still unfolding.

The promise of fusion is enormous: power that is clean, relatively cheap, and virtually limitless. Joining light atoms like hydrogen to produce energy may have few of the drawbacks of splitting heavy atoms like uranium, the fission method used in ordinary nuclear reactors. For nearly four decades, scientists have been

Fusion, the joining of atoms to produce energy, does not produce the dangerous by-products that occur with fission, the splitting of atoms to produce energy. Although a commercially feasible fusion reaction has not yet been devised, researchers haven't stopped trying.

ENERGY   173

By passing electrodes through a chemical solution, physicist Steven E. Jones of Brigham Young University claims to have achieved a modest fusion reaction that produces one neutron per second—hardly enough for commercial development. The pennies shield his instruments from radiation that might distort their readings.

Above: © Eric Lars Bakke; facing page: Princeton Plasma Physics Laboratory

trying to reproduce the mighty fusion processes of the sun at tremendous pressures and temperatures, in huge test reactors like the tokamak at Princeton. If fusion can be achieved at room temperature in simple laboratory apparatuses, and the technique can be made commercially feasible, the impact will be almost incalculable on a world dependent on a diminishing supply of fossil fuels that foul the air as they burn.

The tokamak machine derives its name from a Russian acronym meaning doughnut-shaped chamber. It represents one of the old-fashioned approaches to achieving fusion on a scale needed to run huge power plants. In what is called magnetic confinement, the massive machine compresses hot-fusion fuel into a gaseous plasma that would ignite to produce sustained, controlled energy. General Atomics of San Diego has sunk millions into the process.

KMS Fusion of Ann Arbor, Michigan, is a leader in the other long-standing hot-fusion method, which uses lasers to compress and "burn" tiny glass fuel pellets the size of a grain of salt. While fusion physicists freely use terms like "burn" and "ignition," they don't intend the words in a chemical sense. Fusion burning produces energy a million times more efficiently than any conventional fuel. Conventional fuels burn by combining hydrocarbons with oxygen, and produce oxides of carbon and nitrogen that create smog and damage the ozone layer.

The overwhelming stumbling block is that no one has succeeded in getting a fusion reaction to the critical stage where it throws off more power than it absorbs. At best, with its current fuel, the Princeton tokamak has generated only a fraction of 1 percent of what it consumes. If either hot or cold fusion ever reaches the point where it returns a lot more energy than it uses up, and the process is commercially practicable, the world will have a nearly inexhaustible fuel supply. The main components of fusion fuel are deuterium and tritium, the heavier chemical forms of hydrogen; the first occurs naturally in water, and the second can be readily obtained from conventional nuclear reactors. Oceans, rivers, and lakes all contain enough deuterium to last billions of years. Fusion power could make any country energy-independent.

No Breakthroughs Yet

Critics such as Lawrence M. Lidsky, an MIT nuclear engineer who abandoned fusion after 20 years and now works on gas-cooled fission reactors, have questioned the very essence of the hot-fusion program. Lidsky feels that deuterium and tritium are the wrong fuels to use. They produce neutrons in such superabundance, he says, that they could damage the reactor structure; also, because tritium is radioactive, it could create some of the same contamination problems that are inherent in conventional nukes. Harold P. Furth, director of the Princeton University Plasma Physics Laboratory, home of the largest American tokamak, concedes Lidsky's point in part, but insists that the difficulty can be overcome by developing the proper ceramics and other impervious materials.

While hot fusion shows promise in the laboratory, its future as an economic reality is sufficiently uncertain that many experts now call it the toughest project in the history of science. Says David O. Overskei, a General Atomics senior vice president: "We've made significant progress, but nothing anyone would call a breakthrough." Even if they succeed, scientists don't see fusion reactors in operation before the year 2025, or maybe not until 2050.

Could cold fusion slip into the gap? Small amounts of it may actually be taking place in those supposedly discredited University of Utah experiments. Although scientists trying to duplicate them have seen no sign of familiar fusion reactions, and most have concluded that the Utah process is purely chemical, a rarer type of fusion never before observed could conceivably be happening on a minuscule scale.

So don't give up on cold fusion yet. The furor over fusion-in-a-flask could be just the opening salvo in the battle to harness cold fusion for energy production. Whatever else they may have done, the University of Utah scientists have at least energized physicists and chemists everywhere. If not this time around, then at some future date, cold fusion may become reality.

A Pittance of Neutrons

Jones' discovery, if it proves out, could be one pathway. While his lab and that of the University of Utah team are only 45 miles (72 kilometers) apart, the Utah and BYU scientists had been working independently along strikingly similar lines. Both used electrochemical techniques to pass a current between electrodes in a chemical solution, but neither lab knew what the other was up to—and they arrived at dramatically different results. Jones, quiet-spoken and modest, never claimed to have invented a magic energy box. He says he has produced less than one neutron a second—a pittance in hot-fusion terms. That's partly why his peers welcomed his findings more cordially than they did those of the University of Utah's team.

Scientists strive to achieve fusion at room temperature, and not at the extreme conditions found in tokamak reactors.

CONTROVERSY IN THE PALLADIUM

It was a wonderful dream: a tiny atomic lattice of palladium that could do basically the same thing as the mammoth machines of conventional fusion research. University of Utah chemist B. Stanley Pons and his British collaborator Martin Fleischmann claimed that ions of deuterium—a heavy form of hydrogen—were catalyzed somehow by the palladium and combined to form helium. Cold fusion was generating heat, they said, more energy than their tabletop contraption consumed.

They thought they had made possible unlimited supplies of power. The University of Utah thought it would get rich, and asked for $25 million in federal funds for more research. The world's scientific establishment, largely unable to duplicate the results, thought otherwise. Most investigators said Pons and Fleischmann had erred in their measurements and were merely feeding a chemical reaction—not fusion—with the electrodes. The Utah team stands by its claims.

Drs. B. Stanley Pons (left) and Martin Fleischmann.

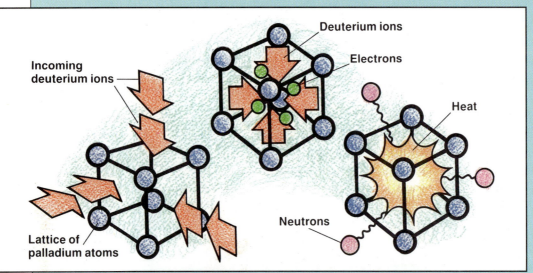

Pons and Fleischmann claim that deuterium inside a cubelike palladium lattice should fuse and emit heat.

Jones is quick to say that as of now, his work is useless for producing energy—and he readily concedes there's a chance he is seeing merely tiny peaks in background neutrons naturally present in the environment, rather than fusion neutrons thrown off when two deuterium atoms join to form an isotope of helium. Johann Rafelski, a theoretical physicist from the University of Arizona who works with Jones, is less guarded. He thinks that if Jones' level of neutron production is real, once the phenomenon is understood, there is a genuine possibility that the reaction can be scaled up to generate useful energy.

One thing is certain: even the most skeptical physicists cannot argue that cold fusion is impossible in principle. One type of room-temperature fusion has been indisputably achieved in a simple chamber containing deuterium and tritium gases; physicist Luis Alvarez of

FUSION WITH MAGNETS

You wouldn't think a company could survive in the never-never land of magnetic-confinement fusion, which is still decades away from any payoff. Privately held—and "highly profitable"—General Atomics of San Diego, once a subsidiary of General Dynamics, is a leader in the field and has been in the business since 1958.

The government has poured $8 billion into fusion research since 1951; this year federal funds will provide the company with some $35 million for that work, about a quarter of its expected revenue. (Most of its sales come from defense contracts and the building of advanced conventional reactors.) The owners, brothers Linden and J. Neal Blue, believe firmly in fusion: since 1980 the company has put about $20 million of its own money into the project. From Japan, it has

David Overskei and Tihiro Ohkawa.

obtained about $30 million in government research funds—and a top physicist, Tihiro Ohkawa, now a vice-chairman of the company, who works with David O. Overskei, senior vice president for fusion.

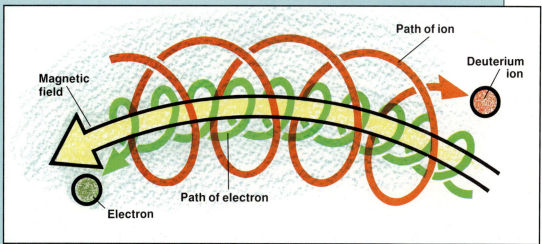

According to Overskei and Ohkawa, a magnetic field confines electron-stripped deuterium, causing fusion.

the University of California at Berkeley, a Nobel prizewinner, first observed it in 1956. Cold fusion has also been shown to work in liquid and solid forms of hydrogen at extremely low temperatures.

This type of cold fusion, originally postulated in the late 1940s, uses heavy subatomic particles called muons (pronounced MEW-ons). Injected into deuterium gas, muons replace the electrons orbiting the deuterium nuclei. Much like lead balls placed on the surface of balloons, the muons squeeze the atomic nuclei together, generating energy in the form of neutrons. The U.S. Department of Energy has been supporting research in muon-catalyzed fusion, as the process is known, since 1982. Most scientists, however, do not see it as a stand-alone source of energy—at least not at this point in time—but rather, as a possible auxiliary to the more familiar hot fusion. Some scientists think muon-

ENERGY FROM A TINY "SUN"

A small, publicly held company pioneers in the second approach to hot fusion, which uses converging laser beams to implode tiny glass spheres filled with fusion fuel, deuterium and tritium. Those are the heavy isotopes of hydrogen that fuse to form helium and generate neutrons for energy production. KMS Fusion of Ann Arbor, Michigan, founded in 1967, got most of its $22 million in 1988 revenues from the U.S. government and made an $800,000 profit.

The company concentrates on the design of the microspheres, which it supplies to federal labs. (In a fusion reactor, each of these spheres would produce as much energy as seven gallons of gasoline.) "With our laser," says senior vice president Timothy M. Henderson, "we verify if the experiments are worth doing in the big national labs."

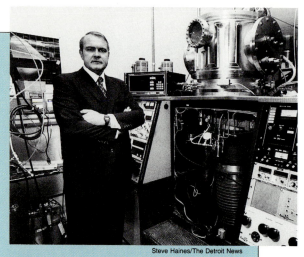

Patrick Long is chairman of KMS Fusion in Michigan.

Says chairman Patrick B. Long: "Now when I go to cocktail parties, people no longer look puzzled and bored after they ask me what I do. They want to know all about fusion."

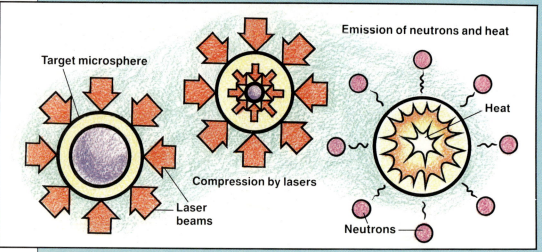

Laser beams set off fusion reactions by crushing tiny glass spheres filled with heavy hydrogen isotopes.

catalyzed fusion could be put to work producing neutrons for use in making fuel for conventional fission reactors.

Nuclear Polygamy?

Even in science, history tends to repeat itself. When Alvarez first observed muon-catalyzed fusion more than 30 years ago, he felt, as he said later, that "we had solved all of the fuel problems of mankind for the rest of time." It turned out, though, that quite apart from the prohibitive cost of producing muons as catalysts, their lifetime is too short to provide a net gain of energy from the reaction. The muons' matchmaking ability is impressive: a single muon can precipitate as many as 150 "marriages" between deuterium nuclei by flitting from one set of atoms to another. But that's not enough to reach the energy break-even point, much less produce more power than the process uses.

Rafelski, the Arizona physicist, thinks that for muon-catalyzed fusion to be practicable, the number of nuclear marriages each muon makes would have to increase "by another factor of two or three"—a result he thinks could be attained eventually. He adds that while no easy and inexpensive way to make muons exists today, such an apparatus could come along in the future. Two years ago he and Jones wrote an article predicting that one day, muon-catalyzed fusion could drive steam turbines.

Researchers digging frantically through the scientific literature recently discovered with amazement yet another type of cold fusion. It may be only a distant relative, and it's probably too weak to serve as an energy source, but it's apparently another example of the phenomenon. Soviet scientists reported in 1986 that they had recorded a few fusion neutrons emanating from cracked crystals of a compound of lithium and deuterium. The Soviets fired bullets at the crystals to fracture them. One theory is that the strong electric field at the spreading cracks in the crystals speeds up deuterium atoms so much that they collide and fuse.

The same effect could occur in the deliberately cracked palladium and titanium crystals the Jones team has used. A reporter visiting Jones' lab recently was startled when Jones picked up a palladium electrode, knelt in a corner, and began hitting the electrode with a decidedly low-tech hammer. Bang! Bang! Bang! Bang! A scientist gone mad? Hardly. Jones was merely following Rafelski's suggestion to create more cracks in the metal, the way the Russians did with bullets, to see if more fusions could take place.

Jones is also looking into achieving fusion under high pressure without using electricity or chemistry. He puts scraps of titanium into a stainless-steel tube, and then forces deuterium into the metal at about 1,000° F (538° C)—lukewarm compared with tokamaks or lasers. He hopes the deuterium will fuse. "I think this process will work better than the other one," he says.

Fusion Fingerprints

He was encouraged in his pursuit of not-so-hot fusion by a 1978 Soviet report that had found in certain metals inexplicably high ratios of the variants of helium produced by hydrogen fusion. Because helium cannot be made chemically, its presence invariably represents "the ashes of fusion," in Rafelski's phrase. That suggested to the Russian scientists that fusion had taken place naturally on earth. But by what means? A Brigham Young colleague further whetted Jones' appetite by speculating that fusion might still be happening deep underground, produced more by high pressures than by high temperatures. The small amounts of helium that volcanoes and hot undersea vents spew out might be fusion's fingerprints.

If Jones' experiment generated any helium, the amounts would have been too small to measure because the experiment produced few fusions. Jones has said that, metaphorically speaking, if his experiment creates $1 worth of energy, then the Pons-Fleischmann scheme, if it really worked, would yield enough to pay off the national debt 10 times over. Jones doesn't think the University of Utah experiments produce any heat by fusion. Most scientists agree.

Near-Term Outlook

Most scientists also think hot fusion is a lot further along as a potential energy source than any other variety. The odds that one or more of these approaches—hot or cold—will prove out are good, though it may not be for half a century. "One fusion neutron does not mean you have a new technology," says Rafelski. "But if you open the door to a new direction, it's almost always true that something comes out of it. It took 200 years to learn how to light up this room with electricity after Luigi Galvani first produced an electric current in a cell full of chemicals and metals."

By one estimate, despite all the hoo-ha, researchers around the world have been spending no more than $1 million or so a week to try to reproduce the disputed Utah results. In the age of big, expensive science, the laboratory equipment used in recent cold-fusion work—simple jars and electrodes—seems refreshingly modest. The mammoth machines required for hot-fusion research give an inflated impression of the size and cost of the fusion effort, but even so, hot fusion will consume only $514 million in U.S. funds this year—close to the projected cost of a single Stealth bomber. Considering the energy independence fusion would make possible, even those inadequate amounts could help buy more security for the U.S. than would a B-2—and more peace for a world that may soon be squabbling with growing violence over who gets what share of the diminishing supply of conventional fuels.

Photos: Jose Lopez/NYT Pictures

Coal Mining Goes HIGH TECH

by Matthew L. Wald

The 4-inch (10-centimeter) carbide-steel claws of a screaming yellow robot tore relentlessly at the coal seam 1,000 feet (300 meters) below the surface of Wana, West Virginia, digging enough in 30 seconds to meet the electric needs of a typical house for a year.

The robot's human master, old enough to remember when miners used picks and shovels and loaded coal into carts drawn by dogs, strolled beside, periodically pushing buttons on support equipment.

Coal mining, an unglamorous business known mostly for accidents and bitter strikes, is going high tech, and the resulting productivity gains are important for a nation that is using up its other fossil fuels.

A New Mining Method

The country's deposits of coal are expected to last more than a century, while oil reserves will last only 10 years at current rates of production, assuming no more oil is found. Gas reserves are also more limited than coal supplies. As oil and gas are used up, finding and extracting them become more expensive; by contrast, because of technical advances like the robot, the cost of digging coal is falling. Thus, over the long term, the price gap between coal and oil and gas will probably grow, and coal could become an even more important energy resource.

The robot here, in the Consolidation Coal Company's Blacksville No. 2 mine, 700 to 1,400 feet (215 to 430 meters) under the Mason-Dixon line, is perhaps the most advanced in the country. It is but one recent improvement in a highly automated technique called longwall mining, in which coal is removed not by tunneling like a worm through an apple, leaving more of the target than is removed, but by methodically shuttling back and forth across the width of the deposit and devouring nearly everything.

Longwall mining has been around for decades, but mine operators have recently made the technique far more efficient. The method

Plentiful, inexpensive coal soon may emerge as the nation's most important fossil fuel, especially now that some mines have improved their productivity through advanced technology. At Blacksville No. 2 mine in West Virginia, workers can even monitor the washing, crushing, and sorting of coal from a computerized control room (left).

can now extract about 75 percent of the available coal, compared with 50 percent for conventional mining, which is done largely with machines that dig tunnels. Moreover, the coal can be recovered far more inexpensively.

Extensive Deposits

While coal has been hailed as the nation's energy savior since the 1973 Arab oil embargo, the cost of oil has not risen enough to make coal a good substitute beyond the power plant, where it is burned to produce electricity. Cutting the cost of coal production is a small step toward wider use. When costs justify it, coal can even substitute for gasoline.

Already, coal is the nation's biggest indigenous energy source, and its role has been growing. Last year, it supplied more than 23.4 percent of all energy consumed in this country, up from 17.5 percent in 1973, the year of the first oil shock, and 19.1 percent in 1979. And it accounted for nearly one-third of all energy produced in this country last year, up from 22.5 percent in 1973. It is the main fuel used to make electricity, providing about 56 percent, and is also the nation's largest energy export by far.

But increased production of coal also raises intense environmental questions. The issues range from the speculative—like the extent of global warming caused by the burning of carbon-rich fuels such as coal—to the spooky, like the collapse of land as robot miners pass beneath. The government is spending more than $5 billion to find cleaner ways to use coal, and the coal companies have tried harder to ameliorate the effects of the collapse, called "subsidence."

Longwall mining is used wherever coal is present in thick, level seams deep under the ground. (Shallow deposits are often exploited by strip mining, in which the surface layers are scraped away and the coal is excavated from an open pit.)

Embracing new technologies and sometimes developing technology on its own, Consolidation Coal, a subsidiary of E. I. du Pont de Nemours and Company, last year mined 55 million tons with 9,500 employees. Fifteen years earlier, it took 23,000 workers to mine 55 million tons. A big part of the change was installing 22 longwall machines like the robot here.

Despite the loss of employment, the United Mine Workers union (UMW) generally favors longwall mines. "Back in the 1950s, John L. Lewis said it would be far better to have highly productive operations that were able to pay our folks good wages and benefits than to have 600,000 shovelers living in poverty," said John Duray, a spokesman for the union.

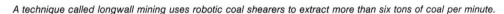
A technique called longwall mining uses robotic coal shearers to extract more than six tons of coal per minute.

SUBSIDENCE SCENARIO

Longwall mining looks worse from the surface in Sherrard, West Virginia.

Toby and Patricia Marsh's two-bedroom retirement home seems as if an earthquake had hit. Walls are cracked, doors will not open, and floors have lifted in some places and sunk in others. The furnace is tilted, the foundation cracked.

Outside, the Marshes say, springs have dried up, and a pond has disappeared. And land is no longer level enough to keep cattle or mow with a tractor; it is going back to brush.

Neighbors in this community south of Wheeling have similar complaints. The reason is that the Consolidation Coal Company's Shoemaker mine lies several hundred feet below, and a longwall machine undermined the area in December 1984. The company promised in advance to make repairs, but the Marshes and their neighbors are still involved in a bitter legal dispute with Consol over the extent of the damage and their compensation for it.

The collapse of land over longwall mines, called subsidence, is prompt. Consol says that conventional mining, which leaves a grid of tunnels, produces subsidence, too, but decades later.

The coal companies bought the rights to millions of acres of underground Appalachian coal around the turn of the century, often in deed amendments that exempt them from liability for surface damage. But faced with tougher state and federal laws, Consol and other longwall mining companies have become increasingly accommodating to the property owners above, promising to install flexible water, sewer, and electric connections while the settling occurs, and, when the settling is complete, to make repairs and "inconvenience payments."

The Marshes, who cashed in their life insurance policies in 1971 to buy 49 acres (20 hectares) where they could raise a few head of cattle and eventually retire, say Consol has not kept its promises, particularly the promise that the subsidence would stop in a few weeks. The Marshes and several neighbors are suing Consol and the state of West Virginia, which they assert has failed to enforce its mining laws.

Other neighbors say Consol gave them money for repairs, and they signed settlements absolving the company of further responsibility. But then the ground shifted again, causing more damage.

Consol executives say that subsidence cannot still be continuing on the Marsh property, and that they can now repair the damage to the house, but they have not been allowed access to the land. Tom Hoffman, a Consol spokesman, said the relationship with the Marshes has been very difficult lately. "No matter what you do, they won't be satisfied," he said, "because what they want is to be left alone."

But the Marshes knowingly exposed themselves to the risk, he said, adding that his own house is over a conventional mine. "I have to live with the uncertainty of not knowing whether it will happen."

Called Safer

Union officials say longwall mines are also inherently safer in their design, and the mines that can afford them are generally the bigger, better-managed ones.

Consol, the company that operates this mine, is by far the biggest longwall operator.

In the longwall system at Blacksville No. 2, miners dig a tunnel the height of the coal seam, and in it they set up a row of floor-to-ceiling steel brackets to hold up the rock and soil above. Sheltered by the shields, the miners install a robot that runs back and forth on a track across the 750-foot (225-meter)-long wall of coal on the open side of the bracket. On each pass, a wheel 4 feet (1.2 meters) in diameter, studded with claws, chews up the coal in a space extending 3 feet (1 meter) high and 30 inches (76 centimeters) in front of it. The coal falls to a conveyor belt. At the end of the wall, the robot miner, called a shearer, reverses direction and makes a pass just below the first.

When the shearer has finished its second pass, devouring almost all of the 6-foot (2-meter) height of the coal seam here, it moves 30 inches (76 centimeters) forward and starts again. Then, in a sequence worthy of a grade-B science fiction film, each of the 153 steel shields in the row takes its turn stepping forward into the space that the shearer just opened, to be in position for the next pass. As the steel shields step forward, the earth behind them collapses.

The longwall moves forward 17 to 20 feet (5 to 6 meters) per eight-hour shift, two or three shifts a day, until it has covered about 1 mile (1.6 kilometers). Then, in a few hours, the equipment is moved to another track 750 feet (225 meters) long, and starts work again.

Fierce Competition

The process can help coal companies achieve enormous efficiencies. Competition among coal producers is vicious, and executives at Consol say they plan to meet it by increasing productivity 3 to 6 percent annually.

Setting up a modern mine can cost $100 million, and the chance of making a killing is very small. Unlike the price of oil, the price of coal has been fairly stable since the early 1980s, and is not likely to go anywhere soon; when inflation is taken into account, the price of coal is down by 44 percent since 1975.

The price varies according to the sulfur content and energy content of the coal. In September 1989, the amount of coal required to make a kilowatt-hour of electricity cost only 54 percent as much as the oil needed for the same job, says the National Coal Association.

Longwall mining has the distinct drawback of causing the land above the mine to collapse, a phenomenon called subsidence. The downgrade above resulted from longwall mining several hundred feet below the surface.

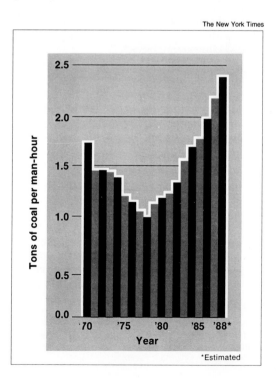

The New York Times

Tons of coal per man-hour vs. Year ('70 to '88*)

*Estimated

Blacksville has two of Consol's 20 longwall machines. There are only about 100 longwall systems running in the 2,000 underground coal mines in the U.S., but they produce about one-third of the coal that comes from such mines, and some experts think that figure will rise to more than half by the year 2000.

About a third of the nation's coal comes from surface mining, mostly in the West. The other third comes from underground mines that use older technologies, either because their owners cannot afford longwall machines or because the coal deposits are unsuitable for them.

Further Refinements

Engineers are now working on additional refinements to longwall mining. For example, operators of the Blacksville No. 2 mine hope that even pushing the buttons to make the 153 shields step forward—now done by a human—will soon be taken over by computer.

High-tech changes are taking place at the surface, too. In a small factory at the mouth of the mine, the coal is washed in water and other liquids, removing some of the sulfur; it is then dried, crushed, and sorted by size.

Above and left: © Phil Schofield/Picture Group

FUNERAL for a REACTOR

by Seth Shulman

The mighty Columbia River carves its way through the Cascade Mountains of the northwestern United States to leave a massive swath of running water dividing Washington and Oregon. Standing on its gorge-ridden banks, one cannot help but feel humbled. Every day the Columbia irrigates 7 million acres (2.8 million hectares) of land and delivers a steady surge of electricity to the entire region, as well as to parts of Los Angeles, nearly 1,000 miles (1,600 kilometers) away.

Today the river's quiet power is especially striking as a tiny dot of a barge called the *Paul Bunyan* peeks over the hazy horizon. It is hard to believe that the barge, flanked by tugboats and dwarfed by its surroundings, totes a thousand-ton payload. Over its 30-million-year history, the river has never known the likes of this barge's cargo. On board, like a huge sepulcher, is the spent reactor vessel from the world's first commercial nuclear power plant.

A Nuclear Milestone

Almost silently, the *Paul Bunyan* navigates the last leg of what must rank as one of the world's longest funereal journeys. Setting off from the

A barge carried the spent reactor vessel (above left) of the world's first commercial nuclear reactor 8,100 miles from Pennsylvania to its burial spot in Washington state. Protesters greeted the barge as it began the final leg of its journey up the Columbia River.

outskirts of Pittsburgh, Pennsylvania, the nuclear remains have traveled down the Ohio and Mississippi rivers, into the Gulf of Mexico, through the Panama Canal, up the Pacific coast, and now inland on the Columbia en route to an earthen burial in a trench on the Hanford Military Reservation (population 30,000) in southeastern Washington.

Beyond even the impressive 8,100 miles (13,000 kilometers) logged, though, the barge voyage represents a significant milestone for the nuclear industry. Shippingport, the nuclear reactor whose spent core now lies atop the barge, was the cornerstone of President Dwight D. Eisenhower's "Atoms for Peace" program. Built in 1957, Shippingport ushered in the age of commercial nuclear-power generation. Now—fittingly, perhaps—the Shippingport reactor vessel's final voyage also hails a new chapter for nuclear power. Its demise is being used by the U.S. Department of Energy (DOE) as a model project to show the world that nuclear reactors can be shut down and torn apart safely.

Until recently, this was not a pressing issue, given the more urgent (and as yet unresolved) problem of how to dispose of a nuclear reactor's fuel rods, which can power a reactor for only about three years. The spent fuel rods—the most highly radioactive entities known on the planet—are still the major problem on the industry's hands, and it is worth a moment's pause to contemplate their nature.

The fuel for a nuclear reactor is made up of small pellets of uranium poured into thin metal rods 12 feet (3.6 meters) long. Each full-size reactor holds roughly 40,000 to 50,000 of these pencil-thin rods, wrapped in groups called fuel assemblies. Every year, approximately a third of a reactor's extremely radioactive rods need to be disposed of. The rods are said to be "spent," but actually, during their time in the reactor, they have become far more radioactive than when they were inserted.

In the United States, the spent fuel from commercial nuclear-power generation—now totaling some 22,500 tons (20,000 metric tons)—is nearly all stored temporarily in water-filled cooling ponds adjacent to the nation's reactors. Many of these ponds are already filled to capacity, but more waste than ever is currently being generated; its volume is expected to double within the decade. And the proposed high-level waste repository deep inside Yucca Mountain in Nevada is still far from ready to receive it.

A proposed complex deep in Nevada's Yucca Mountain could someday isolate radioactive waste from the environment.

Courtesy U.S. Department of Energy

In 1954, President Eisenhower waved a "radioactive wand" that signaled workers to begin construction of the Shippingport Atomic Power Station in Pennsylvania.

Everyone knew, of course, that in the long run, the reactors themselves, with their roughly 30-year operating life spans, would eventually face the retirement quandary: at least 50 nuclear power plants in the Western world will reach retirement age within the next decade. In the United States alone, about a dozen nuclear reactors are ready now for decommissioning, a process that formally terminates a nuclear power plant's operating license and is supposed to clean up residual radioactivity, leaving the site safe for other uses.

Debris Dilemma
Indeed, the problem of disposing of all wastes from nuclear reactors, military as well as commercial, has lately become a matter of nationwide planning and controversy. Currently, federal law calls on states to create regional or individual dumps for "low-level" waste (which can run the gamut from relatively harmless to absolutely lethal) by 1993. The selection process has sparked growing "nimby-ism" ("not-in-my-backyard"), as Americans discover, according to one recent report, "that nuclear waste is coming soon to a dump near them." High-level waste from military sites, except for spent fuel, is destined for burial near Carlsbad, New Mexico, although there is much controversy over the site. And the spent fuel from both military and civilian reactors awaits a tomb tunneled into Yucca Mountain, a site scientists are also questioning (it lies between two prominent earthquake faults and is 12 miles—19 kilometers—from a still-young volcano).

Today, however, 32 years into the age of commercial nuclear power, Shippingport provides a firsthand glimpse of what is actually entailed in the process of decommissioning a nuclear plant. It signals, too, that many daunting challenges still lie ahead as we face the prospect of eventually dismantling the more than 500 reactors now on-line or under construction worldwide.

Project manager John Schreiber has taken a good deal of criticism for some of the choices made by the Department of Energy in the Shippingport project. Some critics see the reactor vessel's lengthy voyage to a military dump site as an embarrassment, only underscoring the siting obstacle facing the disposal of future commercial reactors. At present no civilian site in the United States is willing to take such nuclear debris, and Hanford was available only because of the government's oversight of this unique pilot project.

Many also question the Energy Department's decision to ship the nuclear reactor vessel in one piece. Shippingport, at 72 megawatts, was a relatively small reactor. Decommissioning a full-size, 1,000-megawatt, modern commercial reactor, according to many experts, will almost certainly require carving the vessel into smaller pieces, a problematic and high-radiation-producing step that the department notably managed to avoid in Shippingport's decommissioning—taking some of the shine off Schreiber's sweeping claims for the project.

But now is a time when Schreiber understandably takes a good deal of pride in his work. Despite everything, the job is on schedule for its 1990 completion, and is even slightly under its $100 million budget. The project has managed to avoid any large-scale catastrophes, and has weathered countless minor debacles. "We do wear a belt and suspenders," Schreiber says, "although no one gives us credit for it."

Only a few bright streaks against the tarp covering the vessel mar Schreiber's message—markings left by several dozen protesters on nearby bridges, who, as the barge passed, pelted it with balloons filled with fluorescent green

Hanford workers and their families gave the deactivated reactor its first, and perhaps only, warm welcome.

paint. The streaks of paint reflect the opposition of many activists and residents in Washington and Oregon to the disposal plan for the Shippingport reactor. They also make it clear how much times have changed since the reactor was born not all that long ago.

"Radioactive Wand"

Shippingport's tale is really nothing less than the story of commercial nuclear-power generation. The story began in September 1954, in a much-publicized media event from a Denver television station where President Eisenhower waved a "radioactive wand"—a neutron source—in front of a small detector. This device sent a radio signal that triggered the ground-breaking for the Shippingport Atomic Power Station. These were the heady days of nuclear power. In the year following Shippingport's completion, Lewis Strauss, the second chairman of the Atomic Energy Commission (AEC), made his now-famous prediction that nuclear energy would soon be "too cheap to meter."

It is clear now that the Shippingport reactor was more a key component of a public-relations campaign than a technical or economic breakthrough. At the time, the Eisenhower administration was preoccupied with the fact that the Soviet Union had made its own nuclear weapons, and had just brought on-line its first weapons-producing and electricity-generating reactor at Obninsk. Shippingport offered a way to offset the Soviet Union's newfound nuclear limelight—making the United States the world leader in the peaceful use of the atom.

Ironically, the "civilian" Shippingport reactor was actually a military prototype originally planned as a propulsion system to power a ship in Admiral Hyman Rickover's still-youthful nuclear Navy. The reactor was a scaled-up version of a design the Navy had already used successfully in its submarines, the so-called pressurized water reactor. In this type of reactor design, still the most commonly used throughout the world, the nuclear fuel heats water under high pressure in the reactor vessel. The pressurized water is pumped through pipes that get hot enough to boil coolant water into steam, which in turn is captured to drive turbines that generate electricity.

Shippingport actually began generating commercial electricity for a local utility company by the end of 1957. In truth, though, like the feeble man behind the curtain who was operating the visage of the Wizard of Oz, the Shippingport reactor's record shrank beside its own image. Shippingport was acknowledged early on to be an economic failure, producing electricity more than 10 times as expensive as that generated by conventional coal-fired plants of the

After inspecting the reactor vessel for radiation leaks, workers used cutting torches to free it from the barge.

day. But to early proponents of nuclear energy, it didn't matter. Shippingport was a symbol of a new age. And the utility didn't mind, because Shippingport was a heavily subsidized government project. Part of the arrangement dictated that the government would be responsible for the project from cradle to grave, which explains the Energy Department's involvement today. In decommissioning other large-scale commercial nuclear power plants in the years ahead, however, the industry will be left pretty much on its own.

Shippingport's symbolic mystique continues to overshadow reality even in its demise. The reactor vessel's burial isn't triggered by a "radioactive wand," but it is a media event just the same, a symbol being used to benefit the nuclear industry. Little matter that the Hanford Reservation is already home to six similar reactor vessels removed from decommissioned nuclear submarines and quietly buried by the Navy, or that the government is overseeing the project's every detail, depositing a government-built reactor on a government-built reservation.

This, we are told repeatedly, is a prototype of the first commercial disposal effort of its kind, a demonstration that promises many successful efforts from the nuclear-power industry in the future.

On Thursday morning at the dock, the task ahead is to haul the mammoth cylinder on its 320-wheel transport trailer to its designated burial site at Hanford. Three huge trucks with names such as "Mary Ann" and "Big Jim" painted above their front fenders spew diesel fumes as they strain to lug the reactor vessel off the barge. But on the twisting gravel grade away from the dock, they lose their traction and slip back down. Eventually, two additional enormous vehicles must be recruited to aid the cause, underscoring the uncertainties involved in the entire decommissioning process. Finally the ungainly caravan manages to take the first hill leading away from the dock, but even with all this horsepower, the trucks can move only about 3 miles (5 kilometers) per hour, and even that pace seems difficult.

Meanwhile, in Shippingport, Pennsylvania, 25 miles (40 kilometers) outside Pittsburgh, the reactor vessel's former home is quiet. It still looks like a messy construction site, but not much is left here. It will take almost a year to complete the project's mandate of returning the site to the Duquesne Light Company (the Pittsburgh utility that ran the Shippingport plant) for unrestricted use. But when it is all over, John Schreiber stresses, "the place will be as safe as my backyard."

There is little sign now at the Shippingport site of the years of demolition work that have taken place, but a clue to the project's magnitude is seen in the fragments of several dense concrete foundations marking the muddy earth, 6 feet (1.8 meters) thick in places.

With this kind of bulky composition, nuclear power plants would be tough to dismantle even if they were not radioactive. The same massive structure touted for its safety by the nuclear industry during the plant's construction and operation has now returned to haunt the industry in the decommissioning process.

About 15 percent of all the materials used in the construction of an average reactor will emit a discernible amount of radiation at decommissioning time. While most of these materials will contain only low levels of radioactivity, they make up a hefty volume: some 160,000 cubic feet (4,480 cubic meters), or enough to cover a football field 12 feet (3.6 meters) deep in debris. Significantly, though, these figures do not include the nuclear fuel rods that drive the fission reaction in a power plant. While comparatively small in size, these fuel rods account for the overwhelming majority of radioactivity at a nuclear power plant.

A nuclear reactor's fuel is not normally considered a part of the decommissioning process because it is assumed that the fuel will be removed prior to decommissioning. In fact, no country in the world has yet settled upon exactly what to do with these extremely radioactive spent fuel rods. For the decommissioning of Shippingport, though, the problem of where to store its used fuel was bypassed. The fuel assemblies were moved, temporarily, to a military facility in Idaho.

Lifting a Reactor Vessel

Tearing down a nuclear reactor makes for an odd kind of demolition work, combining large and complex tools with some painstaking hand labor. Pat Coughlin, the principal engineer for the Shippingport decommissioning project, becomes animated when he talks about the operation he helped design and oversee to lift the 1,000-ton (900-metric-ton) reactor vessel from its underground housing. For this key step, a huge frame of steel girders was built, replete with four huge hydraulic jacks and more than 30 steel cables. The jacks, something like those used at a garage to lift a car, were strong enough to lift more than 6,000 tons (5,400 metric tons), or the equivalent of several shopping-center parking lots worth of cars. When the steel frame was complete, the jacks hoisted the reactor vessel 77 feet (23 meters) into the air, moved it half again as far horizontally along a track, and lowered it onto the transport trailer. "It worked like a charm," says Coughlin proudly, "but that baby was heavy."

The Energy Department's decision to remove the reactor vessel whole, which avoided cutting apart the radioactive structure, saved $7 million and dramatically lessened worker exposure to radiation, Schreiber stresses. This is true, but critics like Cynthia Pollock Shea, a senior researcher at the Worldwatch Institute, and Michael Pasqualetti, a decommissioning expert at Arizona State University, warn that such a procedure will not be possible for larger reactors. Because of this, they say, the Department of Energy missed an important opportunity with Shippingport to test the remote-control technologies that will be needed to carve up reactor vessels in future efforts, when government resources will not be on tap.

Even at Shippingport, handling the reactor vessel required caution. Dozens of workers in protective suits with radiation detectors began work in 1985, tearing apart the outside emergency-cooling-water tanks with picks, shovels, and drills. Then, by hand, they scrubbed down the radioactive residue from the walls inside the plant. Later the job required bulldozers and backhoes, blowtorches and larger plasma torches, explosives, and even wrecking balls. During the height of activity in tearing down the main containment building, more than 200 workers attacked various aspects of the project. All this labor generated vast amounts of debris, which workers sorted into piles of radioactive and nonradioactive material. Over the next few years, aside from the barge

Powerful trucks strained to tow the 1,000-ton reactor 30 miles overland to its final resting place on the Hanford Reservation.

Below and facing page: © Phil Schofield/Picture Group

that would eventually transport Shippingport's reactor vessel, more than a hundred large truckloads of radioactive waste were driven across the country to the dump at Hanford.

The sheer quantity of low-level radioactive debris generated by dismantling a reactor presents the final major dilemma of the decommissioning process. Currently none of the three operating disposal facilities for low-level commercial waste will accept the volume of debris a decommissioned reactor produces. The prospect of creating the numerous repositories needed around the country to bury the remains of existing nuclear power plants is so politically volatile that renewed consideration is being given by industry and environmental groups alike to leaving the reactors standing indefinitely. The simplest plan would entail posting guards around the plants, at least until some other solution is found. A more elaborate version would be a kind of mothballing, or "entombment," encasing the reactors in concrete. These plans, though, are unsettling in that many of the reactors that will eventually be retired are in or near populated areas. Because potentially dangerous levels of radiation will be present for tens, or even hundreds, of thousands of years, the prospect of leaving a nuclear reactor, emitting low levels of radioactivity, to decay and possibly even be forgotten centuries into the future is not a comforting one. These concerns would remain even if the spent fuel rods were stored elsewhere, as DOE experts contemplate.

For Shippingport, decommissioning was possible because of the availability of the Hanford Military Reservation. The choice of Hanford as the site for Shippingport's remains is rife with historical and political irony. In many ways, Hanford and Shippingport are like matched halves of the nuclear age. What Shippingport was to civilian nuclear-power generation, Hanford was to weapons production. The Hanford Reservation was established as a top-secret location during World War II to produce weapons material for the Manhattan Project. It was here on this sagebrush-desert reservation half the size of Rhode Island that the nation's first significant quantities of plutonium were made, including the plutonium used for the bomb dropped on Nagasaki, Japan. After the Army Corps of Engineers selected the remote site in 1943, work began here at a furious pace. Almost overnight a city of 30,000 workers was imported—virtually none of whom had an inkling of what they were working to build. A "company town" sprang up next to Hanford, and in the intervening 40 years, it has become the sprawling tri-city area of Richland, Pasco, and Kennewick.

Landscape of Hidden Contamination

It's Friday now, and we are inside the gates of the vast Hanford Reservation, well past the guardhouse at the entrance nearest the town. Bill Klink, my escort, reports that a specially trained police force here is armed with submachine guns and attack helicopters to protect against terrorists, but the place seems fairly quiet nonetheless. The day is hot, and the mood has dimmed a bit. Inside the guarded reservation, the reporters are gone, as are the casual onlookers. But there is still quite a crowd of managers and other personnel on the scene. And there is a good contingent from the site in Pennsylvania. Pat Coughlin came with several other colleagues from Shippingport. They have brought a cooler of soft drinks in the back of their rented car, and they are squinting from a bluff at the vessel being hauled to the western section of the reservation. The bluff is one of only a few elevated vantage points in the entire reservation, and the view is filled with gray-green scrubby flatland—miles and miles of it.

From here the land displays a barren beauty that belies the truth about the hidden contamination it holds. Just as Shippingport symbolized the public's hopes about nuclear power, Hanford has come to represent the nuclear age's worst environmental nightmare. During its 40-year history, the Hanford Reservation, with its nine plutonium-producing reactors, has released untold millions of curies of radioactivity into the ground, water, and air. In just one example, according to information uncovered by local reporters sifting through some 19,000 pages of government documents, more than half a million curies of radioactive iodine 131 were released into the atmosphere at Hanford between 1944 and 1957. The accident at Three Mile Island released only 15 to 24 curies.

This spring, Governor Booth Gardner of Washington and officials from the Department of Energy signed a landmark cleanup agreement for Hanford. The agreement is laudable in its intent, but, if anything, it dramatized for the world the extent of Hanford's environmental woes. The cost of the 30-year cleanup is estimated at a staggering $57 billion, more than four times greater than the Energy Department's entire budget for fiscal year 1989. The enor-

Even in its burial trench, the reactor will continue to emit low but discernible levels of radiation for centuries.

mous cost of the project reflects the amount of waste material involved—an estimated 30 million cubic feet (840,000 cubic meters) of nuclear waste, and perhaps as much as 100 times that amount of contaminated soil. Such facts and figures seem incomprehensible, otherworldly.

It won't be until the next day that the vessel is finally driven into its trench, but, impatient with its pace, Klink and I leave the bluff to drive the last 7 miles (11 kilometers) to the final resting place. Here, adjacent to Shippingport's assigned trench, are row upon row of thousands of black drums of "transuranic" waste, long-lived but slightly less radioactive than high-level waste material. And just 100 yards (91 meters) away lies Hanford's most notorious site, where hundreds of large underground storage tanks have leaked more than 500,000 gallons (1.9 million liters) of high-level liquid waste into the ground. That the majority of the hundreds of single-shell tanks built between the 1950s and the 1970s have breached is a tragedy of immeasurable proportion. And despite the recent cleanup plan, no one has any idea how to clean this facet of the site's contamination. The entire area surrounding the underground tank farm is dangerous. And the tanks are still filled with corroding, highly radioactive sediment.

Of all the travesties that have recently come to light about the handling of radioactive materials at the Energy Department's facilities around the country, Hanford's underground tanks may be the most egregious. But they have much competition. Serious radiation leaks have been found at every one of the Energy Department's sixteen major nuclear-weapons-production facilities, leaks that have forced many of these facilities to shut down permanently. Yet a wooden sign at the edge of the Shippingport trench gives the entire area an unnervingly euphemistic label. It ominously reads: "Burial Garden."

Early the next morning, the trucks will pull away from the site and leave the dry, gravelly dust to settle around the reactor vessel, perched upon a steel pedestal at the bottom of the trench. At some future date when more waste has been added, the trench will at last be filled.

There, under the earth,. the Shippingport reactor vessel will continue to emit a low, but discernible, level of radiation for 100,000 years. For this reactor vessel from the world's first commercial nuclear power plant, the job is through. For the nuclear industry, as it faces the legacy left by our reliance upon nuclear power, the real work has yet to begin.

ENERGY 191

Review of the Year

THE ENVIRONMENT

The growing global concern for the environment, reflected by ecologically oriented legislation on both the national and international level, has been underscored by the debate surrounding oil spills, toxic-waste dumps, and poisoned food.

by Gladwin Hill

The Earth's Atmosphere in Danger?

A primary environmental concern in 1989 was the possibility of drastic changes in the atmosphere caused by human activities.

Most alarming were the continuing depletion of the stratospheric ozone layer, which serves as a global buffer against excessive ultraviolet radiation; and the "greenhouse effect," an increase in the Earth's surface temperature caused when carbon dioxide and other combustion gases in the atmosphere trap in solar heat. A rise in Earth's temperature would drastically raise ocean levels (by melting glaciers and ice caps), and alter both weather and vegetative patterns. Some scientists, however, challenge the "global warming" predictions. Although they acknowledge the existence of underlying factors, they question the rate of contaminant buildup in light of the Earth's natural capacity to compensate and adjust to change.

Despite the dispute, groundwork was laid by United Nations delegates, U.S. President Bush, and British Prime Minister Margaret Thatcher for an international treaty restricting discharges of substances harmful to the environment.

Also during the year, 81 nations, including the United States, agreed to end the use of chlorofluorocarbons (CFCs), used in refrigerants and foam plastics, by the year 2000; this doubled the pace of a phaseout schedule adopted in 1987. CFCs are thought to be a principal contributor to the ozone problem.

In the United States, significant progress was made in eliminating the emission of CFCs. Many large U.S. industrial concerns renounced future use of the chemicals, and Vermont and Hawaii enacted laws aimed at banning CFCs.

Another likely factor contributing to atmospheric alteration is the destruction of rain forests in the Amazon region of South America and in central Africa. Farmers, loggers, and miners in those areas burn large tracts of forest. However, the burning eliminates vital oxygen-producing vegetation and generates excess carbon dioxide—1 billion tons annually in the Amazon Basin alone.

High levels of acid rain occurred in 1989 over African rain forests, showing more clearly the link between burning of rain forests and environmental damage. International conservation organizations in 1989 pursued a policy of arranging reductions in national debts in exchange for forest preservation in Costa Rica, Ecuador, and Bolivia.

Jim Estrin/NYT Pictures

Diesel exhaust is an important source of urban air pollution. In New York City, an air-pollution monitor on Madison Avenue, a busy bus route, found curbside air quality to be unacceptable. The Madison Avenue monitor is the city's only such device located above ground.

In domestic efforts to protect the air, Congress moved closer to passing a new Clean Air Act imposing stringent limitations on industrial and automobile emissions, with a particular emphasis on reducing acid rain.

George Tames/NYT Pictures

In addition to its damage to forests and farmlands, acid rain eats away at many of the country's most beloved monuments. At Gettysburg National Military Park in Pennsylvania, some markers and plaques can hardly be read anymore.

Waste Management

Management of industrial, nuclear, and municipal wastes loomed ever larger as a national problem in 1989. From New York City to Los Angeles, California, communities turned to mandatory recycling programs. "Degradable" plastics became more widely used, despite the warnings of scientists that such products might cause more problems than they solve.

The Department of Energy (DOE) painted a grim picture of cleanup needs for the nation's nuclear-weapon facilities. Seventeen installations in 12 states were said to have areas so radioactively contaminated that it would take more than a generation to make them safe, at an estimated cost approaching $200 billion. The agency proposed an initial commitment of $21 billion over five years to begin the task.

A cleanup program at the former nuclear-weapons plant in Oak Ridge, Tennessee, has emerged as a testing ground for similar projects underway elsewhere.

Michael Patrick/NYT Pictures

The eight-year-old federal Superfund program for cleaning up civilian toxic-waste sites came under increasing criticism in 1989 for its slowness. The number of sites needing cleanup is estimated at 30,000; some 1,200 sites are on the Environmental Protection Agency's (EPA's) high-priority list. An independent study found that, of $10 billion authorized by Congress for the Superfund, only $4.5 billion had been appropriated, and only $2.6 billion actually spent. Work had begun at only 160 of the high-priority sites. The EPA announced a series of measures to speed up the program.

Food Hysteria

Public concern about pesticides and additives used in food reached an unprecedented level in 1989. The alarm was reflected in drastic reactions to two instances of presumed food contamination during the year.

Early in 1989 federal inspectors, alerted by a telephoned poison threat, found minute amounts of cyanide in two grapes imported from Chile. A public warning against eating Chilean produce, and the removal of it from most store shelves, caused widespread apprehension. No more cyanide was found, however, and after the warning was rescinded, Chilean produce was restocked.

At about the same time, it was announced that unacceptably high levels of Alar, a pesticide thought to cause cancer, had been found in apples. New York, Los Angeles, and other cities promptly removed apples from school menus, and many households stopped using apple products altogether. A week later, federal regulatory agencies pronounced the boycott unwarranted. However, the EPA subsequently announced a ban on Alar, to become effective in 1991. Tests indicated that protracted ingestion of the compound could be harmful.

Problems from Oil Spills Continue

In Alaska the Exxon Corporation in September 1989 suspended its 2-billion-dollar, 10,000-worker cleanup of the 11-million-gallon (41.6-million-liter) March 1989 oil spill from the *Exxon Valdez*. The company said that most of the 1,300 miles (2,067 kilometers) of the shoreline of Prince William Sound had been "environmentally stabilized." It claimed that 250,000 gallons (945,000 liters) of oil and 50,000 tons of oily debris had been removed.

Federal and state officials challenged these figures, and charged that the cleanup was less than one-third complete. In February 1990, Exxon announced that it would resume cleanup operations in May. Although officials agreed that winter storms had aided in the cleanup, much work remained to be done.

On February 7, 1990, a fully loaded oil tanker struck its own anchor and gashed its hull while maneuvering toward Huntington Beach, California, spilling nearly 300,000 gallons (1.1 million liters) of light crude oil into the Pacific Ocean. The accident, one of the largest spills off the California coast in years, damaged wildlife in the area when winds pushed deposits of oily sludge close to a nearby wildlife reserve.

The seemingly endless series of oil spills continues unabated. Last February, a tanker spilled oil off California, causing much environmental damage.

AP/Wide World Pictures

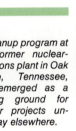

Tourists in Antarctica

by Yvette Cardozo and Bill Hirsch

Not long ago the idea of shipping tourists by the hundreds to the Antarctic seemed as much science fiction as the thought of sending them to outer space. But today the Soviets are selling seats on their space shuttles, and 93-year-old tourists are visiting penguin colonies.

It recently became obvious that tourists would one day overwhelm the fragile Antarctic. But no one expected the situation would come to a head so soon.

worse place. Just as scientists and tour companies were coming to an agreement on the future of tourism in Antarctica, the freighter *Bahía Paraíso*, carrying 250,000-plus gallons (946,350 liters) of diesel fuel, rammed itself onto some rocks less than 1.9 miles (3 kilometers) from Palmer Station, the U.S. research base on Anvers Island.

Bottom of the World

For most people, Antarctica is a thin white strip on the bottom of their maps. Actually, it is a 5.4-million-square-mile (13.9-million-square-kilometer) disk, almost completely covered by ice 1.9 miles (3 kilometers) thick, locking up 90 percent of the world's fresh water. The Antarctic Peninsula trails off one side of this disk, reaching to within 620 miles (1,000 kilometers) of South America and shadowed by a scattering of islands. It is here, in a comparatively "balmy" climate where summer temperatures can reach 60° F (15° C), that most of the Antarctic wildlife gathers.

It is these very features—plus a concentration of wildlife and relatively easy access to the rest of the world via South America—that attract research stations. And for these same reasons, it is where tourism concentrates.

Organized mass tourism in the Antarctic dates back to 1958, when Argentina began sending cruise ships with 400 to 800 passengers to the Antarctic, but this program was eventually discontinued. In 1966 Lars-Eric Lindblad started sending groups of tourists aboard the same kind of government supply vessels as the *Bahía Paraíso*.

Three years later Lindblad had a tourist ship custom-built with a reinforced hull to withstand ice. The MV *Explorer*, 2,500 tons (2,270 metric tons), 250 feet (76 meters) long, capable of carrying 100 passengers in relative comfort, has been visiting the Antarctic on a regular basis since then.

In 1978 the *Explorer* was joined by the MV *World Discoverer*, 3,153 tons (2,860 metric tons), 285 feet (87 meters) long, capable of carrying 136 passengers. Both ships are now used exclusively by Society Expeditions, Inc., of Seattle, Washington. Lindblad Travel, meanwhile, does its Antarctic trips aboard the 90-passenger MS *Antonina Nezhdanova*, while Travel Dynamics uses the MV *Illiria*.

The Lindblad and Society ships have not been, as one company representative put it, "love-boat cruises." Rather, they are billed as

© Yvette Cardozo

What pushed the issue was the grounding last year of an Argentine supply vessel carrying more than 80 tourists. Considering that most Antarctic tourists are from the United States, it could not have happened at a worse time in a

The ever-growing numbers of tourists flocking to Antarctica may ultimately cause irreversible damage to the continent's delicate ecosystems. Even the most inaccessible areas stand threatened by the tourists, who board small inflatable boats to get a firsthand look.

THE ENVIRONMENT 197

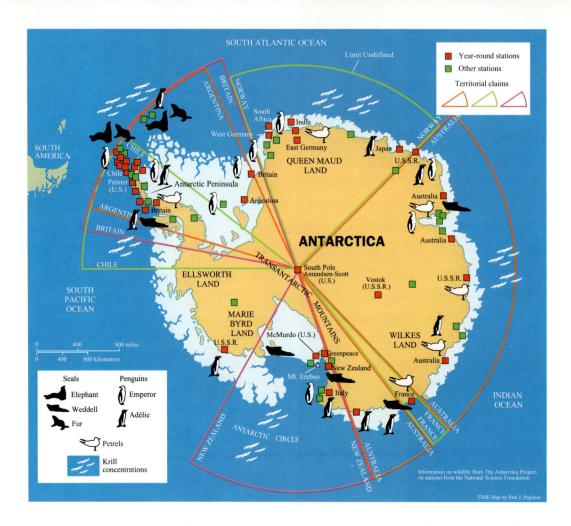

expeditions and have the feel of a university field trip. Passengers are supplied with thick information texts and polar-style gear. They get daily lectures from onboard scientists who are veterans of the Antarctic and often are carrying on research while serving as cruise staff.

From the earliest years of serious scientific research, various governments took tourists sporadically for visits. Mostly they deposited their visitors at research stations, where the people got to meet working scientists and see whatever penguin and seal colonies happened to be in the area.

But not all tourism has gone smoothly. Two years of jet overflights culminated in a disaster when an Air New Zealand flight crashed into Mount Erebus in November 1979, killing all 257 people aboard.

Land tourism, meanwhile, grew slowly. Even five years ago, the total number of tourists visiting Antarctica each year was hardly 500. Then, suddenly, there were companies seemingly everywhere, carrying tourists by ship, flying them to research stations and even to the South Pole, guiding them up mountains, and, in one spectacular case in January 1989, leading them overland 685 miles (1,100 kilometers) on skis all the way to the Pole. By 1989 there were seven companies running trips.

Widespread Visits

Visits to some countries' Antarctic Peninsula stations—especially Great Wall (China), Teniente Rodolfo Marsh (Chile), Bellingshausen (U.S.S.R.), Arctowski (Poland), Palmer (U.S.), Faraday (Britain), and Esperanza (Argentina)—became so commonplace that what started as the informal trading of souvenirs turned into a thriving little business. Polish scientists sell patches, stamps, pennants, shirts, pins, and maps. The Chileans have a tourist store and are renting rooms in a trailerlike building that serves as the local hotel. Palmer staffers sell U.S. Antarctic watches for $30.

But, as one veteran Antarctic scientist pointed out, "More people have been to a single football game in one afternoon than have ever been to the Antarctic." That includes tourists, explorers, and scientists.

The Antarctic is not, however, a Temperate Zone field of grass. It is considered a pristine environment, free of the pollutants that might skew research elsewhere. It is incredibly fragile, a place where a few millimeters of moss may take years to grow. For studies of this fragile environment, there were, at last count, 57 year-round scientific stations in Antarctica representing 18 countries.

Recently, U.S. National Science Foundation (NSF) scientists—and those from other nations—began to assess the tourist problem.

"In the 1986–87 season, the level [of tourists] we had been used to jumped from 400, 500 a season to 1,300," explained Jack Talmadge of the NSF Division of Polar Programs.

"The scientists said, 'Look, these visits are getting to the point where we cannot conduct our research efficiently and effectively.' We were responsive to that because our feeling is that taxpayer money has gone into paying for the research down there, and we are responsible for making sure that research can be done without outside interference.

"As reports came to us," Talmadge continued, "we got early indications of what was likely to hit the following season. And we were shocked, because we were looking at 2,500 visitors. Obviously, if 1,300 is a problem, 2,500 is going to be more of a problem."

Both photos: © Yvette Cardozo

After centuries of virtual isolation, Antarctica's seals and penguins have not yet grown wary of humans. Tour companies that cater to their clients' desire to see Antarctic wildlife close up may unwittingly endanger the animals by exposing them to excessive noise and confusion.

What may have added to the alarm was a short-lived plan to bring a standard cruise ship to Antarctica, packed with 600 passengers per trip. So, even before the *Bahía Paraíso* incident, steps were being taken to control tourism.

"Our concerns about Antarctic tourism are threefold," Talmadge said. "Prospects that it will interfere with science, that there will be an impact on the Antarctic environment, and that there might be a need for search and rescue, which our program is not set up to provide."

Until the February 28, 1989, wreck, the most immediate concern was the interference with science. The U.S. station limited visits to four per ship during the 1987–88 season, and kept tourists out of laboratories, instead showing "sample" tanks filled with swimming krill (euphausiids). Tourists who could not go ashore got a visit on their ship by station personnel who lugged souvenir T-shirts aboard.

But tourists visiting the Antarctic tend to be wealthy and influential. Disturbed by the fact that they were greeted warmly by, for example, the Polish scientists, while getting the cold shoulder from Americans, they wrote to their congressional representatives. And they got NSF's attention.

Sun-bleached whale bones (left) attest to past abuses of Antarctic resources. Beginning early in this century, hunters slaughtered thousands of whales in nearby waters and brought the carcasses ashore for processing. Only the bones and abandoned stations (below) remain; the whale population has not recovered.

The tourist invasion has also disrupted life at the scientific research stations scattered across the continent.

In July 1988, NSF held a hearing and workshop in Washington, D.C., to bring the two sides (scientists and tourism representatives) together. "We had people from tour companies, scientists, lawyers who defended the U.S. government in the Mount Erebus crash, the chairman of the Antarctic Safety Panel, people who had been guides on board ships, and people who had been guides in other, similar areas such as the Galápagos," Talmadge said.

Disruptive Visits

In response to one tour guide's complaint that people do not want a visitor center at Palmer—they want to see "live animals and live scientists," NSF explained that visits have become extremely disruptive.

During a recent tourist-ship visit, Frank Todd, an onboard scientist with extensive experience at stations, and creator of the Penguin Encounter exhibit for Sea World, gave a detailed picture of what was happening.

"In the old days, when there were just one or two ships a year, any visit to a station was greatly looked forward to by station personnel," he said. "But over the years the type of projects has changed. There's a lot of laboratory work now, with greater emphasis, at least at Palmer, on controlled-temperature work inside buildings. Sometimes, by simply opening a door, you may wreck an experiment. Some of these experiments may be sensitive to vibrations, which is a problem if somebody accidentally bumps a sensitive instrument.

"Timing is very important in the Antarctic," Todd added. "It sometimes takes years to get our programs going, and the austral summer is very short. A visit once or twice a month was O.K., but when you're talking every other day, you've got a problem."

The outcome of the Washington hearing was better understanding on both sides. Tour companies such as Society Expeditions set up voluntary restrictions on station visits. Palmer, meanwhile, reopened to visitors on a limited basis: letting people in to see living quarters; setting up a trail, complete with signs; and assigning Talmadge as more or less the Antarctic tourism officer.

Even before the *Bahía Paraíso* incident, NSF was discussing further measures. "We're talking to the National Park Service about sending down an interpretive planner to help us learn how to handle visitors in environmentally fragile areas," Talmadge said. "And we're considering putting people on board as observers to

Above and facing page: © Yvette Cardozo

report back infringements of the law [the Antarctic Conservation Act]," he said, prophetically, a full month before the freighter *Bahía Paraíso* rammed the rocks.

While observers in Alaska become part of the tourist experience, giving lectures and handing out pamphlets, Talmadge thought such people should stay out of the active lecture picture in the Antarctic. "Besides," he added, "those ships already have their own extremely fine lecturers."

Meanwhile, it is sometimes difficult to draw a sharp line between tourism and science in the Antarctic. The tourist exploration cruises all carry working scientists, and many of them, especially the ornithologists, continue their work on board ship.

"[The cruise ship] is essentially a mobile science platform," Todd explained. "It enables us to do more in a shotgun way than people who spend an entire season at one location."

On one recent Society cruise, Todd spent much of his time ashore updating information. Much of the information he relied upon while setting up not only Sea World's Penguin Encounter, but also an earlier experimental penguin colony for NSF, came from observations during tourist trips.

Ships often ferry scientists back and forth between South America and the stations, and, in February 1989, the renamed *Society Explorer* not only carried NSF and Coast Guard personnel working on the oil spill, but, with the *Illiria*, helped rescue the *Bahía Paraíso* tourists and carried them to the Chilean station for airlift.

None of this changes the fact that the *Bahía Paraíso* incident showed just what can happen in the Antarctic. The *Bahía Paraíso* was leaving Palmer Station after a tourist visit, when she ran into submerged rocks in the harbor. While British, Chilean, and American charts show rocks there, the Argentine charts apparently did not, and the ship had used this route successfully before.

The main gash in the ship's hull, said Tom Forhan, an NSF representative, was "the size of a dance floor." Hundreds of thousands of gallons of diesel and jet fuel leaked. In addition, 450 sealed drums of oil bobbed off, but were quickly recovered.

Subtle but Insidious Damage

Fortunately, diesel is a much lighter oil than crude and dissipates more rapidly. So, unlike the *Exxon Valdez* disaster in Alaska, Antarctica was not turned into a scene of pathetic, flapping birds covered with tarry goo.

The damage, say Forhan and scientists such as Todd, is far more subtle and perhaps insidious. There was a tremendous loss of limpets, amphipods, and krill—and that, Forhan said, will affect the birds that normally feed on them. Cormorants apparently ingested oil while preening themselves, and many died when their stomachs hemorrhaged. There were concerns that penguin chicks, smelling of oil but apparently stable at the moment, might die or be weakened eventually.

Equally troubling, years of scientific research have been seriously disrupted. "Let's say you've got a five-year project on kelp gulls or penguins, and you've been banding a certain percentage of young," Todd said. "All of a sudden, these birds disappear. You don't have any direct proof it's oil, but you've got some interesting problems with your study.

"Or let's say next year, adults are doing something a bit differently, or their timing is off, or productivity is down," he said. "It could be a natural swing in their cycle, but it could also be a result of the oil." Todd said a major human-caused event such as the *Bahía Paraíso* spill puts many unknowns into the equation, and it will be four or five years before scientists really know what the ramifications will be.

Meanwhile, talks on tourism and the presence of people in the Antarctic in general continue. What the *Bahía Paraíso* has done is escalate events. Talk at NSF centers on control of ships, most likely requiring a pilot to guide ships into Palmer's Arthur Harbor, and the use of wildlife rangers as onboard observers.

Other people, meanwhile, are equally concerned. Bruce Manheim of the Environmental Defense Fund says it is time someone laid down specific guidelines as to precisely what constitutes pollution and precisely what amounts to harassment of wildlife. "Does that mean," he asks, "that a U.S. citizen can't go within 5 feet or 10 or 20 feet of, say, an Adélie penguin?" Manheim, who did a study on the impact of scientists on the Antarctic, is now doing a similar study on the impact of tourism.

No one—not the NSF, working scientists, or even Manheim—suggests an end to tourism. It is unlikely such a ban could be enforced, anyway. And tourists are not the only threat. Hardly weeks after the *Bahía Paraíso* incident, a Peruvian ship doing strictly research-base work also rammed into a rock, spreading more oil.

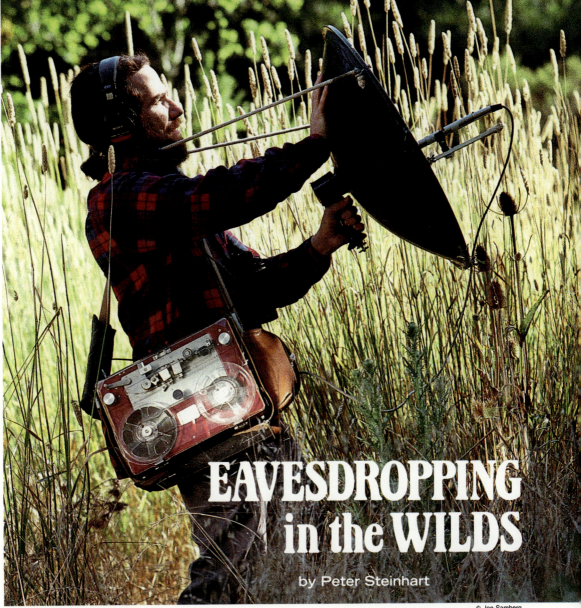

EAVESDROPPING in the WILDS

by Peter Steinhart

We are a visual species. We process more information through our eyes than through our ears. A picture, we say, is worth a thousand words. Most of us find sounds indescribable. Modern culture makes us ever more empirical because it values what we can count and describe. That makes us engineers and athletes rather than philosophers and storytellers. It makes us careless about what we bray into the airwaves. It makes us deaf to the voices of the natural world.

So it shocks us when music wells up out of the silence. Bernie Krause, a California bioacoustician, plunged a microphone into a desert spring to record insect larvae making a wicking sound reminiscent of hedge clippers. He placed a microphone in an ants' nest and heard their tiny feet crackling on the microphone, a rush of whispering, then a loud rhythmic rasping, like that of an agitated jay.

In recent years we have begun to listen to the sounds of nature. The growth of wildlife on television, with its needs for accurate sound, has contributed to the trend. So has the develop-

Birds, mammals, and even insects subtly vary the frequency, tempo, or pattern of their vocalizations to convey specific messages. Scientists called bioacousticians have begun to record the sounds of nature before the din of modern civilization drowns them out forever.

THE ENVIRONMENT 203

Animals evolved their particular ways of making sounds to match the needs of their individual environments. The low-frequency roar of the lion (left) apparently developed to travel well over the open land on which it lives. The gregarious zebra (near right), with no need to project its voice over long distances, makes a barking sound. The rhinoceros (far right) has a curious high-pitched voice.

ment, since the early 1960s, of high-quality transistorized recording equipment. And with the growing sense that wildlife is vanishing, there has been a desire to memorialize species on tape before they disappear. Today there are bioacoustic libraries, archives of the recorded utterances of birds and insects, whales, and popping shrimps. The largest, Cornell's Library of Natural Sounds, holds over 65,000 recordings. There are societies for nature-sound recordists in the United States, England, France, and the Netherlands. According to Paul Matzner of the Nature Sounds Society, there are at least five regular journals for nature-sound enthusiasts. And hobbyists in increasing numbers have been taking microphones and tape recorders out to eavesdrop in the wilds.

They have discovered a world full of mysterious and entrancing voices. A walrus grunts, moans, burps, and then makes a delicate metallic sound, like the ringing of a small Oriental gong. Bearded seals make weird spiraling moans, ghostly sounds that suggest lost souls tumbling down the depths of a bottomless sea. A rhinoceros has a high-pitched, quizzical voice that resembles that of a young child talking into a bottle. Zebras bark. Elephants make a low rumbling sound that is beneath human ability to hear, but which may convene gatherings of elephants over many miles.

Signature Calls

Those who listen discover that the world of natural sound is as rich and complex as the world of natural shape and color. Donald Borrow of Ohio State University found that every chipping sparrow he recorded has a slightly different song. Luis Baptista of the California Academy of Sciences observes that white-crowned sparrows on three different sides of the academy's building sing three distinct variations of a dialect. Diana Reiss, who is investigating the vocalizations of dolphins at Marine World-Africa U.S.A. in Vallejo, California, observes that each dolphin has its own unique signature call, which announces its identity to all other dolphins. (See also *The Day of the Dolphins,* p. 356.)

It is increasingly clear that bird and mammal vocalizations have precise meanings. Subtle differences in frequency or tempo or pattern may convey very specific messages. A killdeer makes one sound when it is courting, another when it is building a nest, another when it is trying to decoy an intruder away from the nest, and another when it is calling young. A territorial male white-crowned sparrow ignores the songs of immediate neighbors, but if a stranger moves in and sings an unfamiliar song, he challenges that bird. There are grammars and rules of discourse we don't understand, says Marie Mans, an Oregon recorder who calls birds into view with her recordings. If you play a recording to a hermit thrush, she says, it upsets the bird. "It starts to sing stridently. It sounds like a bad recording."

Auditory Nitwits?

Humans are, by comparison with most animals, auditory nitwits. What we hear as a senseless burp from a dolphin or a colorless twit from a

sparrow may be a rich and complex expression to a creature equipped to hear it. When we slow down the song of a winter wren, what we hear as a single note becomes immensely complex and melodious. Birds have been shown to have more sophisticated powers of processing song than we have. Masakazu Konishi of the California Institute of Technology (Caltech) has found that the auditory neurons of barn owls lock onto sound at higher frequencies than the auditory neurons of mammals, giving owls the ability to localize sound. Says Konishi: "Birds can detect very fast changes in sound. You listen to a tape recording of birdsong and slow it down by half, and you begin to hear some of the details. But birds can hear these details without slowing them down."

Until recently, we assumed each animal had its own pip or squeak, and perhaps only one of them. That may be true of flycatchers, whose single calls may be encoded in their genes. But most bird species learn their calls, and they may learn a great variety of them. A marsh wren in San Francisco Bay may sing 200 different songs in an hour. A brown thrasher may sing more than two thousand. The languages they thus develop are constantly changing. At a Nature Sounds Society meeting, Catherine Thexton, who records birds on her Manitoba farm, plays a recording of "my million-dollar song," a lilting, springtime meadowlark tune she taped 15 years ago. Today none of the meadowlarks in her neighborhood sing that same song. Baptista says some of the original white-crowned sparrow songs heard 15 years ago are extinct.

Most humans listen to such voices only for the oddest of reasons. For years, for example, the National Aeronautics and Space Administration's (NASA's) Search for Extraterrestrial Intelligence (SETI) program has probed the universe with large radiotelescopes, listening for the sounds of other intelligences. It was thought that our radiotelescopes would recognize the impeccable mathematical order of distant radio broadcasts. But what of understanding extraterrestrial voices themselves? Given that we understand almost nothing of wolf howl or gorilla hoot, the odds are slim that we'd understand anything a singer from Aldebaran had to say. Realizing that they are unprepared to make sense of intergalactic voices, NASA researchers are now studying films and tapes of dolphins, hoping thereby to learn how to listen.

Acoustical Niches

There are more mundane reasons to listen. Animal sounds are adapted to the terrain. A lion roars loudly at a low frequency designed to be heard over open territory. Birdcalls are high-pitched in small, dense spaces and low-pitched in wide-open areas. Every creature finds its frequency. Robert Bowman of San Francisco State University has shown that the voices of Galápagos finches are adapted to the particular terrain of the island on which they evolved, so that the local vegetation does not absorb the sound frequencies of their calls. The call of one finch will travel farther when uttered on its own island than when uttered on a neighboring island. Krause found that hyenas utter one cry at night, when dew on leaves and ground makes sounds reverberate, and another by day, when the air is deadened.

Each creature's sound fits into the warp and weave of the others'. Bioacoustician Krause believes birds and animals fill acoustical niches in their environment. He displays a spectrogram that shows ten seconds of ambient sound he recorded in Kenya. Two broad bands of black cut across the page, showing the chirp of insects and the police-whistle chorus of frogs. In a white band in between, there are rhythmic lines. He plays the tape that produces the graph. A hyena and a bird are heard between the two black bands. "Those guys fit right smack in the middle," says Krause. He adds: "North America's white-crowned sparrows will slot right down into a part of the audio spectrum that has very little sound. It may be the only habitat in which they can survive."

The human voice may have adapted to openings in the acoustical landscape of the African plains millions of years ago. And speech may be shaped by the baffles and silences of plant cover, open space, and competing noise. Farmers tend to speak volubly, to reach with voices across unplowed fields. City slickers and forest dwellers speak softly, so as not to arouse strangers.

Every landscape has its own acoustical ecology. The music of a place is as invitingly distinctive as the aromas of a household kitchen. The Bioacoustics Research Program being run by Dr. Christopher Clark at the Cornell Laboratory of Ornithology's Library of Natural Sound is now perfecting a computer program that will enable researchers to tape-record the ambient sound of a rain forest or a desert and play it into a computer that will recognize and identify which species are present. The music of a place may have more meaning than any of its individual voices. Migrating birds may be guided back to their breeding grounds by the varied music that they passed through on their journeys.

The Drone of Civilization

The music of a place may also be important to our sense of well-being. Just as a voice is shaped by the landscape, our minds may be tuned to the ambient sound of a natural world. We vacation at the seashore or in the mountains in part to get away from the noise of the city. It is not just silence we seek, but the sounds of things our bodies and minds wrapped around as they evolved: birdsong, cricket chant, the hiss of surf on sand, the gurgling of brooks. Until the past two centuries, those were the meters to which our hearts and minds were calibrated.

Appreciating the complexity of natural sound may lead us to worry about its delicacy. Nature-sound recorders report that they are less and less able to find places free from the drone of civilization. There is no place on Earth that is free of the noise of helicopters and airplanes.

Herb Melchior of the Alaska Department of Fish and Game reports, "You can be in the middle of the Arctic National Wildlife Refuge, but overhead the over-the-Pole direct flights go by every hour." Krause keeps a detailed journal of his field recording. He says that more than 20 years ago, "about 22 hours of recording would yield 15 minutes of usable material." But recording in Rwanda in 1988, he had to invest 200 hours to get 15 minutes. He spent 200 hours recording in northwestern Washington that year, but could not get a single recording worth saving, because of the din of chain saws or passing trucks, day and night, on a highway 7 miles (11 kilometers) away. In the Amazon the racket of chain saws prevented recording. In 1989, when Cornell University held a field seminar in nature-sound recording, it held it at California's Yuba Pass, because it could not find an area in the eastern states that offered enough silence.

The din gets louder and louder. More than half the low-altitude airspace in the United States is allocated to military flights, and the traffic in both civilian and military airspaces grows. In Goose Bay, Labrador, the culture of the Innu Indians, one of the last hunting and fishing peoples in North America, is about to vanish. The airspace over nearly 40,000 square miles (104,000 square kilometers) of forest inhabited by the Innu is used by German, Dutch, and British military planes flying low-level training missions. The jets fly down the river valleys, where the Indians fish and hunt for caribou and moose. They scream overhead at treetop level, and the noise leaves the Indians deafened, dizzy, and demoralized. There are currently 8,000 such flights a year, but by the year 1992, there are expected to be 40,000. Says John Olthuis, attorney for the Innu: "They will be wiped out as a people."

Losing Acoustical Integrity

What all that noise does to the rest of creation is a matter of conjecture. Says Karl Kryter, a California acoustician: "Most of the problem with noise is that it interferes with other signals we

Human-generated noise is by no means limited to urban areas. No place on Earth remains completely free from screaming airplane engines (above). Sound barriers (right) that reduce highway noise bothersome to nearby residents may unintentionally create an insurmountable obstacle to animal communication.

Top: © Billy E. Barnes/Uniphoto; above: Connecticut Department of Transportation

are trying to hear. The message that you want to hear just can't get into the brain." It's rather like a radio broadcasting on a similar frequency. If there is a continual noise at a certain frequency, says Kryter, "the animals that depend on that frequency to communicate can't do it. I suspect the noise might drive them out of their environment."

Krause has shown that killer whales in Puget Sound alter their vocalizations at the frequencies of cavitation noise caused by fishing vessels. But we have no evidence that any species has suffered acoustical extinction. The effects of noise on wildlife have been little studied. Most funding for noise-related research has gone to projects investigating whether aircraft noise reduced the value of poultry and livestock farms. Some studies suggest that when an airport arrives, the birds depart, but gradually return and adapt to the noise. But it is not clear whether the same species return.

The changes we make in the physical environment may be even more important. Clearly humans have, by laying down walls of concrete and glass and carpets of exotic vegetation, altered the acoustical environment to which bird and insect voices were tuned. We have changed the music. That may be what happened to Manitoba farmer Thexton's "million-dollar song."

Most of us are unaware of the loss of the land's acoustical integrity. Few of us know the voices of walrus or rhinoceros, or even backyard birdsong. In part, that is because we don't listen. In part, it is because, even if we do listen, our neighborhoods are so rattled with the sounds of automobiles and airplanes, leaf-blowers and amplified music, that we can't hear crickets or mockingbirds. We roll up our windows and turn up the stereo, borrowing noise to screen out noise. We are the only creature that tries not to hear.

Our environmental-impact statements talk about noise pollution. But noise pollution is defined only as sound that annoys people. We have no notion of our own needs for the music of places. We have no standard of acoustical purity for wild creatures, which may have far more serious needs for it than humans. Natural sound may be as important to the land as are its water and air.

The DWINDLING DRY FOREST

by Thomas A. Lewis

The dean of tropical biologists stands shirtless, graying hair flying in the wind. Daniel H. Janzen glares northeastward over his black-rimmed reading glasses into the hot, dry wind shoving down from the Costa Rican Cordillera. He looks past the littered yard of a long-abandoned ranch house, out across nearly 10 miles (16 kilometers) of parched savanna, to where the twin volcanoes Cacao and Orosi jut into the sky.

What Janzen is surveying is much more than a failed, overgrown ranch. The baked landscape is in the process of being reclaimed from near ecological ruin. Janzen is leading the effort—more accurately, the crusade—to make sure this resurrection really comes to pass. His mission is to return what was once a tropical dry forest to its former vitality.

© Phyllis Nunn/National Wildlife Magazine

Restoration Ecology

Janzen and a brigade of scientist colleagues are part of a growing global movement known as restoration ecology. Already projects are underway to re-create natural ecological systems in places as diverse as the grassland of the North American Midwest, the wetlands and rain forest of southern Brazil, and the arid grazing lands of Kenya. But few ecological restorations match the scale of the experiment here in northwestern Costa Rica, where a small existing park, Santa Rosa, is the nucleus for creating an enormous additional park, Guanacaste. And few pioneers of the restoration ecology movement approach the energy, achievements, or just plain stubbornness of Dan Janzen, whose dream to bring about a détente between civilization and the natural world is meeting with increasing success.

The once-expansive Central American dry forest has dwindled to only 2 percent of its original size. Restoration ecologist Daniel Janzen (above) leads a private crusade to revive the dry forest in a tiny corner of Costa Rica (see map). The careful nurturing of small trees in an old cattle pasture (left) represents a step in the right direction, although full restoration may take hundreds of years. Fire poses the biggest threat to Janzen's dream.

Above and below: © Bob Krist; upper right: © Peter C. Poulides

Reviving the dry forest will expand the habitats of such local wildlife as the white-faced monkey (left) and the ocelot (above). By burying acorns in deforested areas, the agouti (top) indirectly helps the forest to grow. The coati (right) aids the cause by defecating fruit seeds in pastureland.

But today more than victory is in the air. Janzen's eyes narrow and flash. He shouts a fervent expletive. Two park employees and a visitor follow his gaze to a wispy ribbon of gray rising from the flanks of Orosi. By the time the visitor sees and comprehends, the other men are dashing for a vehicle. Fire, the tool of tropical agriculture, is the deadly enemy of the dream.

As his Toyota Land Cruiser hurtles across the field toward the Inter-American Highway, a grim-faced Janzen barks answers to his visitor's questions. Who is going to fight this fire? We are. With what? With a bundle of straw brooms that appear to have all the fire-suppression qualities of gasoline. Just us? "Sure," says Janzen. "One guy can stop one of those fires if he knows what he's doing." In that wind? In those thickets of waist-high, dry grass? Visions arise of roasted reporter.

Mercifully, a short sprint north on the highway (toward the Nicaraguan border, 25 miles [40 kilometers] away) reveals that the fire is not in the park, but across another highway, on private land. Back to headquarters.

210 THE ENVIRONMENT

Top: © Winnie Hallwachs; others: © Ola Jennersten

Project Control Center

A tin-roofed, cinder-block shed measuring perhaps 20 by 24 feet (6 by 7 meters) in the administrative area for Santa Rosa National Park, it is also a biological field station, kitchen, and visitor's dormitory. Additionally, it is the nerve center for the giant Guanacaste National Park Project (GNPP), now in full swing. The shed seems ready to collapse under pressure of the people, clutter, and crises it contains.

Inside on a late-spring day, with a troop of monkeys screeching in the trees overhead and a beefy, iguanalike lizard baking indolently in the sun nearby, GNPP administrator Randall Garcia struggles with the logistical needs of 20 employees working on the larger Guanacaste project. Garcia responds with unruffled cheer to telephone calls, shouted questions, and a constant jostle of humanity. Meanwhile, a woman whose impeccable dress and professional manner reveal nothing of her life on a hardscrabble Costa Rican farm methodically pins freshly collected moths to wooden display frames, helping catalog the area's biological diversity. Nearby, a newly arrived Venezuelan biologist makes arrangements for building a research station at the edge of the rain forest; his young children acclimate themselves to life in the field ("Hey, there's a snake coming into the kitchen, wanna see?").

Janzen plunges into the crush, now in consultation with Garcia, now diving into his office (a screen tent pitched *inside* one of the shed's minuscule rooms) to whack out a fund-raising letter on his computer, now weighing and examining a couple of dozen trapped mice whose family trees, habits, and home life he has scrutinized for eight years.

Habitat Fragments

The ecosystem that so intrigues him is a mere remnant of what was. Almost five centuries ago, Christopher Columbus was looking at the rain forest on the Atlantic side of what is now Central America when he dubbed the place *costa rica,* or rich coast. But the real wealth was across the central spine of mountains, on the Pacific side. There rain fell only half the year, from May through October. Green during the rainy season, leafless and brown the rest of the year, this primeval dry forest occupied more than 200,000 square miles (520,000 square kilo-

Janzen conducts his ecological research from a makeshift laboratory that doubles as his home.

THE ENVIRONMENT 211

meters)—all of western Central America from Mexico to central Panama. Larger in extent than the rain forest, it was also more vulnerable. It was easier to clear and more hospitable to people and livestock, and it stood on better soils. It went fast.

A favorite technique of early Spanish settlers was to chop a field out of the forest and encourage grass growth. Once each year, to prevent uneaten grass and unwanted saplings from impeding new grass, the ranchers set fire to their pastures. Each year new pastures were opened, the grass edged farther into forest margins, and more forest went up in smoke. After 400 years of this, a mere 2 percent of Central America's dry forest remains standing. No large expanses survive, says Janzen, only "habitat fragments and degraded patches, all of which are shrinking, if not in conserved areas."

The consequences for dry-forest dwellers, from white-faced monkeys to white-tailed deer, from the raccoonlike coati to scarlet macaws, are obvious. As habitat boundaries tighten like a noose, some of the larger mammals—the exceedingly shy jaguars and mountain lions and the endangered Baird's tapir—are making a last stand in the remnants of dry forest.

Man with a Mission

Yet, until 1985, Daniel Janzen was not overly concerned about the fate of the dry forest. He had been at work in it for three decades, and he had made himself a name. An award-winning professor at the University of Pennsylvania, he was a prolific author of well-received scientific papers, most of them on coevolution (as Janzen explains it, coevolution is a process whereby insects or larger animals adapt over a long period to specific species or communities of plants; the plants, at the same time, adapt to the animals—until neither can survive without the other).

Janzen had come a long way from his days as an engineering student in his native Minnesota, where he discovered only belatedly that one could make a living doing what he loved best—contemplating insects. But he admits he had not come so far from the days when "I used to chain-saw big trees just to count rings for my research with no more thought than you'd flick an insect off your sleeve." Then, in 1985, the Australian government called.

The Australians wanted advice on how to preserve their dwindling dry forest. Janzen went to have a look, but did not have much to recommend. The surviving parcels of forest were too small and remote, and the surrounding expanses of grassland too prone to fires, to allow for restoration.

He returned to Costa Rica. Where there were a few dwindling areas of dry forest, but where fire was not so intractable an enemy. Where the government, though small and strapped for money, was renowned for its commitment to conservation. An idea began to grip Dan Janzen.

What if one added about 250 square miles (650 square kilometers) to Santa Rosa National Park's 42 square miles (108 square kilometers)—creating a preserve that stretched from the Pacific Coast across the seasonally arid plains, up the slopes of the volcanoes and right into the rain forest? In fewer than 30 miles (48 kilometers), Janzen realized, it would be "as if Arizona were on one side and the Florida Everglades on the other, with the whole spectrum of flora and fauna in between."

Biological Rembrandts

The size of the dream was a product of biology, not hubris. Small parks—Janzen calls them "biological Rembrandts"—give the illusion of preserving a forest. But the forest gradually withers unless there is room for all the little fragments of habitat—the diverse combinations of soil types, rainfall and drainage patterns, elevations, wind exposures, and slope gradients—that each make minute but vital contributions to the complex mosaic.

Animals, as well, need room in the dry forest. Room to make seasonal migrations upslope and down to the seaside and back, or into the rain forest during the dry season and back out to the dry forest in the rainy season (a recently discovered connection, vital to many species).

To create a viable dry-forest park, Janzen calculated, one would need $8.8 million to "buy the dirt," and another $3 million for an endowment muscular enough to fund management forever. One would also need the total sup-

Restoring Costa Rica's dry forest depends in large part on letting nature take its course. Birds feeding in the existing forest on the mountainside deposit their seed-filled droppings in pastures. A cud-chewing deer spits out seeds. Such predators as the coyote later excrete the seeds eaten by their prey. Isolated trees attract wildlife by providing food and shade; the congregating animals spread seeds outward from the tree's base. From any of these sources, new trees can germinate. Planting further encourages the restoration process.

port of the Costa Rican government, help from area residents, and a willingness on the part of some 40 landowners to sell their ranches.

A Flair for Fund-raising

Did Janzen wonder whether he, who owned neither a home nor a business suit, was the man to raise $12 million? Whether he, with the scholar's usual distaste for politics, could achieve the delicacy of maneuver such an undertaking required? Whether a man who had spent much of his life probing the secrets of animal excrement, for clues to the distribution of seeds, was prepared for truly distasteful tasks such as asking for money and dealing with journalists? "You see something that needs doing," he shrugs, "and you do it."

Janzen undertook a metamorphosis worthy of one of his beloved moths: from reclusive biologist to garrulous after-dinner speaker. He wheedled money for the Guanacaste National Park Project from individuals, corporations, and foundations, and he made use of a relatively new financial strategy called debt-swapping. By the spring of 1988, he had raised $9 million. Of the 285 square miles (740 square kilometers) he had gone after, only 37 square miles (96 square kilometers) remained to be purchased. He was going to succeed.

At the same time Janzen was revealing a flair for fund-raising, he was taking on the job of fire chief. He hired local men to clear broad lanes within and around the project property; he organized lookouts and broom squads. "The fire crew is deadly," Janzen bragged at the end of the 1988 dry season. "Only 6 percent of the GNPP burned this year, as compared with the traditional 30 percent barbecue."

Forest Emergent

In terms of fire effects, the difference is not easy for a novice to spot. But saplings are emerging in strength from the abandoned pastures of tall, highly flammable jaragua grass, introduced by ranchers. For the first time in 400 years, the dry-forest canopy is edging outward instead of shrinking inward. Even so, recovery will not come rapidly. According to Janzen, it will take about 20 years to convert an unused pasture into closed-canopy forest; in 200 years the forest will *look* like the original dry forest; in perhaps 1,000 years, it will *be* the original dry forest, for all practical purposes, with all the myriad dry-forest interrelationships of insect and animal and plant in place.

The dry forest has two natural allies in its struggle to regain lost ground: wind and animals. The wind lofts seeds into new areas. Animals eat fruits and nuts, then defecate seeds elsewhere. But Costa Rica's winds have strong seasonal patterns and tend to spread seeds only in certain directions. Likewise, most of the creatures that inhabit existing dry forest seldom venture into pasture areas. Left to their own devices, these fruit eaters and seed swallowers contribute little to the restoration of the very habitat they need.

Thus, tree-planting will be required in order to nudge the forest composition in the right directions. But restoration ecologists apply uncertain methods to systems of incalculable complexity in order to achieve unpredictable results. Much remains to be done before they really know the "right" compositions to choose.

Nearly as challenging is Janzen's campaign to involve local residents, from touring schoolchildren to apprentice research assistants to chamber of commerce and provincial-government officials, with the excitement of restoring the park.

According to Janzen, there is a fundamental reason why Costa Rica has preserved 11 percent of its land area in national parks, compared to the 1 percent thus protected in the United States. It is, he says, "the biology courses taught 20 years ago in Costa Rican schools," courses that instilled an appreciation among the country's decision makers for the biological wealth possessed by their homeland.

Janzen makes countless economic and biological arguments for restoring the dry forest. His main theme is that Guanacaste Park's greatest offering—he goes further, and says it is biology's greatest contribution to humanity—is stories. Stories?

To Janzen, experiencing the staggering biological diversity, energy, and inventiveness of an ecosystem such as the dry forest is like browsing through a library. We cannot predict the worth of every story its volumes contain. But as we pursue our interests, we gain wisdom, often from unexpected sources, as well as a reconnection to the natural world. "To lose the abundance of the tropical dry forest to the demands of agriculture," he fumes, "is comparable to processing the books in the Library of Congress to relieve a temporary paper shortage." It would be, Janzen says with customary passion, "the greatest criminal act that life on earth ever could or will sustain."

HUGO'S ENVIRONMENTAL AFTERMATH

by Frank Graham, Jr.

Matchsticks! That old cliché was heard again and again in South Carolina after Hurricane Hugo had devastated the Low Country's forests—"trees snapped like matchsticks," or "big pines scattered every which way, like handfuls of matchsticks." But the mundane little word hardly reflects the horror of the reality: mile after mile of shattered wood, decapitated and dismembered giants standing over oceans of debris.

A thousand journalists have chronicled the human misery. Homes in splinters, the roofs of others heaved aslant, like the caps of tipsy soldiers. Twenty-six lives were lost. But what was nature's impact on nature itself? It will be months, even years, before all the consequences reveal themselves.

Hugo Hits

Norman Brunswig, manager of the Audubon Society's Francis Beidler Forest Sanctuary, some 40 miles (64 kilometers) northwest of Charleston, had kept track of Hugo's approach for a week as it smashed through the Caribbean.

"It wasn't clear what it was going to do as it approached South Carolina," Brunswig said. "We kept hoping it would change direction—not to sandbag somebody else, but just turn out to sea, or maybe lose force."

When Hurricane Hugo struck South Carolina in September 1989, it took 26 lives and leveled countless homes and businesses. Also ravaged were miles of forests and nature preserves. The Cape Romain National Wildlife Refuge (above) was destroyed by a 19-foot storm surge.

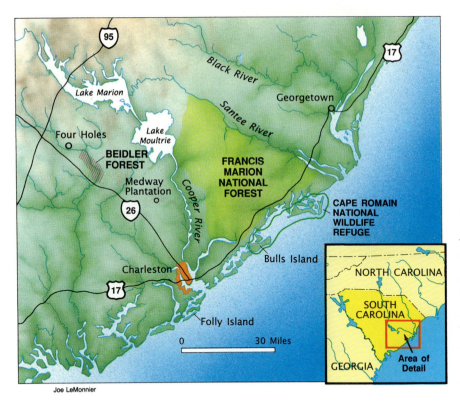

Hugo's winds intensified to 135 miles per hour just before the storm slammed almost head-long into Charleston, South Carolina. The hurricane's landfall coincided with high tide, swelling the already enormous wall of water, or storm surge, that devastated coastal areas. Several environmentally delicate areas lay within the storm's path of greatest destruction.

But those hopes were dashed on Thursday morning, September 21, 1989, when Hugo not only headed straight for Charleston, but also picked up speed and power. Winds accelerated to 135 miles (217 kilometers) per hour. Before closing the Beidler visitors' center, Brunswig supervised the taping of windows and the filling of the sanctuary's canoes with water to hold them in place. Then he left to board up the windows at home, casting a worried glance at the tumbling menace of the sky.

"We had bought our house because of the large pines that surrounded it," he said. "Now the trees were a real danger, and late in the day, we accepted our neighbor's invitation to sweat out the storm at their place, which is in the middle of a field."

Hugo struck at 11:30 that night. No one slept as the storm tore at the building, shaking it almost without letup, and drowning out with its shrill keening the smashing of trees in the nearby forest. At dawn, and almost reluctantly, Brunswig and his wife, Beverly, returned to their house. Twenty-six of the big pines in their yard had fallen, three of them hitting their house. But a new roof and some touching-up where water had seeped in undid most of the damage, meaning that the Brunswigs had fared better than many of their Low Country neighbors. The scene at the sanctuary, however, was to depress Brunswig for weeks.

A World in Tatters

The 5,819-acre (2,357-hectare) Beidler Forest, in the rambling Four Holes Swamp, holds earth's largest remaining virgin stand of bald cypress and tupelo. (Some of the cypresses are believed to be more than 600 years old.) It is an aqueous forest, the big trees standing in three feet of black swamp water. River otters, bobcats, water moccasins, alligators, turtles, white ibis, and prothonotary warblers are among the wild creatures that inhabit the sanctuary. Before Hugo hit, visitors walked into the swamp on a boardwalk 6,500 feet (1,980 meters) long, immersing themselves in the shaded grandeur of another world.

Norm Brunswig found that world in tatters once the staff had cut a way into the visitors' center through a jumble of fallen trees. Miraculously, the center itself was intact. The wind had taken down dozens of trees just behind it, but they had fallen into the swamp or across the boardwalk. The boardwalk was in ruins for most of its length. Oaks, with their shallow root systems, were toppled, the muck clinging to the

Facing page, left: © Terry Richardson; right: © Larry C. Cameron roots where trees lay in windrows on their sides. Many of the tall, slender pines were snapped like—well, like matchsticks. But the huge cypresses, supported by their buttresses, lived to fight again.

There was little satisfaction even of that kind at the Medway Plantation, a 6,800-acre (2,750-hectare) showplace 20 miles (32 kilometers) closer to the coast. Brunswig (who is also assistant director of National Audubon's Sanctuary Department) has worked closely with the Medway staff for some time, and 700 acres (280 hectares) of the plantation have already been donated to the society by the land's owner. There the ruin was on a greater scale than that suffered at the Beidler Forest.

"The lights went out when the storm came on," Robert Hortman, Medway's manager, told Brunswig afterward. "Even through the boarded windows, we could hear the loblolly pines snapping: *Tsss-chuck! Tsss-chuck!* When the eye of the storm passed over and everything was strangely quiet for a while, I went out and shined my flashlight up into the trees, and there was nothing there!"

Hortman and his staff were trapped in the interior of the plantation for five days while emergency crews worked to open the tree-clogged access roads. The old plantation house somehow stood up to the storm, the uprooted live oaks and other ancient trees piled around it, like a precious egg packed in its nest.

Profound Consequences

What made the privately owned plantations like Medway special was that they weren't under the commercial gun. There was no need to squeeze every penny from the forest. Their managers could be selective, in the old-fashioned European way, letting the pines grow for 90 years or more, producing "perfect" trees. The most valuable timber is usually found in the first, or bottom, log, but at Medway the pines were snapped off in that lower portion, or, if the tree broke higher, everything below was often structurally damaged.

"In some areas there aren't enough trees left to regenerate the forest, so we'll have to do a lot of planting," Hortman reported. "The wildlife? I've seen foxes wandering around like they were dazed, and wild turkeys limping or with broken feathers. Because the forest canopy is gone, the warblers are confused, and there are birds in the understory I've never seen before."

Brunswig might exult in the survival of the giant cypresses and tupelos in Beidler Forest. But the splendor around them was gone for now, replaced by the stench of rotting vegetation in the water, and by hordes of mosquitoes that hatched after the storm from puddles where their normal predators could not get at them. Many of the standing loblolly and longleaf pines showed ugly scars, their bark ripped away by the glancing blows of their stricken neighbors.

The once-scenic boardwalk at Beidler Forest (left) lay in ruins under oak and pine trees toppled by Hugo (right).

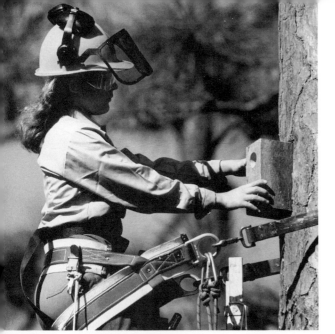

The Francis Marion National Forest, home to a rare growing population of red-cockaded woodpeckers (left), lost many of the pine trees in which the birds nest. The U.S. Forest Service rushed to the rescue, attaching boxlike artificial cavities for the birds to use (above).

Above: © Larry C. Cameron; left: © C. C. Lockwood

Fair Game

As soon as Hugo began its rampage through the Caribbean, wild areas and their inhabitants became fair game:

• The Puerto Rican parrot, down to only 13 birds in the wild in 1971, had responded to affirmative-action programs launched by interagency biologists in the Caribbean National Forest (its only known home), and counts earlier last year turned up 47 individuals. After Hugo demolished large areas of the forest, computer projections estimated that half the wild population had been wiped out. A few days later, preliminary surveys in the field found only 24 parrots. Fortunately, another 54 of them were safe in an aviary maintained for a captive-breeding program.

• Sixty-five miles (105 kilometers) of South Carolina's dunes took a fearful beating when the hurricane made its landfall. Federal, state, and private funds were used to scrape sand into new protective dunes, efforts that were criticized by some scientists, who predicted that the artificially steepened slopes would only cause waves to break farther inland. Under state law, some 200 beach houses carried away in the storm surge cannot be replaced. Their location would now be on the seaward side of the "critical line," where future erosion can be expected.

• Shellfish beds along parts of the South Carolina coastline were closed because Hugo's heavy rainfall caused runoff to overburden sewage plants. The overflow contaminated shellfish, posing "a threat to public health."

• Enormous numbers of fish died when leaves and bark blew into lakes and rivers, rotting there and robbing the water of oxygen. Further oxygen was lost when organic muck, churned up on lake bottoms, decomposed. "The turtles are feasting," an observer said.

The Red Wolves

Bulls Island, just north of Charleston, was one of South Carolina's glories. On its nearly 5,000 acres (2,025 hectares) grew the finest maritime forest along the Atlantic Coast—a green, densely textured tangle of live oaks, pines, and palmettos. The island, a component of the Cape Romain National Wildlife Refuge, also proved eminently suitable for the experimental release of red wolves as a part of the federal program to restore that species to the wild.

A visitor last fall had ample warning of what to expect on reaching the island. Mile after mile of forest along Route 17, the coastal high-

"This spring the fresh greenery in the swamp will hide part of the destruction," Brunswig said, knowing well that the effects of Hugo will be around for years to come. Brunswig continues: "There may be profound consequences of the storm here. Pine bark beetles find scars to be inviting points of entry, so we could have a beetle epidemic that will finish off a lot of the trees that Hugo didn't get. And it looks as if we lost almost all our oaks and hickories, the trees that deer, squirrels, and our big wild-turkey population depend on for mast [nuts and acorns that serve as food]. They'll have a feast this fall, perhaps, but in future years there's going to be famine."

way, had simply ceased to exist, now resembling isolated fence posts with chewed-off tops. Refuge headquarters at Moore's Landing, in the mainland village of Awendaw, had also ceased to exist. The scene put one in mind of a wartime outpost, after the bombs have gone off and the survivors have crawled from the mud to try to carry on as before.

Men and women wearing U.S. Fish and Wildlife Service insignia walked among the trailers and campers hastily installed on the headquarters site. Waiting on the remains of a dock for a boat to take them to Bulls Island were George Garris, the refuge manager, and Warren Parker, leader of the red-wolf-recovery team.

"It wasn't the wind that wrecked us," Garris was saying. "A surge of water 19 feet [5.8 meters] high came ashore at this spot, picking up trees and mud and smashing everything in its path. The visitors' center and headquarters was built well—it would take a million dollars to put up one like it today—but it's gone. So are 50 years of refuge records, our books, our color slides. One of the staff hasn't even found the remains of his desk."

Parker expressed his amazement that the wolves had somehow lived through the storm on Bulls Island.

"We had a pair of wolves out there that produced a litter of five pups last spring," he said. "An alligator got the female and one of the pups—we found the radio the female was carrying in some gator droppings! When we flew over the island after the storm, we saw the male and the four remaining pups still together. We dropped carnivore logs—they're big chunks of horsemeat like the ones they feed to tigers in zoos. The wolves may not be able to move around enough now to capture prey. I'm going out there today to see if we ought to take them off the island."

Scientists believe the red wolf evolved here in the New World. (The gray wolf is a larger European animal that apparently came to North America over the Bering land bridge.) Restricted to what became the southern United States in recent times, the red wolf declined precipitously after the arrival of European settlers. By the 1970s the remnants of the population were hybridizing with coyotes. Only a handful

The critically endangered red wolf (below) has flourished since being reintroduced into the wild on South Carolina's Bulls Island several years ago. Wildlife experts, fearing the worst when the hurricane devastated the island, rejoiced when radio tracking (right) confirmed that the wolves had miraculously survived the storm.

Below left: Tom & Pat Leeson; below: © Larry C. Cameron

of purebred red wolves were left, prompting the Fish and Wildlife Service to capture all of them. The red wolf, like the California condor, became extinct in the wild.

"This is the first time an 'extinct' animal has been reintroduced," Parker said. "We now have a small population living free on the Alligator River National Wildlife Refuge in eastern North Carolina. Wolves that are born in captivity don't do well when they're released. They usually live only six or eight months, at best. But they are valuable if they can produce and rear pups during that time, as they've done on Bulls Island. The pups learn from them and can make it in the wild."

Garris and Parker joined several other men in a small boat that headed through the offshore marshes toward the island.

"These marshes were full of clapper rails," Garris shouted above the roar of the twin outboards. "Some of them survived that big surge of water and got carried inland. I've seen rails up there, headed back toward the marshes."

The men landed at a battered dock on Bulls Island. Its maritime forest lay in untidy layers, the horizontal trunks forming an almost impassable barrier. The palmettos, deprived of their lush neighbors, were nothing but frayed and upended broomsticks. Huge army-surplus frontloaders and other equipment, operated by emergency crews sent from refuges all over the Southeast, were opening rudimentary roads through the debris.

"Here are some tracks!" called Parker, who was standing only a few feet from the dock. "It's the big male, and the tracks are fresh."

Garris nodded. "Before the storm the male came here every night," he said. "This is where he landed, and maybe this is where he thinks he's going to get off."

"We use Bulls Island to acclimatize wolves to life in the wild," Parker said. "There isn't enough room here to sustain a growing population, but it gives a young wolf a taste of what freedom is like."

As Garris and Parker followed a path opened by the cleanup crews, they found more distinct tracks in the muck. The wolves were moving, and so were raccoon, squirrels, even deer. Parker, climbing a metal observation tower left standing by the storm, wasn't able to pick up signals from the radios carried by the wolves, but he was more than satisfied with his inspection tour.

"The wolves are probably up at the north end of the island, and we know now that both the wolves and their prey can get around," he said. "We'll leave them on the island until late winter, then take them off and either release them at Alligator River or use them in the captive-breeding program. The male is a marvelous animal. This is the second year in a row he has lost the female to gators, but in both years he reared the pups by himself. We'll bring out another female in February [1990]."

Bulls Island's maritime forest may take a century to recover, but life goes on in the ruins.

Making Hollows

A Forest Service worker, wearing a hard hat and protective eyewear, climbed a metal ladder that was propped against a pine in the Francis Marion National Forest. At the foot of the ladder, another forester started a chain saw, then hoisted the clattering machine to her on a light rope. In a moment she was cutting into the tree, hollowing out a deep, rectangular cavity.

"We can't afford to wait," said Robert G. Hooper, a Forest Service research biologist who was standing nearby. "The biggest threat to the red-cockaded woodpecker is habitat destruction, and Hugo destroyed a large part of the habitat here. The birds nest in old pines. They seldom accept nest boxes, and so we're trying to give them a focus for establishing new colonies." Few areas of South Carolina were ransacked quite as thoroughly by Hugo as the 250,000-acre (101,000-hectare) Marion National Forest. Perhaps 50 percent of the trees were knocked down, a loss in lumbermen's terms amounting to anywhere from 700 million to 1 billion board feet. (A forester calculated that the downed timber was the equivalent of a board, 1 by 12 inches, long enough to circle the globe seven times.) The blow was compounded by the fact that this forest had sustained an increasing population of red-cockaded woodpeckers. Under the Endangered Species Act, the little woodpecker is a priority item wherever it breeds on public lands.

"We had about 475 colonies," said Hooper, who is assigned by the Forest Service to red-cockaded protection work. "These woodpeckers live in family groups, from two to nine birds, including the breeding pair, young of the year, and older siblings that hang around as helpers. Each colony may excavate three or four cavities and defend them against other family groups. We're trying to create about 800 artificial cavities now because we estimate that 80 percent of their old nest trees went down.''

Hooper and his colleagues may be optimistic in preparing such a high number of woodpecker homes. Early surveys disclosed that 75 to 95 percent of the birds survived in areas of the forest escaping heavy damage, but only 10 percent were detected elsewhere.

''I've actually found only one dead red-cockaded,'' Hooper said. ''It was in a tree that broke, like so many did, right at the cavity. The tree had hinged over, and I could see the bird's feathers sticking out. We don't know exactly what happened to most of the missing birds. They would probably fly out as the tree went down. They could have been beaten to death in the storm or maybe blown 50 miles [80 kilometers] away.''

The woman on the ladder had finished carving out a hollow in the tree. Hooper held up a rectangular piece of wood that would fit neatly inside the new hole. It was a portable woodpecker nest, coated with a sealant to keep out the tree's resin, and itself hollowed out with a woodpecker-sized cavity. In the event that other birds or mammals enlarge the cavity for their own purposes, biologists have designed a metal ''restrictor'' to fit over the entrance, bringing it back down to red-cockaded size. The forester, after coating the inside of the hollowed tree with glue, pounded in the plug with a mallet. Everyone hoped it would not remain unoccupied for very long.

Outlook

There will be trying times ahead for the Francis Marion National Forest. With sawmills glutted by the unexpected harvest everywhere in the region, much of the fallen timber may rot in the warm climate before markets can be found for it. Threats of fire and insect infestation will grow once spring arrives. Either occurrence, by prematurely removing more trees, would further cloud the woodpecker's future.

''It will take a very long time for the forest to recover, and at least several decades before the red-cockaded population can return to pre-hurricane levels,'' Bob Hooper said. ''But the woodpecker will be a driving factor in whatever decisions are made affecting this forest in the future.''

The Montana conservationist Les Pengelly once described selective logging as the practice of selecting a forest and then logging it. In the aftermath of Hurricane Hugo, one might be ex-

© Walt Johnson/Picture Group

Hugo wreaked environmental havoc in residential areas throughout the state. In Charleston, the graceful oaks that shaded many city streets snapped in the high winds generated by the storm, compounding the destruction and rendering roads impassable.

cused for wondering what the fuss made by forest preservationists is all about. Perhaps the advocates of clear-cutting are right: Why not cut the trees as soon as they are ''harvestable,'' rather than let them rot where they stand, or be reduced to rubble in a violent storm?

Such thoughts may have troubled Norm Brunswig when he saw the trees strewn across Francis Beidler Forest. But if they did, they soon flickered out, submerged by his plans for rebuilding the sanctuary's boardwalk so that it will be ready to lead curious and wondering visitors into the swamp again this spring.

''Our job at the sanctuary is to preserve natural systems,'' Brunswig told a friend several weeks after the hurricane. ''The species of plants and animals at Beidler evolved in a disaster-prone ecosystem. We're not here to log the forest or salvage the downed timber. We're here to let events play themselves out in good times and bad, and tell the story to the public.''

HEALTH
AND
DISEASE

Review of the Year

HEALTH AND DISEASE

People afflicted with AIDS and hepatitis were given new hope in 1989. But the incidence of these and other diseases continued to grow, as did the chaos and despair caused by the ever-widening drug epidemic.

by Jenny Tesar

The Nightmare of Killer Drugs

Although a survey of drug users found a significant decrease in casual drug use, addiction to cocaine and crack (a smokable form of cocaine) continued to grow. And deadly new drugs appeared. Among the most worrisome was a mixture of crack and heroin, called "moonrock," that combines the intense euphoria of crack with the physical addiction of heroin.

A drug whose use has reached epidemic proportions in Hawaii began to spread to the mainland U.S. "Ice" is a smokable form of methamphetamine. It is cheap to produce, and can be made in clandestine laboratories in the U.S. Like crack, it is highly addictive and produces bouts of severe depression. However, ice produces a much longer period of euphoria than crack, and users exhibit potentially violent paranoia and other disturbing psychological reactions for up to 14 hours after taking the drug. "The ice problem is so bad that crack pales by comparison," warns Douglas G. Gibb, Honolulu's police chief.

Illicit drugs damage the body in myriad ways. Crack smoking causes a lung disorder with symptoms similar to those of pneumonia. Nicknamed "cracklung" by physicians, the ailment does not respond to antibiotics or other standard pneumonia treatments. Rather, it must be treated with corticosteroids.

Women who trade sex with multiple partners for cocaine and other drugs play a major role in the transmission of syphilis, not only to their partners but to their fetuses. There has been a major increase in babies born with congenital syphilis.

The craving for crack, a smokable form of cocaine, can linger long after an addict kicks the habit. Acupuncture may help block the craving in some former users.

Many are stillborn. Those who do survive to birth may die soon after or suffer brain damage, hearing disorders, and other problems.

Women who use cocaine during their pregnancies also are more likely to give birth to premature babies and babies that weigh less, are shorter, and have smaller heads than average. Their babies are at greater risk of having heart defects and other abnormalities.

The cost of illicit drug use in the U.S. is estimated to be more than $60 billion annually. Medical bills for drug abusers, their children, and the victims of their crimes comprise a significant percentage of these costs. For example, it costs $80,000 to detoxify a baby born addicted to crack.

"Ice," a smokable form of methamphetamine, is more powerful than crack cocaine and at least as addictive.

"Still Out of Control"

The AIDS epidemic continued to spread, with 6 million to 8 million people believed to be infected worldwide. AIDS (acquired immune deficiency syndrome) had claimed at least 67,000 lives in the United States by the end of 1989.

In the U.S., unlike some other nations, the AIDS virus has been transmitted primarily through homosexual sex or intravenous drug use, with the number of infected men far exceeding the number of infected women. In 1989, however, researchers reported an alarming spread of the virus among teenagers, with equal numbers of boys and girls infected. Some studies indicated that as many as 1 percent of teenagers in cities like New York were already infected. The virus may lie dormant for years before symptoms of the disease appear. Thus, although there was a 40 percent increase in reported AIDS cases in teenagers between 1987 and 1989, most infected adolescents will not have symptoms until they are in their twenties.

The time it takes for AIDS to develop is longer than previously believed. A study by epidemiologists at the University of California in San Francisco found that only half of the people studied had AIDS symptoms 9.8 years after they were infected with the virus. Other studies found that the older people are when they become infected, the faster the disease develops. People who were under 20 when infected had a 2 to 5 percent chance of developing AIDS within seven years. Those age 20 to 40 had a 20 to 30 percent chance, and those over age 40 had a 30 to 40 percent chance of developing AIDS within the same time period.

AZT (azidothymidine), the only drug licensed by the U.S. Food and Drug Administration (FDA) for combating the AIDS virus, had been approved for use only in the sickest patients. But research sponsored by the National Institute of Allergy and Infectious Disease showed that AZT can significantly delay the onset of AIDS and slow the early stages of the disease. Hundreds of thousands of people in the U.S. who have the virus but show no symptoms could be helped, but early intervention is costly. AZT is one of the most expensive drugs on the market, costing up to $8,000 a year per patient. "This announcement is great news for those people with good insurance," commented U.S. Congressman Henry A. Waxman (D-Calif.). But many infected people do not have adequate insurance, and federal health programs are designed to help only those people who are already sick or disabled.

© Chester Higgins/NYT Pictures

Researchers have made little progress toward developing a cure for AIDS. The demand grows daily for better treatment and services for victims of the disease.

Smoking and Drinking

Illicit drugs aren't the only ones grabbing headlines. The ways in which smoking tobacco and drinking alcohol affect people continue to raise concern.

Findings announced in early 1990 report that women become drunk more quickly than men because their stomachs have lower levels of an enzyme that neutralizes alcohol. Researchers found that women end up with 30 percent more alcohol in their blood than do men of similar size after drinking the same amount. The higher blood-alcohol levels, in turn, make the women more prone to liver damage. The same researchers found that the ulcer drug cimetidine (Tagamet) made the same stomach enzyme less active, thereby doubling the blood-alcohol levels in men who take the drug and drink. It is suspected that cimetidine has a similar effect in women.

The first direct evidence linking cigarettes and heart disease came from a study that showed that heart-disease patients who smoke are three times more likely to have chest pains and restricted blood flow to their hearts than are patients who do not smoke. Another study indicated that intake of cadmium—at levels roughly equal to that found in the blood of smokers—promotes osteoporosis. A disease characterized by the loss of calcium from bone, osteoporosis is a leading cause of bone fractures among older people.

A new study indicated that smoking is also hazardous to pets. Researchers found that dogs whose owners smoke have a 50 percent greater risk of developing lung cancer than do dogs whose owners are nonsmokers.

Ophthalmology Update

Tobacco smoking was reported to increase the risk of prematurely developing nuclear cataracts, a clouding that forms in the center of the eye lens. A study of 838 men found that those who smoked were twice as likely to develop nuclear cataracts before age 70 as those men who had quit smoking more than ten years previously.

A study conducted at the University of Western Ontario suggested that vitamin C and vitamin E supplements may significantly reduce the risk of getting cataracts. People who took at least 400 international units (one regular capsule) of vitamin E daily had about 50 percent less risk of getting cataracts. People taking a minimum of 300 milligrams of vitamin C daily reduced their risk of cataracts by 70 percent.

Extended-wear contact lenses were designed to be safely worn for up to 30 days without being removed from the eyes. But two studies indicated that people who wear their lenses while they sleep are 10 to 15 times more likely than strictly daytime users to have painful and potentially blinding eye ulcers. The risk appeared to increase the longer the lenses were worn before removal. To counteract the negative publicity, manufacturers of extended-wear lenses reduced the maximum period for wearing the lenses before removing them for cleaning from 30 days to seven days.

An estimated 10 million Americans suffer from dry-eye disease, a condition caused by either reduced tear production or increased tear evaporation. Most medications provide only limited relief, and their prolonged use tends to exacerbate the condition. Researchers reported a new electrolyte-based solution that maintains healthy eye lubrication without undesirable side effects.

© Kenneth Murray/Photo Researchers, Inc.

People who leave in extended-wear contact lenses while they sleep run a high risk of developing serious eye ulcers. Eye specialists recommend wearing them for no longer than seven days at a time.

Advances in Organ Transplants

Each year, some 12,000 organ transplants are performed in the U.S. The types of operations continue to grow. In 1989 surgeons at the University of Chicago performed the first U.S. liver transplant from a living donor. A section of liver was removed from a woman and transplanted into her 21-month-old daughter. It is hoped that the technique will save the lives of hundreds of children who die each year while waiting for donor livers from cadavers.

Thomas Starzel of the University of Pittsburgh, the world's largest organ-transplant center, reported the development of a new drug, FK-506, that has proven to be extremely effective in preventing the body's rejection of transplanted organs. The rejection rate was one-sixth the rate in transplant patients taking cyclosporine, the major antirejection drug now in general use. FK-506 also produced fewer side effects, although its long-term effects were still unknown.

AP/Wide World Photos

Alyssa Smith recovered well after becoming the first American to receive a liver transplant from a living donor.

Limiting the Spread of Hepatitis

There are three main types of viral hepatitis, labeled A, B, and C. All three can be transmitted through infected blood. But hepatitis A is commonly transmitted through contact with fecal material from infected people. Hepatitis B is transmitted primarily through sexual contacts, though there has been a dramatic increase in transmission among intravenous drug users who share hypodermic needles. Hepatitis C is a major threat to blood supplies.

Each year, about 150,000 Americans become infected with hepatitis C, many through blood transfusions. More than 20 percent develop chronic liver problems, including cirrhosis and liver cancer. Researchers at the Chiron Corporation announced the development of a screening test that can detect blood tainted with the virus, offering hope that this threat can be greatly diminished.

Two studies found that the drug alpha interferon can control hepatitis C by preventing the virus from destroying liver cells. It was the first drug found to combat the virus. It did not cure the disease, but rather, controlled it. In more than half of the patients who were helped, the disease returned when treatment ended. Further research may find that giving interferon in higher doses or for longer durations of time will decrease the relapse rate.

During the 1980s the number of hepatitis B cases jumped from 200,000 a year to 300,000 a year. The increase was unusual because two hepatitis B vaccines are available in the U.S. The vaccines are well tested and considered safe, but they are comparatively expensive, costing about $120 for adults and $60 for children. Dr. Mark A. Kane of the Centers for Disease Control (CDC) noted that the fastest way to reduce the incidence of hepatitis B would be to vaccinate teenagers before they become sexually active or use intravenous drugs. By the time people enter college or the military, it is too late to protect many of them.

© David Powers/Chiron Corporation

A test that detects the hepatitis C virus in blood reduces the chance of contracting the disease via transfusion.

Measles on the Rise

At least 42 people—mostly young children—died of measles in 1989. It was the most deaths in any year since 1971. The total number of reported measles cases also surged, to approximately 15,000. One outbreak among unvaccinated preschoolers in Houston involved more than 1,700 cases. Other outbreaks occurred at colleges and universities.

The CDC recommended that children receive two vaccinations: one at 15 months of age, the second at five years. This "double-whammy" approach is intended to stop the spread of measles among children who were vaccinated young, but whose vaccination, for some reason, failed to protect them. The CDC also recommended second vaccinations for people born since 1957, because studies have shown that vaccine effectiveness diminishes with time. (In 1957 measles was so widespread that Americans developed some natural immunity.)

Physicians stressed that measles vaccinations should not be given to pregnant women. Nonpregnant women in their childbearing years should be counseled not to conceive within three months of vaccination, because of potential harm to the fetus.

Measles wasn't the only resurgent childhood disease. The percentage of children aged 1 to 4 who were immunized against such common diseases as diphtheria, polio, and mumps fell during the 1980s. The drop possibly may be due to a complacency among parents and physicians, but the rising costs of vaccines is believed to be a factor, too. Between 1982 and 1987, for example, the cost of DTP (diphtheria, tetanus, and whooping cough) vaccine rose more than 5,000 percent—from 15 cents to $7.69 per dose—due in large part to at least $5 billion in lawsuit claims.

An alarming rise in measles cases has led physicians to urge that children receive two vaccinations: one at 15 months and another at five years. Doctors have also suggested that anyone born after 1957 should receive a second vaccination.

© Sam C. Pierson/NYT Pictures

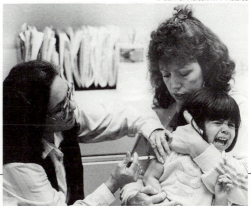

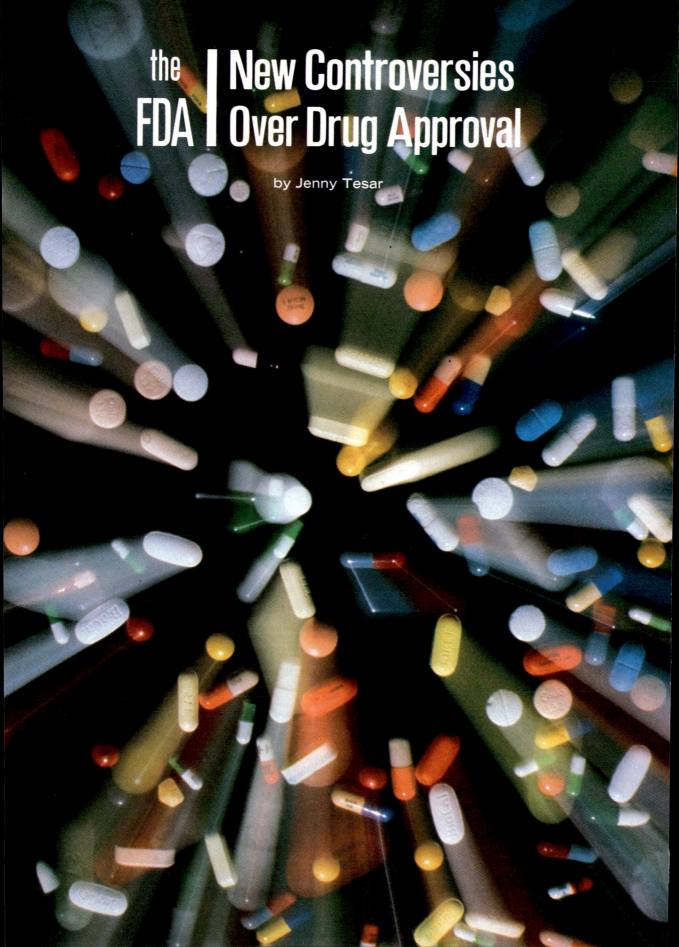

the FDA | New Controversies Over Drug Approval

by Jenny Tesar

Hood's Vegetable Pills: guaranteed to end colds, prevent fever, "regular" the bowels, invigorate the liver, and cure headaches, jaundice, sour stomach, and biliousness. If your druggist didn't carry these pills, order them by mail from C.I. Hood & Company in Lowell, Massachusetts. Price: 25 cents a box; five boxes for $1. Sound too good to be true? It is. But 100 years ago, C.I. Hood and others could make such medical claims about the products they peddled.

We've come a long way in the past century. Today claims of a product's medical merit must be supported by scientific proof. A federal agency, the Food and Drug Administration (FDA), has the authority to ensure that all drugs on the market are both safe and effective.

But regulation of medical products and treatments continues to evolve. In recent years the specter of AIDS has spurred revision of the government's drug-approval process. And a 1989 scandal in the generic-drug industry brought renewed attention to federal review of the lower-priced medications.

An Agency with Increasing Power

The FDA is a unit of the Department of Health and Human Services. Its prime purpose is to protect Americans against harmful, unsanitary, and falsely labeled foods, drugs, cosmetics, and therapeutic devices. Feed and drugs for pets and farm animals also come under the FDA.

The FDA was born early in the 20th century, following passage of the nation's first drug law. Initially, the agency's powers were limited. The Food and Drugs Act of 1906 required only that drugs meet official standards of strength and purity. In order to take a drug off the market, the FDA had to prove that the drug's labeling was false and fraudulent.

It wasn't until 1938—after 107 people died from using an untested, poisonous new drug formulation called Elixir Sulfanilamide—that Congress passed a stronger law. The Federal Food, Drug, and Cosmetic Act of 1938 required a manufacturer to prove the safety of a drug before it could be marketed. The law also provided for tolerances for unavoidable poisonous substances, authorized factory inspections, and added the remedy of court injunction to previous remedies of seizure and prosecution.

The Drug Amendments of 1962 further tightened control over drugs. Now firms also had to prove effectiveness for the product's intended use before they could market a drug. They were required to send adverse-reaction reports to the FDA. And drug advertisements in medical journals were required to provide complete information on both risks and benefits.

The Drug-Approval Process

Before a pharmaceutical firm can begin to test an experimental drug on humans, it must conduct extensive laboratory research to determine the compound's positive and negative effects and its safe- and toxic-dose levels. If test-tube research is positive, indicating at least some desired effects, then the compound is tested on animals. Generally a compound is tested first on mice or other rodents. If results are positive, then tests on higher animals follow. Among things scientists want to learn at this stage are how much of the substance is absorbed into the blood, how it is broken down chemically by the body, what metabolites (breakdown products) are produced, and how rapidly the drug and its metabolites are excreted from the body.

By this point in the testing process, the great majority of potential drugs studied by pharmaceutical firms prove to be unusable for one reason or another. The Upjohn Company estimates that of every 2,000 chemicals studied, only 200 show any potential in early tests, and only 20 of those may be tested in people.

Testing a drug on humans requires FDA approval. The pharmaceutical company must file a report that includes such information as the results of animal testing, plans for the proposed human study, and the qualifications of the researchers who will conduct the study. If the FDA grants permission for testing, the compound is classified as an Investigational New Drug (IND).

All IND tests are conducted on volunteers. Each person is fully informed of the procedures, the potential risks and benefits, and the possible alternative treatments. He or she may withdraw from a study at any time.

Typically, testing on humans consists of three phases:

Phase I. Phase I tests for toxicity and metabolism involve healthy adults. Generally neither children nor women who might possibly be or become pregnant are accepted for this phase. Usually fewer than 100 subjects are involved, and they receive the drug for about a month. Initially the drug is given to the volunteers in very low doses. Researchers look for negative reactions, how the drug is metabolized, and how to best administer the drug.

DRUG APPROVAL: The Standard Process and the Fast Track Approaches

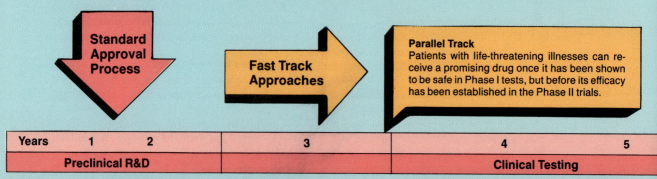

Preclinical R&D (Years 1 and 2). The compound is synthesized (left), then tested for biological activity and safety in animals. The drug company then files an Investigational New Drug (IND) application with the FDA. The IND becomes effective if the FDA does not disapprove it in 30 days.

Phase I (Year 3). The drug is tested in usually no more than 100 healthy volunteers (right). Dosages are slowly increased to gauge the drug's safety, evaluate its effects, and to learn more about the way it is metabolized.

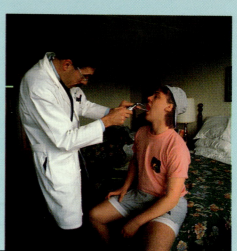

Phase II. In Phase II the drug is tested on volunteer patients with the specific disease or symptom the researchers hope to treat (unless the drug is a vaccine or other preventive medication). Generally the first volunteers chosen are people who have not responded to other treatments and may benefit from the experimental drug. In most Phase II studies, some of the patients receive the drug while others act as a control group, receiving either a standard treatment or a placebo (an inactive substance). Phase II usually involves several hundred patients, and it lasts from several months to two years.

Phase III. In Phase III the drug is tested on a large group of volunteers—often thousands of patients—to further determine safety and effectiveness, and to determine optimum dose rates and schedules. Phase III involves several hundred to several thousand patients who have the condition against which the drug is intended to be effective. These tests may take four years or more to complete.

Only when one phase has been completed satisfactorily may the pharmaceutical firm proceed to the next phase. The FDA oversees the progress of each phase, and it must be informed promptly of any negative effects of the drug. At any point in the process, the agency can stop or modify testing to protect or benefit the patients.

A new drug may not be commercially marketed in the United States, imported, or exported from the United States unless it has been approved by the FDA. Such approval is based on a New Drug Application submitted by the sponsor of the drug, containing the results of tests to evaluate its safety, plus substantial evidence of effectiveness for the conditions for which the drug is to be offered. It is not uncommon for such an application to be 100,000 pages long, with the bulk (80 percent or more) consisting of data gathered during the clinical tests.

Even after receiving approval to market a drug, the manufacturer must report regularly to

One IND, aerosol pentamidine, is given to some AIDS patients to prevent pneumocystic pneumonia (right).

Treatment IND
Under special circumstances, an investigational new drug (IND) can be administered to patients not taking part in Phase III clinical trials and prior to FDA approval. Drug companies may charge patients for the drug.

6	7	8	9	10
			Approval	

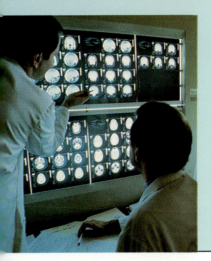

Phase II (Years 4 and 5). The drug is given to about 200 to 300 patients to determine its side effects and to measure its effectiveness against existing treatments or a placebo. Trials are conducted by a university or government medical center.

Phase III (Years 6, 7, and 8). The effectiveness of the drug is tested in 1,000 to 3,000 patients in clinics and hospitals; the results are compared with existing treatments (left). Trials must confirm Phase II studies and yield a low incidence of adverse reactions.

Approval (Years 9 and 10). The drug company files a New Drug Application (NDA) with the FDA, containing reams of laboratory, animal, and clinical data. After extensive review, the FDA may approve the application, in which case full production of the new drug begins (right). The FDA may, however, seek more data before granting approval.

the FDA on the drug's effectiveness. These data are obtained from physicians and other health professionals who use the medication. In addition, the manufacturer may sponsor Phase IV studies to develop new indications (uses) for the drug. One such example is SoluMedrol sterile powder, a corticosteroid medication developed by the Upjohn Company. It had 14 indications when it was approved in 1959; today it has more than 60 approved uses.

Speeding Up the Process

It takes an average of seven to 10 years to bring a drug to market in the United States. Much of this is a result of the rigorous premarketing trials required by U.S. laws. Another factor is the lengthy time taken by the FDA for reviewing new-drug applications. The law allows the FDA six months, but at present it takes an average of 31 months to process a new-drug application. At least part of the delay has been caused by budget constraints and insufficient staff. During the 1980s the FDA had to begin to deal with a deadly new disease, AIDS. It took on the added responsibility of regulating the biotechnology industry. And 24 new laws were passed that had to be enforced by its employees. Yet during that decade the FDA's staff was cut, and its budget remained almost constant.

Many European nations have much simpler review procedures and approve drugs much more quickly. Thus, effective medications may reach European consumers years before they become available to U.S. consumers. For example, calcium blockers were used to treat Europeans with heart disease for at least eight years before receiving approval in the United States.

But European patients pay a price for this faster availability of drugs, experiencing more adverse drug reactions. Perhaps the worst case of premature drug approval occurred with thalidomide, which was developed as a tranquilizer to depress the nervous system and quiet the patient without inducing sleep. The FDA delayed approval of this drug because of doubts about its safety, but it was widely prescribed elsewhere. In 1961 thalidomide was found to cause fetus malformations when taken during pregnancy. These abnormalities resulted in babies born with some or all arms and legs missing. About 10,000 such children were born worldwide, with 5,000 in Germany alone.

Major drug disasters have also occurred in the U.S. as well. Oraflex, an arthritis drug, was removed from the market within three months of its introduction in 1982 after being linked to 72 deaths in the U.S. and Britain.

Nonetheless, at times the FDA has been accused of undue strictness in approving new drugs. Criticism of the lengthy approval procedure became particularly vociferous among AIDS patients and their advocates, who claimed it was cruel and unethical to withhold promising experimental drugs from people with life-threatening conditions for which there is no other hope.

In 1987 the FDA responded to the critics by changing its drug regulations. ''These new procedures are proposed to apply to immediately life-threatening conditions, recognizing that in those cases, patients are willing to accept a greater risk, since there may be no other hope,'' explained then FDA Commissioner Frank E. Young. ''FDA anticipates that approvals for expanded studies of drugs for immediately life-threatening diseases can be given near the end of the second phase of clinical testing; that is, after the drug's safety testing has been done and the proper dose determined, and after some evidence of therapeutic benefit is available. Approval for expanded uses of experimental drugs for serious but not immediately life-threatening diseases would ordinarily occur at the middle of the third and final phase of testing. . . . We trust that these procedures will allow clinical trials to continue unfettered while permitting physicians and patients—with their informed consent—to obtain access to breakthrough drugs earlier than usual.''

Among the first to benefit from the new regulations were AIDS patients. After the experimental drug AZT (azidothymidine) showed very promising results in early studies, the FDA took only one week to approve its broader use. This enabled more than 4,000 patients to receive the drug while it underwent final review and approval—a process that took less than four months, making it one of the shortest approval actions on record.

An even further departure from the agency's traditionally cautious approach occurred in September 1989, when the FDA permitted another promising AIDS drug, DDI (dideoxynosine), to be widely distributed after the manufacturer, Bristol-Myers Company, had finished only Phase I tests on 26 patients. Under

The FDA does much more than oversee the approval of drugs. The agency's rules and regulations also help ensure that foods and cosmetics sold in the United States have been produced in a sanitary manner, are not fraudulently advertised, and carry the correct labels.

IT'S A FRAUD!

The days of traveling medicine shows have faded into history. No longer do enterprising peddlers tout such panaceas as Kickapoo Indian Worm Killer, Stephens' Magical Egyptian Oriental Eye Ointment, and Violet Blossom's Tiger Fat (made from petroleum jelly and various oils, and advertised as "effective for all ailments of the flesh").

Though the colorful hustlers are gone, medical quackery remains very much a part of our lives. Each year, an estimated 38 million Americans use fraudulent health products and treatments, spending some $27 billion for worthless and often dangerous "cures."

In reality the cost is even higher, for the victims may lose precious time in obtaining effective medical help. And in many cases the purported cure actually is dangerous. Chelation therapy—promoted as a preventive of heart attacks, strokes, and other circulatory diseases—involves an injection or tablet of the amino acid EDTA (ethylenediaminetetraacetic acid). There is no scientific evidence that chelation therapy works, but lots of evidence that EDTA can cause convulsions, kidney failure, and bone-marrow depression.

According to the FDA, these are the ten major health frauds:
- Fraudulent arthritis cures;
- Phony cancer clinics;
- False and unproven AIDS cures and treatments;
- Instant weight-loss plans;
- Aphrodisiacs and other fraudulent sexual aids;
- Baldness remedies, wrinkle removers, and other "fountain of youth" products;
- Phony nutritional schemes;
- Chelation treatments;
- Unproven use of muscle stimulators;
- Treatments for unproven hypersensitivity to *candida* fungus.

To avoid being ripped off by medical quacks, check the credentials and affiliation of individuals before accepting their advice and treatment. Be wary of secret remedies and quick, miracle cures for complex health problems. Accept only scientifically documented information. And don't believe that pills and gadgets can provide quick, effortless ways of accomplishing weight loss and other difficult goals.

---

the new regulation, Bristol-Myers could distribute DDI to people with AIDS who were not eligible to take part in clinical trials. It soon became apparent, however, that the procedure was threatening clinical trials of the drug.

Bristol-Myers needed 1,540 patients for two trials that would compare DDI with AZT. Two-thirds of the patients in the trial would receive DDI, in differing dosages, and one-third would receive AZT. (Using a comparison group that receives another medication or no medication is the standard method used to determine the effectiveness of a substance under study.) But AIDS patients, wanting a better than 67 percent chance of getting DDI, which they believed was more effective and less dangerous than AZT, applied to receive the drug outside the trials. In mid-November, it was reported that only 75 patients had enrolled in the trials, while more than 1,300 had enrolled in the expanded-access program. Although the latter group is being monitored for signs of the drug's efficacy, clinical trials are needed to clearly demonstrate if DDI's benefits outweigh its risks.

The Generic-Drug Controversy

Developing new drugs is an expensive procedure. It can cost up to $200 million to develop and bring a drug to market. To recoup the costs and to make a profit, manufacturers depend on patent protection, giving them a period of monopoly.

When the patent of a highly profitable drug expires, however, competing manufacturers quickly enter the market with cheaper generic versions of the drug. To win approval of a generic drug, a company is not required to undertake large-scale clinical trials. It simply has to provide the FDA with data showing that its version contains the same active ingredients as the original product, and that the medication is absorbed into the bloodstream of 18 to 24 healthy adult volunteers in a similar fashion to the brand-name drug. FDA rules allow generics to use different additives—fillers, coloring, preservatives, etc.—as long as these ingredients do not significantly change the way the drug is absorbed into the bloodstream.

On average, generics cost about half as much as brand-name drugs. Due to their lower cost, generics have captured a growing share of the $20-billion-a-year market for prescription drugs. Between 1982 and 1989, generics' share climbed from 2 percent to 33 percent. And as brand-name makers look ahead, they see a gloomy picture: between 1991 and 1995, patents are scheduled to expire on brand-name prescription drugs with an estimated $10 billion in annual sales.

Brand-name manufacturers have used various arguments in an effort to fight the advance of generics. For instance, they claim that variations in the rate of absorption make it risky for many patients, especially senior citizens, to switch drugs. According to the FDA, there is no scientific evidence to support this contention. The agency noted that some 430 million prescriptions were filled with generics in 1988, with no evidence of systematic problems.

At the same time that they are criticizing generics, however, some large brand-name manufacturers are producing generic versions of their rivals' brand-name drugs. Ciba-Geigy, for example, owns Geneva Generics, one of the largest generic manufacturers. It is estimated that at least 50 percent of all generic drugs are made by brand-name companies and their subsidiaries. But the brand-name drugs have higher profit margins, and so are more important to the companies than are their generic lines.

Criticism by brand-name firms of the safety of generic drugs and of FDA standards for approving generics took an unexpected turn in 1989, when investigations disclosed that several generics manufacturers had bribed federal officials, while others cheated on applications to win federal approval of their drugs.

Three FDA employees pleaded guilty to charges of accepting illegal gratuities from generics makers, presumably because the companies hoped to speed the approval process for their new products. Then Vitarine Pharmaceuticals was found to have won approval of its substitute for Dyazide (a popular hypertension drug made by SmithKline Beecham PLC) by filling test capsules with the brand-name drug instead of with its own product, thereby assuring that its product would pass equivalency tests.

Prompted by congressional inquiries, the FDA began a major investigation of the industry. It found manufacturing and record-keeping irregularities at 11 companies. It withdrew from the market more than 25 generics from several manufacturers, citing unclear or incomplete data. And it analyzed the 30 most-prescribed generics to check if they were equivalent to their name-brand counterparts.

"These inspections raise very serious questions about whether the industry can be trusted to supply accurate information to the FDA in their ANDAs [Abbreviated New Drug Applications], and whether that information remains accurate for those generic drugs which have been approved for sale to American consumers," said Congressman John D. Dingell of Michigan, who chairs the House Energy and Commerce Committee's Oversight and Investigations Subcommittee.

The FDA reorganized its generic-drug division and took other steps to avoid a recurrence of the problems, and most generics makers claimed that in the long run, the investigations will lead to stronger regulations and positive changes.

"I think the FDA needs more power," said Bill Haddad, chairman of the Generic Pharmaceutical Industry Association. "Companies that are tainted by corruption, even though it didn't affect health and safety, should be barred from doing business."

© Dan McCoy/Rainbow

New Advances in the
CANCER WAR

by Elisabeth Rosenthal

In September of 1984, Dr. Ronald H. Spiro told Robert A. Pfeiffer that he had a large tumor at the base of his tongue, right next to the vocal cords in his larynx, that had already spread to some lymph nodes in his neck. The standard treatment for such tumors was an extensive operation that involved removing the entire larynx, or voice box. Even with a total laryngectomy, his chance of being cured was less than 20 percent. Pfeiffer, who had spent 30 years as a broadcast journalist for CBS and the United Nations, accepted the news of the tumor as a challenge to be overcome, but the talk of laryngectomy terrified him. "I thought, 'Oh my God, they're going to take away my precious instrument, my voice.' It was like saying you're going to amputate the fingers of a violinist. My first reaction was: 'Don't take out the larynx under any circumstances.'"

Robert Pfeiffer said that in a recent interview—and he was using his own voice. Instead of proceeding with conventional therapy, Dr. Spiro sent him to Dr. George J. Bosl, a medical oncologist at Memorial Sloan-Kettering Cancer Center in New York City who had started an experimental "larynx-preservation" program. Dr. Bosl used a combination of chemotherapy and radiation as a first strike against certain can-

Through advanced imaging techniques, physicians frequently can pinpoint the exact location and extent of tumors. Such foreknowledge allows surgeons to better plan operations and, if needed, to construct the appropriate prosthetic devices before surgery begins.

HEALTH AND DISEASE 235

cers of the mouth and throat in an attempt to eliminate the need for surgery that ultimately leaves the patient deformed, disfigured, or otherwise handicapped.

Today, five years later, Pfeiffer is considered cured, and larynx preservation, though still experimental, is generating widespread interest among cancer specialists; it may soon be standard care.

Preservation of Function

The effort to save the voices of people like Robert Pfeiffer reflects a new concern among cancer specialists with what they call "preservation of function." Until recently, removal or regression of a tumor was the only measure of successful treatment. Surgeons practiced their art with broad brushstrokes, cutting out large margins of visibly normal tissue around a tumor in hopes of preventing a recurrence. The cancer would sometimes be cured, but neither the patient nor the surgeon ended up feeling very good about the process or the results. Now physicians are trying to modify their treatment in an effort to enable survivors to lead relatively normal lives.

"Ten, 20 years ago, mutilating and destructive operations were the only options for survival," says Dr. Murray F. Brennan, chairman of the surgery department at Memorial Sloan-Kettering. "We used to say, 'To hell with the ovary or the leg or whatever; let's treat the cancer.' Now we have the luxury of thinking about quality of life." Major advances in chemotherapy, radiation oncology, pathology, and surgery have made the crusade for function preservation possible.

The crusade has also been fueled by the dramatic increase in the number of long-term cancer survivors. Half of all newly diagnosed patients are now expected to live disease-free for at least five years, the length of survival traditionally defined as a "cure." As recently as 1976, only about 41 percent were cured. Although lung cancer, the most common cancer in the United States, still has a dismal 13 percent five-year survival rate, improved screening and treatment have nudged the rates for most other cancers slowly upward. Seventy-five percent of breast-cancer patients and 54 percent of colon-cancer patients, for example, can now expect to live for five years. And current cure rates of greater than 90 percent for patients with testicular cancer and most types of bone cancer confound the popular notion of the disease as inevitably terminal.

THE FIGHT FOR SURVIVAL

1930s: *Bombardment by X-ray radiation offers the only nonsurgical method of eradicating cancerous tumors.*

1950s: *Chemotherapy, the use of drugs to kill cancer cells or inhibit their growth, emerges as a weapon.*

Clockwise from above: J. J. Kath/Merck & Company; NYT Pictures; AP/Wide World Photos; © Mark Ferri; chart by Hank Iken/The New York Times

1970s: Ted Kennedy, Jr., endures high doses of chemotherapy and a leg amputation to treat a malignant bone tumor. He beats the 1-in-4 odds for a cure.

1990s: Dan Rosenthal walks on his own two feet, thanks to limb-sparing surgery performed in 1987 that removed a cancerous bone tumor from behind his left knee. Since the 1950s, medical science has seen a sharp rise in the five-year survival rates of all patients with cancer of the testis, larynx, and bone (see chart below). Lung-cancer survival rates remain very low.

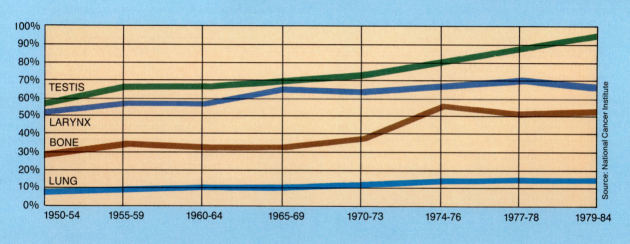

Source: National Cancer Institute

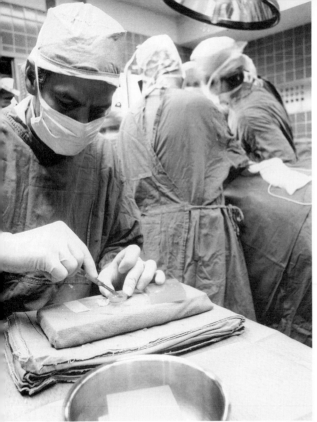

For many bone-cancer patients, surgeons can use a silicone bone-replacement material (above), prosthetic devices, or bone from a donor to salvage a functional limb. Today, 80 percent of bone-cancer victims retain their limbs.

Tumor Reduction

Each follow-up visit to the head-and-neck clinic must remind larynx-preservation patients of their good fortune. In the waiting room, they see survivors of conventional larynx surgery, each breathing through a small hole in the neck. These patients can no longer talk normally, and most carry devices that enable them to communicate: pads and pencils and small sensors that are held against the neck to convert vibrations from inside the throat into an electronic voice.

Until recently, head and neck cancers—90 percent of which are related to tobacco or alcohol use—were the domain of surgeons working in association with radiologists. But in the late 1970s, Dr. Waun Ki Hong, an oncologist then at the Boston Veterans Administration Hospital and now at the M. D. Anderson Cancer Center in Houston, made a discovery that kindled hope for nonsurgical treatment. Dr. Hong and his colleagues found that a new cancer medication called cisplatin, used in combination with older chemotherapeutic drugs, significantly reduced the size of tumors in 70 to 80 percent of patients.

Larynx-preservation treatment today consists of two to three months of chemotherapy followed by six to seven weeks of radiation. Patients whose tumors do not shrink within two months are whisked to the operating room for radical laryngectomies. After two years, according to Dr. Hong, patients treated with chemotherapy and radiation have the same survival rate as those treated immediately with surgery—66 percent. And about 70 percent of the patients who elect larynx-preservation treatment retain their voices.

Robert Pfeiffer will never forget the day when, after one week of chemotherapy, Dr. Spiro looked into the back of his throat and said only, "Sensational. Amazing." The tumor had shrunk so much that it could no longer be seen.

Thelmo Castillo, another larynx-preservation patient, was also fortunate. His hoarseness —caused by a large tumor on the vocal cords —cleared after one chemotherapy treatment.

Chemotherapy does have its costs, however. Cisplatin, for example, often produces terrible nausea and profound fatigue, as well as transient hair loss. Some patients who received very high doses of the drug were left with long-standing nerve damage that resulted in varying degrees of hearing loss and numbness of the extremities. Castillo recalls dropping a subway token on the way to a clinic visit about a year after he had finished his chemotherapy treatment, and having to call his wife in helpless frustration because he didn't have enough sensation in his fingertips to pick the token up.

Survivors are philosophical. "Look, I'm a little clumsy," says Pfeiffer. "But I'm alive."

Larynx-preservation patients today need not worry about hearing loss or nerve damage, because the doses of chemotherapy they receive are much lower—and just as effective.

Saving Limbs

Many people assume that breast surgeons were the first to develop less drastic means of treating cancer. Mainly because of the number of people involved—approximately 150,000 new cases each year—changes in breast-cancer treatment have received a tremendous amount of publicity. Indeed, in the past 10 years, most breast surgeons have stopped performing disfiguring radical mastectomies—in which the patient loses the normal contour of the upper arm and chest wall—in favor of modified-radical mastectomies or, in certain selected cases, breast-conserving lumpectomies.

But many doctors within the cancer community regard orthopedists, in their treatment of bone and connective-tissue tumors, as the true founding fathers of modified surgical techniques. Orthopedists have come a long way since 1973, when newspapers reported the story of 12-year-old Ted Kennedy, Jr. He heroically endured high doses of chemotherapy with methotrexate, a particularly toxic drug, in addition to the amputation of his leg from above the knee, to treat a malignant tumor. His chance for a cure was 25 percent. But he beat the odds.

Although bone cancer is relatively uncommon—there are fewer than 7,000 new cases each year—most of the people who get it are between the ages of 15 and 30. "When I got out of residency in 1960, everybody with bone tumors got an amputation, and no one lived longer than a year or two," recalls Dr. Henry J. Mankin, chief of the orthopedic service at Massachusetts General Hospital in Boston. "By 1980 we had a survival rate in the 60 percent range, and half the kids kept their limbs. Now almost everybody lives, and at least 80 percent keep their limbs."

Regular mammograms help detect breast cancer in its early stages. Surgeons rarely perform radical mastectomies anymore, opting instead for the less-disfiguring modified mastectomy or breast-conserving lumpectomies.

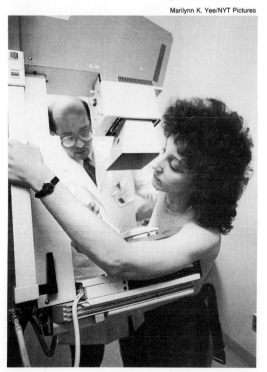

Marilynn K. Yee/NYT Pictures

Bone tumors originally had a dismal prognosis because, unlike head and neck tumors, they tend to shed cancerous cells into the blood. These cells then take up residence at distant sites, such as the lungs. Removing an afflicted limb, therefore, often leaves behind disease elsewhere in the body that later proves fatal.

Surgery will probably always be the only way to completely remove bone tumors, but in the 1970s concomitant treatments of either chemotherapy or radiation were found to shrink the tumors and kill cancer cells in the surrounding tissue and elsewhere in the body, allowing orthopedists to try less extensive operations. Instead of amputating an entire limb, they began to remove only the cancerous part of the bone and replace it with a synthetic prosthesis or bone from a human donor—with the goal of salvaging a functional limb. In 1971 Dr. Mankin performed his first limb-sparing procedure on a 16-year-old girl, replacing a section of her thighbone with bone from a cadaver. Today she is well and lives an active life.

With experience and advances in medicine, orthopedists have been able to get good results with progressively less extensive surgery. To salvage a limb successfully, surgeons must remove the tumor without damaging certain muscles or major nerves and arteries. Modern chemotherapeutic agents can sometimes tease the tumor edge away from these vital structures, and advanced technologies for looking inside the body—such as computerized axial tomography (CAT) scans and magnetic resonance imaging (MRI)—show surgeons exactly where a tumor ends, allowing them to better plan operations and build prostheses before surgery.

Living with Limitations

Dan E. Rosenthal, a 20-year-old Lehigh University sophomore, opted for a limb-sparing procedure instead of amputation in 1987, when a two-year-old painless lump behind his left knee proved to be an osteosarcoma. Dr. Joseph M. Lane, chief of the orthopedic-surgery service at Memorial Sloan-Kettering, warned him that limb-sparing surgery was the more arduous route, often complicated by infections around the prosthesis, fractures of the salvaged limb, and cracks or breaks in the hardware.

During seven hours of surgery, Dr. Lane removed 10 inches (26 centimeters) of thighbone, the knee, and the top of the shinbone. In their place he inserted a custom-made titanium prosthesis that consisted of an artificial knee and rods that slipped into what remained of Rosenthal's own thighbone and shinbone. Since the tumor had not invaded many of the muscles that stabilize the knee, they were preserved; these muscles are able to control the movement of the artificial joint.

Compared with many patients whose limbs have been spared, Dan Rosenthal is lucky. The slab of metal inside his leg fits well, his leg never became infected, and he limps only imperceptibly when he walks and climbs stairs. By contrast, a girl he met when they were both having limb-sparing surgery spent the next two years in and out of the hospital, fighting an infection of the tissue surrounding her prosthesis. As a last resort, her doctors replaced the bacteria-laden prosthesis with a new one, and the infection finally went away.

The old debate about whether limb-sparing surgery provides good cancer treatment is over. It does. The new focus is on which kind of surgery enables a patient to function best. Dan Rosenthal can swim, and he rides a stationary bicycle several times a week. But a salvaged limb is a salvaged limb, and most patients live with some limitations. Rosenthal, like all patients whose prosthesis includes a joint, cannot run or play tennis. Only the relatively small number of patients who receive pure-bone prostheses that do not include a joint have the potential to return to full activity.

When limb salvage is uncomplicated, patients are usually pleased. "Sure I have to ask for help with certain things," says Rosenthal. "So what? Friends who are short ask me to reach things for them because I'm tall. At least now I look down and I'm whole. It's my foot. They just took part of my bone."

Some patients have such stormy courses with limb-sparing surgery that they throw in the towel. Four years ago, after spending three months confined to a hospital bed nursing a sal-

Both photos: American Cancer Society

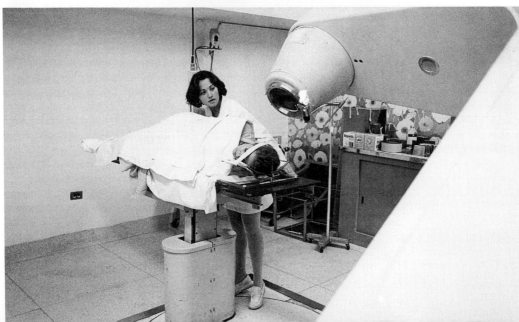

Patients who have undergone conventional surgery for cancer of the larynx typically must hold a small sensor against their necks to convert vibrations into an electronic voice (left). Specialists have now developed a larynx-preservation treatment that entails two to three months of chemotherapy followed by six to seven weeks of radiation (below). Seven out of ten patients who choose the preservation therapy retain their voices.

vaged limb that had become infected, Shawn Dempsey, then 13, decided on amputation. Dr. Lane removed Dempsey's leg from just above the knee, and then, in an ingenious operation called a turnoplasty, transformed the amputated ankle into a knee by turning it backward and attaching it to the thighbone. Dempsey was out of the hospital in a week, wearing only a lower-leg prosthesis. Now, at 17, he plays ice hockey, baseball, and tennis.

The dramatic advances in prostheses for amputees have led Dr. Lane and other surgeons to temper their enthusiasm for saving limbs. "Just because a mountain's there doesn't mean you have to climb it," Dr. Lane says. He believes that in terms of function, many spared limbs cannot compare with good prostheses. By way of example, he points to 20-year-old Todd Schaffhauser, whose left leg was amputated above the knee five years ago (he was not a candidate for limb sparing because the tumor had wrapped around one of the main arteries in the leg). Schaffhauser, who trains with the Aspire athletic program at the Hospital for Special Surgery in New York City, won a gold medal in the 1988 Paralympics in South Korea with a time of 15.77 seconds in the 100-meter dash.

Addressing Private Burdens

Missing limbs attract public attention, but the side effects of many cancer treatments have often been a more private burden. It was long assumed that sexual function and fertility were small sacrifices for a cancer cure. Dr. Brennan of Memorial Sloan-Kettering attributes this attitude, in part, to prudery: "Twenty years ago we weren't willing to talk about certain things that are very important to quality of life. Men wouldn't say, 'I've had my operation, thank you, but unfortunately I'm impotent.'"

Standard operations for cancer of the prostate and bladder leave men infertile and often impotent, and the surgery for testicular cancer usually causes infertility. Dr. Paul H. Lange, chairman of the urology department at the University of Washington School of Medicine, has seen marriages that had weathered the initial stress caused by these diseases later run aground on the fertility issue. Most men with testicular tumors are between 15 and 35.

One very real barrier to retaining sexual function in men who had cancer surgery in the pelvic region was that no one knew exactly where the nerves that control erection were located. After discovering these nerves in 1981,

Dr. Patrick C. Walsh, chief of urology at Johns Hopkins University, devised a modification of the conventional surgery for prostate cancer. The new procedure allows the nerves to be spared in many patients whose cancer is detected early. Seventy-four percent of the patients who undergo this modified surgery remain potent. Dr. Lange and other surgeons developed a similar operation for testicular cancer that leaves 85 to 90 percent of patients fertile.

For many types of cancer, the optimal method for both preserving function and ensuring a cure has not yet been determined. The traditional treatment for invasive bladder cancer, for example, has long been the removal of the bladder. "I remember watching radical surgery for a bladder cancer when I was in training, and thinking, 'This has got to be one of the worst damn operations known to man,'" says Dr. George R. Prout, Jr., a consultant in urology at Massachusetts General Hospital. Today Dr. Prout is reporting good preliminary results with chemotherapy and radiation treatment, obviating the need for any surgery.

Dr. Lange is not convinced that bladder-sparing treatment is as effective as surgery, so he continues to remove the organ. But for male patients he makes a "neobladder" from part of the bowel, and attaches it to what is left of the urethra, so patients can void relatively normally. (For anatomical reasons, the procedure does not work on women.)

Stay Informed

Although many function-preserving treatments show promise, oncologists are embracing them cautiously, awaiting final statistics about their capacity to cure. But Dr. Brennan of Memorial Sloan-Kettering has found that most patients are "more than willing to take a slightly higher risk to save a body part." Indeed, several clinical trials, such as the larynx-preservation studies, started because so many patients refused radical surgery, even when they were told that the alternative treatments might not offer as great a chance for survival.

With so many new options available, and others on the horizon, it is becoming increasingly important for patients to be informed consumers. "Each person has to seek out the best form of care for himself," says Dr. Bernard Gardner, secretary of the Society of Surgical Oncology. "It can't be like it was in the old days, when one surgeon took one patient to the operating room and just cut out the cancer."

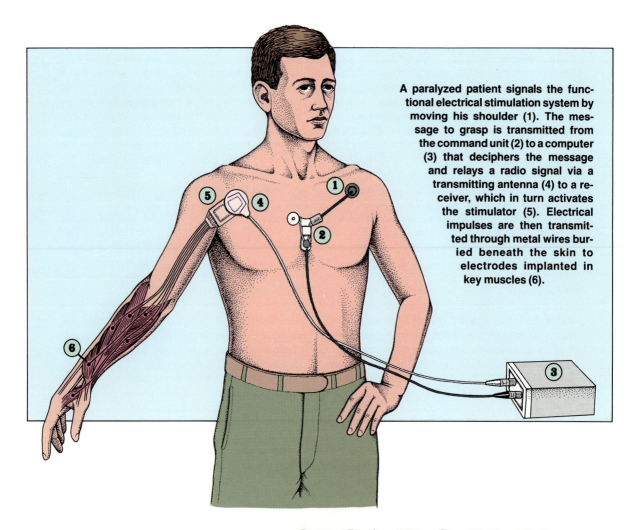

A paralyzed patient signals the functional electrical stimulation system by moving his shoulder (1). The message to grasp is transmitted from the command unit (2) to a computer (3) that deciphers the message and relays a radio signal via a transmitting antenna (4) to a receiver, which in turn activates the stimulator (5). Electrical impulses are then transmitted through metal wires buried beneath the skin to electrodes implanted in key muscles (6).

Waiting for the BIONIC MAN

by Ruth Simon

A 1977 diving accident paralyzed James Jatich, leaving him able to move only his arms and shoulders and raise his left wrist. But today Jatich can write, feed himself, and even drive, thanks to an experimental technology known as functional electrical stimulation, or FES, which uses low levels of electrical current to stimulate movement in paralyzed arms and legs. The results have a truly uplifting effect on a paraplegic's psyche.

"This is a big step toward independence for me," says Jatich, 40, a participant in an FES program at Cleveland's MetroHealth Medical Center. "My dad can put the system on me in the morning and go off to work."

FES first captured the public's imagination in 1982, when CBS's "60 Minutes" reported that a researcher at Wright State University in Ohio had used electrical charges to make a paralyzed woman walk. Researchers who have been experimenting with FES for more than two decades have suggested that the technique could ultimately be used not only to move paralyzed arms and legs, but also to provide bladder control, regulate breathing, prevent bedsores, and treat scoliosis.

But years of experimentation have yet to produce a system that can live up to the early promise of replacing a damaged central nervous system with sensors, electrodes, and computers. Nerves are in a different league from hips and teeth. Researchers are beginning to look at FES as an incremental improvement in a paraplegic's capabilities, not as a replacement for an attendant or a wheelchair. "The objective is to make them more independent," says P. Hunter Peckham, director of MetroHealth's Rehabilitation Engineering Center.

FES could prove useful to only a fraction of the disabled. "Between 5 percent and 10 percent of all spinal-cord-injured patients might be candidates for our system," says Dr. Robert Jaeger, a professor at the Pritzker Institute and the Rehabilitation Institute in Chicago, who has used FES to enable about two dozen paraplegics to stand. Regrettably, many more potential users will not benefit from the technology. Says Jaeger: "People with only one [functional] hand will never be able to use FES to stand or walk because of problems with balance."

Sending Signals

Nevertheless, FES is about to take its first hesitant steps outside the lab. Next spring the Veterans Administration and the National Aeronautics and Space Administration (NASA) will award a contract worth up to several million dollars to develop and manufacture 200 FES prototype systems for patients with paralyzed upper bodies. Modeled after the system now used by Jim Jatich, the prototypes are expected to be delivered in 1994, and could become commercially available in as few as five years if the Food and Drug Administration (FDA) gives quick approval to the devices. Four small biotechnology companies—Life Systems, Inc.; MiniMed Technologies; Thomas Medical; and Cochlear Corporation—are competing to produce the FES prototypes, which are anticipated to carry a price tag upwards of $40,000 apiece. Unfortunately, lower-body systems are more complex and much further away from commercialization.

How does FES work? The artificial brain of the Cleveland system is a pair of microproces-

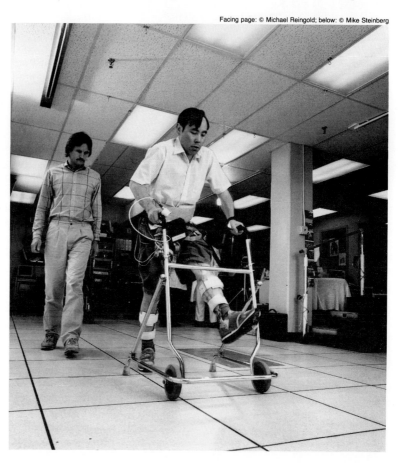

Dr. Ronald Lew, a nuclear-medicine specialist paralyzed below the waist, walks thanks to an advanced FES system that uses 48 electrodes and leg braces. For the most part, however, such lower-body FES systems have yet to emerge from the laboratory.

Facing page: © Michael Reingold; below: © Mike Steinberg

sors that can be attached to a patient's wheelchair or belt. The first microprocessor receives and deciphers a patient's signal to move. Jim Jatich sends a signal to open his hand by shrugging his shoulder, but other systems could respond to blinks or nods. A second microprocessor relays the signal to a radio transmitter, which sends an electric pulse down wires to electrodes implanted in key muscles. The pulses produce muscle contractions that crudely duplicate a normal grasp.

But controlling even basic movements like writing or stepping requires keeping track of massive amounts of information. Writing involves about two dozen individual muscles. Coordinating the movements of all those muscles is a computer programmer's nightmare. "Even with today's electronics, you can't duplicate the number of messages the body receives," says Cynthia Brogan of Ohio's Edison Biotechnology Center, which is helping to commercialize the work of Cleveland researchers.

A major challenge is providing the devices with a "natural" sensing ability. Several sensors may be needed for lower-body systems to adjust automatically to changes in speed and knee angle. "If the person starts to trip, the system has to react," explains Dr. Donald McNeal, director of the Rehabilitation Engineering Program at Rancho Los Amigos Medical Center in Downey, California.

Surviving the Body's Environment

An unusually advanced FES system developed by the Cleveland VA Medical Center team, using 48 electrodes and leg braces, enabled Dr. Ronald Lew, a nuclear-medicine specialist paralyzed below the waist in a 1979 auto accident, to visit Disneyland and try several rides recently. Lew's trip aside, lower-body FES systems have been used outside the lab only for demonstration purposes.

Researchers intend to adapt for human use sensors developed by the aerospace industry to determine a rocket's position and speed. The FES team at Cleveland's Veterans Administration hospital recently turned for help to NASA's Jet Propulsion Laboratory (JPL). Researchers are also hoping that more advanced microprocessors will process all the data without being connected to a large computer.

Another major technical hurdle is coming up with implantable sensors and wires that can survive within the body's saltwater environment for long periods, an increasing concern lately for makers of implantable body parts. Implantable electrodes provide more precise control than those applied to the skin, which can shift around and must be reset daily. "There's still quite a lot of research that has to be done on how these will interact with the body," admits Douglas Ritchie, senior vice president at MiniMed, a $15 million (1988 sales) maker of drug-infusion pumps, based in Sylmar, California.

Despite the obstacles, there's reason to be hopeful. "We're just now getting to the point where the impact on patients is going to be significant," says William Heetderks, a medical officer at the National Institutes of Health (NIH). That doesn't mean FES will produce bionic people, but it could improve some patients' lives.

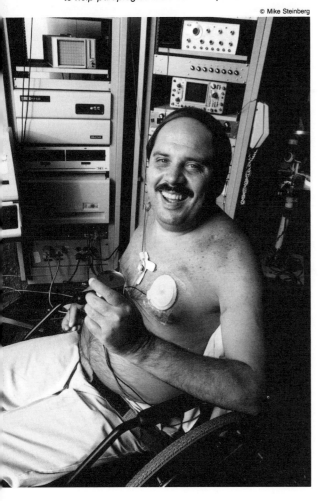

The low-level electrical currents stimulating James Jatich's paralyzed limbs allow him to write, feed himself, and even drive. The principal objective of FES systems is to help paraplegics lead more independent lives.

© Mike Steinberg

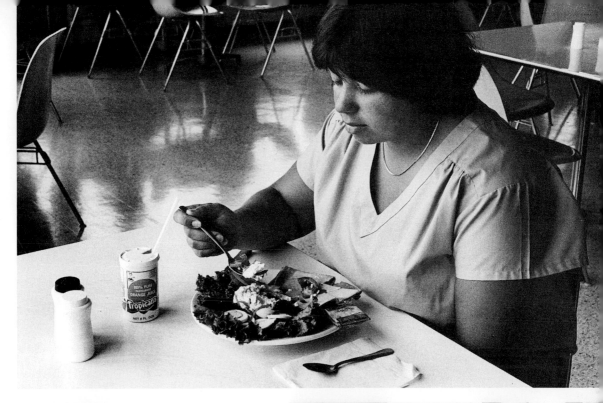

CRYING the WEIGHT-LOSS BLUES

by Nancy Henderson

O prah Winfrey took it off in front of millions of people. And many of us would like to lose weight, too—at least in moderation.

Winfrey trimmed her weight using one of the most drastic of the national weight-loss programs, Optifast, a highly structured regimen that relies on a liquid diet. That's only one of a multitude of paths to a svelte silhouette. You also can cut back on your portions, as we've all been told too many times, or step up your exercise quotient. But what about when you just can't do it on your own? Plenty of programs tout the benefits of their particular approach.

Weight-loss programs like the one Winfrey tried—and other major ones such as Nutri/

Today's diets go far beyond the low-cal meal (top). Many people turn to structured programs that promise rapid weight loss; some even volunteer for experiments that use advanced technology (center) to track the fate of each calorie consumed in a precisely measured meal (bottom).

Top: © Teri Leigh Stratford/Photo Researchers, Inc.; center and bottom: © Fred Ward/Agricultural Service/USDA

System, Diet Center, Weight Watchers, TOPS, and Overeaters Anonymous—range from informal to highly structured, from low-cost, self-help groups to large commercial enterprises that promise to take off pounds for hundreds, even thousands, of dollars. Do they work? Unfortunately, success rates are often a corporate secret that even the most persistent consumer won't be able to unlock.

Generally, the results are discouraging. Numerous studies have shown that about 90 percent of those who lose weight—no matter how they do it—eventually regain it. Programs with the highest success rates tend to be hospital- and university-based research clinics, where 10 percent to 20 percent of graduates have managed to keep their weight off for at least five years.

And some programs have a high dropout rate. Even in the relatively inexpensive Weight Watchers, half of the participants quit within six weeks. More than 70 percent quit in the first 12 weeks, according to a University of Pennsylvania study.

The highest success rates are among people who lose weight on their own, says Dr. John Foreyt, director of the Nutrition Research Clinic at Baylor College of Medicine. But if you find you need outside help, Foreyt advises a scaled approach based on cost. First try a nonprofit, self-help group, he says. In addition to Overeaters Anonymous and TOPS, community health centers and groups such as the Seventh-Day Adventists offer local programs. Hospitals and universities should be your next alternative. Hospital programs usually have a staff physician and a registered dietitian (R.D.). The best also include experts in exercise physiology and behavior. University programs generally have an R.D., an advanced-degree nutritionist, and other professionals. As a last resort, you may need the highly structured approach of a liquid diet.

How to Pick a Diet Program
Look for a regimen that matches your style. Do you feel more comfortable with group meetings or individual counseling? Neither approach has proved more effective than the other, says Kelly Brownell, a professor in the psychiatry department at the University of Pennsylvania. The more regimented approach of Diet Center or a liquid diet may work best for people who need guidance and direction. Prepackaged meals, such as Nutri/System's, may be helpful if you live alone and hate to cook.

Scheduling and location are important. When you first join, you'll probably feel enthusiastic enough to go out of your way to get to meetings. But eventually you'll feel like dropping out if meetings are hard to get to.

If you have special medical, dietary, or emotional problems, a hospital or university program with the right expertise on staff could be more economical than seeing a physician, dietitian, or psychotherapist on your own. Hospital and university programs often have a more supervised approach to exercise than do the big-name programs.

TOPS (TAKE OFF POUNDS SENSIBLY)

© TOPS Club, Inc.

TOPS' upbeat approach to weight loss emphasizes motivation and positive reinforcement.

Camaraderie, competition, and recognition help motivate members of **TOPS**, a loosely structured network of local groups with nearly 12,000 chapters in the U.S. and other countries. Weekly meetings consist of a weigh-in, followed by an upbeat program that stresses motivation and positive reinforcement. Unpaid volunteers direct the meetings. Each chapter has a different approach and should be evaluated individually. Members must see their own physicians to set weight-loss goals and a dietary regimen.

"Before" and "after" photos of TOPS weight-loss contest winners appear annually in the club's national magazine, which also prints news tips and simple recipes.

About 13 percent of members manage to maintain their weight goal for 13 weeks, which makes them members of KOPS (Keep Off Pounds Sensibly).

OVEREATERS ANONYMOUS

The focus of OA is not on weight, calories, or dieting, but on learning to stop eating compulsively. Members follow a 12-step plan patterned after that of Alcoholics Anonymous, which focuses on their emotional and spiritual life. There are no rules, but members are encouraged to see a physician for medical and dietary guidance. An OA survey found that 78 percent of members lose at least 10 pounds, with an average loss of 50 pounds. About 10 percent of those members have maintained their weight loss for five years. Most OA members are 50 to 100 pounds overweight when they join. They're usually feeling desperate, says Carol Hilliard, OA's publications and communications manager. Many have tried other diets and weight-loss programs without lasting success. OA can be a great source of emotional support, says Evelyn Tribole, a registered dietitian and representative of the American Dietetic Association. She recommends OA to clients who need a low-cost alternative to counseling. Members are encouraged to call each other, and it's not unusual for everyone to join hands at the end of a meeting and say a prayer. Chapters differ in their emotional tone, so Tribole recommends trying different groups to find one you can identify with.

© James Wilson/Woodfin Camp & Associates

Overeaters Anonymous offers emotional support to people trying to overcome compulsive eating habits.

WEIGHT WATCHERS

Nutrition specialists count Weight Watchers among the best programs because the good habits you acquire while losing weight can also help you keep the pounds off. Its flexible food plan uses ordinary store-bought foods, which you must weigh according to Weight Watchers guidelines. The plan includes drinking at least eight glasses of water a day. Weight Watchers–brand foods are available but not required. They are, however, implicitly endorsed through discount coupons distributed at meetings. H. J. Heinz, the company which owns Weight Watchers International, has no specific dietary guidelines for products that carry the Weight Watchers label; many of the Weight Watchers desserts contain a high amount of saturated fat.

Dietitians tout the fact that Weight Watchers lets you run your own program. Each day, you keep track of your meals in a booklet that sets out what you're allowed from each food group. Your choices and calorie intake expand each week for the first five weeks. You're also allowed "optional calories." You could lose as much as 10 pounds in the first week, but the rate should taper off to less than 2 pounds per week.

The exercise plan is optional. Like the food plan, it presents choices and provides suggestions, but leaves the initiative up to the program participants.

© Mark Bolster

Many nondieting, weight-conscious people use Weight Watchers products to help control caloric intake.

NUTRI/SYSTEM

The company supplies your meals each week in preportioned packages. Many of the entrées—and the desserts—are flavored with NutraSweet.

The program is based on a theory that overweight people feel deprived of flavor, so they overeat. "It's very expensive, but it's the easiest way I've ever lost weight," says Robin Levey, a New York graduate of the program who went from 140 pounds to 125. Because she lives alone and hates to cook, Levey liked the convenience of prepackaged food. But it reduced her flavor cravings in an odd way. "Some of it tasted so bad, it taught me not to feel I have to finish everything on my plate," she says. Once you reach your desired weight, weekly visits taper off to once a month during the one-year maintenance program. If you gain back no more than 5 pounds in a year, you get back 50 per-

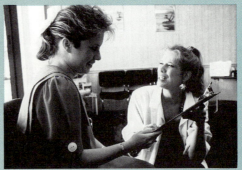

Nutri/System dieters eat special prepackaged foods and attend nutritional- and behavioral-therapy sessions on a weekly or monthly basis.

cent of the program fee. Nutri/System claims that nearly everyone who keeps coming for a year keeps the weight off, but the company won't divulge the dropout rate.

How to Evaluate the Plans

"Commercial weight-loss programs are very persuasive with sales," warns Joan Horbiak, a registered dietitian who evaluates programs for corporations. "They try not to give information on the phone. They want to get you in the door and sign you up." Instead of rushing into a costly commitment to a food-based diet plan, evaluate a program this way:

• What's the recommended calorie level? Less than 1,000 calories a day may be medically risky.

• How will the program help you maintain weight loss? The weight-maintenance part of

DIET CENTER

Private counseling six days a week is central to this highly structured approach. The plan eliminates dairy products and adds a food supplement of soy protein, sugar, and B-complex vitamins. You'll also be offered other Diet Center vitamins and food products; they're a major source of profit.

Upon reaching your weight goal, you stop taking the food supplement and begin a "stabilization" period of up to three weeks that includes counseling every two days and gradual reintroduction of normal foods. You continue seeing your counselor at weekly sessions for a year. Throughout the program, you keep an exercise diary and attend weekly group sessions to view videotaped lessons on meal planning, stress management, and exercise.

Some dietitians, including Tribole, question the balance of Diet Center's eating plan. "Eliminating dairy products is unnecessary and doesn't promote normal eating," says Tribole, who also argues that the diet's emphasis on protein, and especially meat, makes it harder to maintain weight loss.

The Diet Center features protein supplements, frequent weigh-ins, private counseling six days a week, and a yearlong maintenance program.

OPTIFAST

Optifast is a very-low-calorie liquid diet for moderate to severe obesity. You must be at least 30 percent or 50 pounds overweight to sign up. Medical and psychological evaluations are required because severe obesity is usually accompanied by medical problems, such as diabetes or hypertension.

You get powdered meals to mix with water—420 calories a day. After 12 weeks, you begin adding fish and poultry. Aerobic exercise is encouraged, and the supplements are phased out as you approach goal weight. About 60 to 70 percent of participants complete the six-month program, according to Sandoz Nutrition, which makes the meals.

Participants are examined regularly by a physician and attend classes run by a behavioral psychologist and a registered dietitian. For an additional $500, you can continue to attend group meetings and have your diet and exercise monitored for six months after you reach goal weight.

Like all liquid diets, Optifast, while rigorous and expensive, may be ineffective in the long term. Nutritionists caution against trying any liquid diet unless the plan is administered under close medical supervision and includes an exercise program and behavioral counseling.

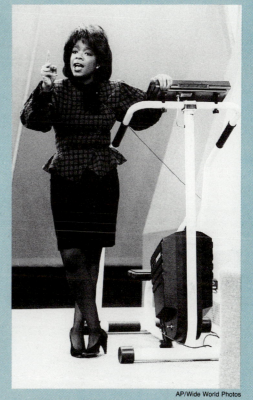

TV's Oprah Winfrey lost 67 pounds by combining the Optifast liquid diet with a regular exercise program.

commercial programs is often very superficial and unstructured, Horbiak believes, because companies make most of their profit from weight loss, not maintenance.
• Are detailed descriptions of the program and its fees readily available? Some programs are reluctant to reveal the total cost at the outset. Press them. Also check whether your health insurance will cover a program prescribed by a physician.
• Are required foods available from supermarkets? Foods that are expensive, sold only by the program, or hard to get don't encourage good habits that are likely to last.
• Are certain vitamins, minerals, herbs, or drugs required? They aren't necessary for weight loss, but some programs make a big profit on them.
• What tests or assessments will be required? A blood test, nutritional assessment, and tests of the percentage of body fat are common. Stay away from clinics that do enzyme tests, hair and nail analysis, or live-cell testing to show vitamin and mineral deficiencies. Such tests, often performed to show a need for the company's products, are unnecessary and often inaccurate.
• What are the staff's qualifications? Anyone can use the title "nutritionist," but a registered dietitian must have a nutrition degree, pass a professional exam, and take continuing-education courses. Check out the school; degrees from unaccredited or improperly accredited institutions flourish in this industry, says Horbiak.

Your state dietetic association, public health department, or American Heart Association office can help you locate good programs. Some dietitians run group sessions for $10 to $15 per hour—much less than you'd pay to see a dietitian individually. Also call the department of psychology or medicine at universities and the nutrition outpatient clinic at hospitals.

Review of the Year

PAST, PRESENT, AND FUTURE

New discoveries and reevaluation of existing finds have led to debate over human evolution and prehistoric settlement of the Americas. Debate rages over hunger and the inequitable food distribution that plague modern-day humans.

ANTHROPOLOGY AND ARCHAEOLOGY
by Peter S. Wells

In London, archaeological work continues on the remains of the Globe Theatre, where many of Shakespeare's plays were first performed in public.

Elizabethan Theaters Discovered

In London, two famous theaters were uncovered during the course of construction work in 1989. The Rose and the Globe theaters, situated on the south bank of the Thames River near Southwark Bridge, were partially excavated by construction crews. The Rose was built in 1587 and was likely the site of the first staging of some of Shakespeare's plays. The Globe, only 100 yards (92 meters) away from the Rose, was constructed in 1599 and was the home theater of an acting company to which Shakespeare belonged. Substantial remains of the foundations of both theaters were discovered and partly excavated, providing unique information about the size and character of the buildings. Little is known about Elizabethan theaters, and this find may further our understanding of the theaters and their relation to the society of that time.

Understanding the Neanderthals

Archaeological finds in Israel have shed new light on the role of the Neanderthal in human evolution. Remains of some 400 Neanderthal individuals have been found throughout Europe and in western Asia, dating from between 120,000 and 40,000 years ago. For many years, scientists had believed that either modern humans replaced Neanderthals, outcompeting them because of superior intelligence, or that Neanderthals evolved into modern humans. The evidence from the Israel site suggests that modern-type humans coexisted with Neanderthals, possibly for as long as 50,000 years. The evidence also raises questions about interaction, cooperation, competition, and communication between these anatomically different humans.

Another important new discovery about Neanderthals was also made in Israel. At Kebara Cave was found the first well-preserved specimen of a Neanderthal hyoid bone—part of the structure of the throat used for speech. Scientists long had debated whether or not Neanderthals could speak, with part of the argument resting on anatomical differences between Neanderthals and *Homo sapiens sapiens*. The Neanderthal hyoid's similarity to the hyoid in modern humans would seem to imply a Neanderthal capability to make sounds similar to those of modern humans.

Human Evolution

The generally accepted theory—that modern humans (*Homo sapiens sapiens*) evolved in just one place, most likely Africa—received further support from recent discoveries. After studying DNA patterns in modern populations, anthropologist Allan C. Wilson and his colleagues suggested that the genetic evidence points to Africa as the location of the common ancestors of all modern peoples.

Fossil finds in Israel add even more support to this supposition. Fossils of modern-looking humans were found recently at a site believed to date from about 90,000 years ago, much earlier than the numerous *Homo sapiens sapiens* fossils, which are dated from less than 40,000 years ago. If the evidence holds up to further scrutiny, these finds will revolutionize our understanding of the evolution of modern humans.

Anthropologists now believe that Neanderthals and modern-type humans might have coexisted, and that Neanderthals had the capability for speech.

Courtesy of the American Museum of Natural History

Archaeology in the Americas

Archaeologists still debate over when humans first arrived in the New World, and how and why they came. There is substantial evidence for the presence of humans in the Americas by 12,000 B.C., but some scientists argue for much earlier dates. Sites recently investigated in both North and South America may demonstrate the presence of humans well before 12,000 B.C., but the evidence for dating the sites is still in question.

Between 25,000 and 12,000 B.C., Siberia and Alaska were connected by a landmass, and humans could have crossed anytime during that interval. Traditionally, archaeologists have envisioned groups of hunters following big game across the land bridge and arriving in America. However, recent excavations in Alaska and Siberia are yielding evidence that the first peoples to come to the New World may have arrived along the coast in boats and been primarily hunters of sea mammals rather than of large land animals.

In South and Central America, new evidence suggests that complex societies with large urban and ceremonial centers emerged earlier than has been believed. On Peru's coast, excavations show that huge settlement complexes (called cities by some scientists), featuring great temples and towering pyramids, date back as far as 3,000 B.C. According to this new evidence, civilization developed along the Pacific coast of South America well before it did in Mesoamerica.

As archaeological research progresses in the jungles of Mesoamerica, dates for the origins of complex societies there are being pushed back also. Recent research in Guatemala suggests a date as early as 600 B.C. for the beginnings of large Mayan ceremonial centers. These centers contained temple complexes, public plazas, and residential districts. Many such sites probably still remain to be explored in Central America.

FOOD AND POPULATION
by Martin M. McLaughlin

The United Nations Food and Agriculture Organization (FAO) estimated 1989's total world cereal grain harvest at 1.877 billion metric tons, roughly 7 percent more than 1988's relatively poor crop. Nevertheless, grain consumption was higher than production for the third consecutive year. Global food stocks thus declined further, reaching the minimum needed for world food security.

The world's more than 5 billion people depend mainly upon the current harvest for their food. Although 1989's harvest provided enough to feed the world's total population, the perennial disparity in distribution of food, with the more prosperous nations consuming the great majority, leaves approximately 500 million people malnourished or facing starvation each year.

Fewer nations were at serious risk of famine at the end of 1989 than at the end of 1988; only six face unfavorable crop prospects, while 16 anticipate a shortage of supply. Weather was not a major adverse factor in 1989 and is not expected to become one in 1990. But the availability of seeds, fertilizer, and equipment has been limited by the heavy outlays Third World nations require to service their external debts, which appear to have stabilized at about $1.3 trillion.

The famine in Africa has largely receded, except in Ethiopia. Chronic hunger, however, still affects hundreds of millions of people, particularly among the world's 12 million refugees, many of whom are concentrated in Africa. The real hunger problem lies more in access than in supply. The poor lack the money needed to buy or grow enough food. Even in the United States, evidence indicates that hunger is on the rise, despite

The number of homeless people in U.S. cities has grown at an alarming rate. The problem is even more acute in poor and developing countries.

© Robert Maass/Photoreporters

steady economic growth. Ten million Americans receive food stamps; many more are eligible.

During 1990 basic U.S. food policy will be under intense scrutiny because of the need to renew farm legislation. Three other issues that will affect the global food system also merit close attention:

• The food shortages and distribution problems that figured prominently in the tumultuous changes in Eastern Europe last year. Food aid and agricultural-development assistance to those countries will play a key role.

• The crucial trade negotiations in the General Agreement on Tariffs and Trade (GATT), which may include a possible change in agricultural-trade policies with Third World nations. The United States has proposed cutting out all agricultural subsidies by the year 2000, but other member nations are resisting this recommendation.

• Mounting concern about the environment and farmers' efforts to cut costs by using fewer chemicals and less heavy equipment, which have resulted in an increasing trend toward low-input "organic" agriculture. This method is considered the most effective way to reduce producer costs, improve food quality, and protect the environment.

THE BLACK DEATH

by Charles L. Mee, Jr.

In all likelihood, a flea riding on the hide of a black rat entered the Italian port of Messina in 1347, perhaps down a hawser tying a ship up at the dock. The flea had a gut full of the bacillus *Yersinia pestis*. The flea itself was hardly bigger than the letter *o* on this page, but it could carry several hundred thousand bacilli in its intestine.

Repeated outbreaks of bubonic plague swept through 14th-century Europe, felling large segments of the population without regard to rank or status (left). Art of the period (above) bears grim testimony to the anxiety and horror that the plague aroused in medieval society.

Left and above: The Granger Collection

Scholars today cannot identify with certainty which species of flea (or rat) carried the plague. One candidate among the fleas is *Xenopsylla cheopis*, which looks like a deeply bent, bearded old man with six legs. It is slender and bristly, with almost no neck and no waist, so that it can slip easily through the forest of hair in which it lives. It is outfitted with a daggerlike proboscis for piercing the skin and sucking the blood of its host. And it is cunningly equipped to secrete a substance that prevents coagulation of the host's blood. Although *X. cheopis* can go for weeks without feeding, it will eat every day if it can, taking its blood warm.

One rat on which fleas feed, the black rat (*Rattus rattus*)—also known as the house rat, roof rat, or ship rat—is active mainly at night. A rat can fall 50 feet (15 meters) and land on its feet with no injury. It can scale a brick wall or climb up the inside of a pipe only 1.5 inches (3.8 centimeters) in diameter. It can jump a distance of 2 feet (60 centimeters) straight up and 4 feet (1.2 meters) horizontally, and squeeze through a hole the size of a quarter. Black rats have been found still swimming days after their ship has sunk at sea.

A rat can gnaw its way through almost anything—paper, wood, bone, mortar, half-inch sheet metal. It gnaws constantly. Indeed, it *must* gnaw constantly. Its incisors grow 4 to 5 inches (10 to 13 centimeters) a year: if it were to stop gnawing, its lower incisors would eventually grow—as sometimes happens when a rat loses an opposing tooth—until the incisors push up into the rat's brain, killing it. It prefers grain, if possible, but also eats fish, eggs, fowl, and meat—lambs, piglets, and the flesh of helpless infants or adults. If nothing else is available, a rat will eat manure and drink urine.

Rats prefer to move no more than 100 feet (30 meters) from their nests. But in severe drought or famine, rats can begin to move en masse for great distances, bringing with them any infections they happen to have picked up—infections that may be killing them, but not killing them more rapidly than they breed.

Rats and mice harbor a number of infections that may cause diseases in human beings. A black rat can even tolerate a moderate amount of the ferocious *Yersinia pestis* bacillus in its system without noticeable ill effects. But bacilli breed even more extravagantly than fleas or rats,

Below left: © G. I. Bernard/Animals Animals; photos below: The Granger Collection

A tiny rat flea (below) harbors the bacilli that cause plague. When rats infested with such fleas arrived in Europe, the people lacked the natural immunity, hygiene standards, and medical knowledge needed to combat the infection. The children's game "Ring Around the Rosy" (right) arose during plague time; so did the Pied Piper legend, although historians doubt any plague connection.

Plague victims developed a blackish pustule at the point of the flea bite, followed rapidly by swelling of the lymph nodes. Medieval medicine, mired in astrology and superstition (left), offered physicians no methods by which to treat plague victims, except perhaps to lance painfully inflamed lymph nodes (above).

often in the millions. When a bacillus finally invades the rat's pulmonary or nervous system, it causes a horrible, often convulsive, death, passing on a lethal dose to the bloodsucking fleas that ride on the rat's hide.

The Ultimate Bacillus Breeder

When an afflicted rat dies, its body cools, so that the flea, highly sensitive to changes in temperature, will find another host. The flea can, if need be, survive for weeks at a time without a rat host. It can take refuge anywhere, even in an abandoned rat's nest or a bale of cloth. A dying rat may liberate scores of rat fleas. More than that, a flea's intestine provides ideal breeding conditions for the bacillus, which will eventually multiply so prodigiously as finally to block the gut of the flea entirely. Unable to feed or digest blood, the flea desperately seeks another host. But now, as it sucks blood, it spits some out at the same time. Each time the flea stops sucking for a moment, it can pump thousands of virulent bacilli back into its host. Thus, bacilli are passed from rat to flea to rat, contained, ordinarily, within a closed community.

For millions of years, there has been a reservoir of *Yersinia pestis* living as a permanently settled parasite—passed back and forth among fleas and rodents in warm, moist nests—in the wild-rodent colonies of China, India, the southern part of the Soviet Union, and the western United States. Probably there will always be such reservoirs—ready to be stirred up by sudden climatic change or ecological disaster. Even last year, four authentic cases of bubonic plague were confirmed in New Mexico and Arizona. Limited outbreaks and some fatalities have occurred in the United States for years, in fact, but the disease doesn't spread, partly for reasons we don't understand, partly because patients can now be treated with antibiotics.

And at least from biblical times on, there have been sporadic allusions to plagues, as well as carefully recorded outbreaks. The emperor Justinian's Constantinople, for instance, capital of the Roman Empire in the East, was ravaged by plague in 541 and 542, felling perhaps 40 percent of the city's population. But none of the biblical or Roman plagues seemed so emblematic of horror and devastation as the Black Death that struck Europe in 1347. Rumors of fearful pestilence in China and throughout the East had reached Europe by 1346. "India was depopulated," reported one chronicler, "Tartary, Mesopotamia, Syria, Armenia, were covered with dead bodies; the Kurds fled in vain. In Caramania and Caesarea none were left alive."

Untold millions would die in China and the rest of the East before the plague subsided again. By September of 1345, the *Yersinia pestis* bacillus, probably carried by rats, reached the Crimea, on the northern coast of the Black Sea, where Italian merchants traded.

From the shores of the Black Sea, the bacillus seems to have entered a number of Italian ports. The most famous account has to do with a ship that docked in the Sicilian port of Messina in 1347. According to an Italian chronicler named Gabriele de Mussis, Christian merchants from Genoa and local Muslim residents in the town of Caffa on the Black Sea got into an argument: a serious fight ensued between the merchants and a local army led by a Tatar lord. In the course of an attack on the Christians, the Tatars were stricken by plague. From sheer spitefulness, their leader loaded his catapults with dead bodies and hurled them at the Christian enemy, in hopes of spreading disease among them. Infected with the plague, the Genoese sailed back to Italy, docking at Messina.

Although de Mussis, who never traveled to the Crimea, may be a less-than-reliable source, his underlying assumption seems sound. The plague did spread along established trade routes. (Most likely, though, the pestilence in Caffa resulted from an infected population of local rats, not from the corpses lobbed over the besieged city's walls.)

In any case, given enough dying rats and enough engorged and frantic fleas, it will not be long before the fleas, in their search for new hosts, leap to a human being. When a rat flea senses the presence of an alternate host, it can jump very quickly and as much as 150 times its length. The average for such jumps is about 6 inches (15 centimeters) horizontally and 4 inches (10 centimeters) straight up in the air. Once on human skin, the flea will not travel far before it begins to feed.

Anxiety and Terror

The first symptoms of bubonic plague often appear within several days: headache and a general feeling of weakness, followed by aches and chills in the upper leg and groin, a white coating on the tongue, rapid pulse, slurred speech, confusion, fatigue, apathy, and a staggering gait. A blackish pustule usually will form at the point of the fleabite. By the third day, the lymph nodes begin to swell. Because the bite is commonly in the leg, it is the lymph nodes of the groin that swell, which is how the disease got its name. The Greek word for "groin" is *boubon*—thus, bubonic plague. The swellings will be tender, perhaps as large as an egg. The heart begins to flutter rapidly as it tries to pump blood through swollen, suffocating tissues. Subcutaneous hemorrhaging occurs, causing purplish blotches on the skin. The victim's nervous system begins

Out of plague hysteria came such bizarre cults as the flagellants, who went from town to town scourging themselves.

As the plague's death toll grew more devastating, formal funeral ceremonies (above) gave way to mass burials.

to collapse, causing dreadful pain and bizarre neurological disorders, from which the "Dance of Death" rituals that accompanied the plague may have taken their inspiration. By the fourth or fifth day, wild anxiety and terror overtake the sufferer—and then a sense of resignation, as the skin blackens and the rictus of death descends.

In 1347, when the plague struck in Messina, townspeople realized that it must have come from the sick and dying crews of the ships at their dock. They turned on the sailors and drove them back out to sea—eventually to spread the plague in other ports. Messina panicked. People ran out into the fields and vineyards and neighboring villages, taking the rat fleas with them.

When the citizens of Messina, already ill or just becoming ill, reached the city of Catania, 55 miles (88 kilometers) to the south, they were at first taken in and given beds in the hospital. But as the plague began to infect Catania, the townspeople there cordoned off their town and refused—too late—to admit any outsiders. The sick, turning back, stumbling and delirious, were objects more of disgust than pity; everything about them gave off a terrible stench, it was said, their "sweat, excrement, spittle, breath, so foetid as to be overpowering; urine turbid, thick, black, or red. . . ."

Wherever the plague appeared, the suddenness of death was terrifying. Today, perhaps only those caught in the AIDS epidemic can grasp the strain that the plague put on the physical and spiritual fabric of society. People went to bed perfectly healthy and were found dead in the morning. Priests and doctors who came to minister to the sick, so the wild stories ran, would contract the plague with a single touch and die sooner than the person they had come to help. In his preface to *The Decameron*, a collection of stories told while the plague was raging, Boccaccio reports that he saw two pigs rooting around in the clothes of a man who had just died, and after a few minutes of snuffling, the pigs began to run wildly around and around, then fell dead.

In Florence, everyone grew so frightened of the bodies stacked up in the streets that some men, called *becchini*, put themselves out for hire to fetch and carry the dead to mass graves. Having in this way stepped over the boundary into the land of the dead, and no doubt feeling doomed themselves, the *becchini* became an abandoned, brutal lot. Many roamed the streets, forcing their way into private homes, and threatening to carry people away if they were not paid off in money or sexual favors.

Visiting Men with Pestilence

Some people, shut up in their houses with the doors barred, would scratch a sign of the cross on the front door, sometimes with the inscription "Lord have mercy on us." In one place, two lovers were supposed to have bathed in urine every morning for protection. People hovered over latrines, breathing in the stench. Others swallowed pus from the boils of plague victims. In Avignon, Pope Clement was said to have sat for weeks between two roaring fires.

The plague spread from Sicily all up and down the Atlantic coast, and from the port cities of Venice, Genoa, and Pisa as well as Marseilles, London, and Bristol. A multitude of men and women, as Boccaccio writes, "negligent of all but themselves . . . migrated to the country, as if God, in visiting men with this pestilence in requital of their iniquities, would not pursue them with His wrath wherever they might be. . . ."

Some who were not yet ill but felt doomed indulged in debauchery. Others, seeking protection in lives of moderation, banded together in communities to live a separate and secluded life, walking abroad with flowers to their noses "to ward off the stench and, perhaps, the evil airs that afflicted them."

It was from a time of plague, some scholars speculate, that the nursery rhyme "Ring Around the Rosy" derives: the rose-colored "ring" being an early sign that a blotch was about to appear on the skin; "a pocket full of posies" being a device to ward off stench and (it was hoped) the attendant infection; "ashes, ashes" being a reference to "ashes to ashes, dust to dust," or perhaps to the sneezing "a-choo, a-choo" that afflicted those in whom the infection had invaded the lungs—ending, inevitably, in "all fall down."

In Pistoia the city council enacted nine pages of regulations to keep the plague out—no Pistoian was allowed to leave town to visit anyplace where the plague was raging; if a citizen did visit a plague-infested area, he was not allowed back in the city; no linen or woolen goods were allowed to be imported; no corpses could be brought home from outside the city; attendance at funerals was strictly limited to immediate family. None of these regulations helped save the city.

In Siena, dogs dragged bodies from the shallow graves and left them half-devoured in the streets. Merchants closed their shops. The wool industry was shut down. Clergymen ceased administering last rites. On June 2, 1348, all the civil courts were recessed by the city council. Because so many of the laborers had died, construction of the nave for a great cathedral came to a halt. Work was never resumed: only the smaller cathedral we know today was completed.

In Venice, it was said that 600 were dying every day. In Florence, perhaps half the population died. By the time the plague swept through, as much as one-third of Italy's population had succumbed.

In Milan, when the plague struck, all the occupants of any victim's house, whether sick or well, were walled up inside together and left

Fear of contracting plague ran rampant throughout medieval society. Physicians, lacking even a single clue to explain how the disease spread, resorted to wearing curious bird masks as a means of preventing infection.

The plague's horror, expressed vividly in Brueghel's Triumph of Death, *changed medieval society forever.*

to die. Such draconian measures seemed to have been partially successful—mortality rates were lower in Milan than in other cities.

Contemporary Conjectures

Medieval medicine was at a loss to explain all this, or to do anything about it. Although clinical observation did play some role in medical education, an extensive reliance on ancient and inadequate texts prevailed. Surgeons usually had a good deal of clinical experience, but were considered mainly to be skilled craftsmen, not men of real learning, and their experience was not much incorporated into the body of medical knowledge. In 1300 Pope Boniface VIII had published a bull against the mutilation of corpses, which had the unintended effect of discouraging dissection.

Physicians, priests, and others had theories about the cause of the plague. Earthquakes that released poisonous fumes, for instance. Severe changes in the earth's temperature creating southerly winds that brought the plague. The notion that the plague was somehow the result of a corruption of the air was widely believed. It was this idea that led people to avoid foul odors by holding flowers to their noses, or to try to drive out the infectious foul odors by inhaling the alternate foul odors of a latrine. Some thought that the plague came from the raining down of frogs, toads, and reptiles. Some physicians believed one could catch the plague from "lust with old women."

Both the pope and the king of France sent urgent requests for help to the medical faculty at the University of Paris, then one of the most distinguished medical groups in the Western world. The faculty responded that the plague was the result of a conjunction of the planets Saturn, Mars, and Jupiter at 1:00 P.M. on March 20, 1345, an event that caused the corruption of the surrounding atmosphere.

Ultimately, of course, most Christians believed the cause of the plague was God's wrath at sinful Man. And in those terms, to be sure, the best preventives were prayer, the wearing of crosses, and participation in other religious activities. In Orvieto the town fathers added 50 new religious observances to the municipal calendar. Even so, within five months of the appearance of the plague, Orvieto lost every second person in the town.

There was also some agreement about preventive measures one might take to avoid the wrath of God. Flight was best: away from lowlands, marshy areas, stagnant waters, southern exposures, and coastal areas; toward high, dry, cool, mountainous places. It was thought wise to stay indoors all day, to stay cool, and to cover any windows that admitted bright sunlight. In addition to keeping flowers nearby, one might burn such aromatic woods as juniper and ash.

The retreat to the mountains, where the density of the rat population was not as great as in urban areas, and where the weather was inimical to rats and fleas, was probably a good idea—as well as perhaps proof, of a kind, of the value of empirical observation. But any useful notion was always mixed in with such wild ideas that it got lost in a flurry of desperate (and often contrary) stratagems. One should avoid bathing, because that opened the pores to attack from the corrupt atmosphere, but one should wash face and feet, and sprinkle them with rose water and vinegar. In the morning, one might eat a couple of figs with rue and filberts. One expert advised eating 10-year-old treacle mixed with several dozen items, including chopped-up snake. Rhubarb was recommended, too, along with onions, leeks, and garlic. The best spices were myrrh, saffron, and pepper, to be taken late in the day. Meat should be roasted, not boiled. Eggs should not be eaten hard-boiled. A certain Gentile di Foligno commended lettuce; the faculty of medicine at the University of Paris advised against it. Desserts were forbidden. One should not sleep during the day. One should sleep first on the right side, then on the left. Exercise was to be avoided, because it introduced more air into the body; if one needed to move, one ought to move slowly.

Three Forms of Plague

By the fall of 1348, the plague began to abate. But then, just as hopes were rising that it had passed, the plague broke out again in the spring and summer of 1349 in different parts of Europe. This recurrence seemed to prove that the warm weather, and people bathing in warm weather, caused the pores of the skin to open and admit the corrupted air. In other respects, however, the plague remained inexplicable. Why did some people get it and recover, while others seemed not to have got it at all—or at least showed none of its symptoms—yet died suddenly anyway? Some people died in four or five days; others died at once. Some seemed to have contracted the plague from a friend or relative who had it; others had never been near a sick person. The sheer unpredictability of it was terrifying.

In fact, though no one would know for several centuries, there were three different forms of the plague, which ran three different courses. The first was simple bubonic plague, transmitted from rat to person by the bite of the rat flea. The second, and likely most common, form was pneumonic, which occurred when the bacillus invaded the lungs. After a two- or three-day incubation period, anyone with pneumonic plague would have a severe, bloody cough; the sputa cast into the air would contain *Yersinia pestis*. Transmitted through the air from person to person, pneumonic plague was fatal in 95 to 100 percent of all cases.

The third form of the plague was septocemic, and its precise etiology is not entirely understood even by modern medicine. In essence, however, it appears that in cases of septocemic plague, the bacillus entered the bloodstream, perhaps at the moment of the fleabite. A rash formed, and death occurred within a day, or even within hours, before any swellings appeared. Septocemic plague always turned out to be fatal.

Some people did imagine that the disease might be coming from some animal, and they killed dogs and cats—though never rats. But fleas were so much a part of everyday life that no one seems to have given them a second thought. Upright citizens also killed gravediggers, strangers from other countries, gypsies, drunks, beggars, cripples, lepers, and Jews. Through the winter and into early spring of 1349 they were burned in Landsberg, Burren, Memmingen, Lindau, Freiburg, Ulm, Speyer, Gotha, Eisenach, Dresden, Worms, Baden, and Erfurt. Sixteen thousand were murdered in Strasbourg. In other cities Jews were walled up inside their houses to starve to death. That the Jews were also dying of the plague was not taken as proof that they were not causing it.

The recurrence of the plague after people thought the worst was over may have been the most devastating development of all. In short, Europe was swept not only by a bacillus, but also by a widespread psychic breakdown—by abject terror, panic, rage, vengefulness, cringing remorse, selfishness, hysteria, and, above all, by an overwhelming sense of utter powerlessness in the face of an inescapable horror.

After a decade's respite, just as Europeans began to recover their feeling of well-being, the plague struck again in 1361, and again in 1369, and at least once in each decade down to the end of the century. Why the plague faded away is still a mystery that, in the short run, apparently had little to do with improvements in medicine or cleanliness, and more to do with some adjustment of equilibrium among the population of rats and fleas. In any case, as agents for Pope Clement estimated in 1351, perhaps 24 million people had died in the first onslaught of the plague; perhaps another 20 million died by the end of the century—in all, it is estimated, one-third of the total population of Europe.

Very rarely does a single event change history by itself. Yet an event of the magnitude of the Black Death could not fail to have an enormous impact. Ironically, some of the changes brought by the plague were for the good. Not surprisingly, medicine changed—since medicine had so signally failed to be of any help in the hour of greatest need for it. First of all, a great many doctors died—and some simply ran away. "It has pleased God," wrote one Venetian-born physician, "by this terrible mortality to leave our native place so destitute of upright and capable doctors that it may be said not one has been left." By 1349, at the University of Padua, there were vacancies in every single chair of medicine and surgery. All this, of course, created room for new people with new ideas. Ordinary people began wanting to get their hands on medical guides and to take command of their own health. And gradually more medical texts began to appear in the vernacular instead of in Latin.

An Old Order Besieged

Because of the death of so many people, the relationship between agricultural supply and demand changed radically, too. Agricultural prices dropped precipitously, endangering the fortunes and power of the aristocracy, whose wealth and dominance were based on land. At the same time, the deaths of so many people caused wages to rise dramatically, giving laborers some chance of improving their own conditions of employment. Increasing numbers of people had more money to buy what could be called luxury goods, which affected the nature of business and trade, and even of private well-being. As old relationships, usages, and laws broke down, expanding secular concerns and intensifying the struggle between faith and reason, there was a rise in religious, social, and political unrest. Religious reformers such as John Wycliffe in England and Jan Hus in Bohemia were among many leaders of sects that challenged Church behavior and Church doctrine all over Europe. Such complaints eventually led to the Protestant Reformation, and the assertion that Man stood in direct relation to God, without need to benefit from intercession by layers of clergy.

Indeed, the entire structure of feudal society, which had been under stress for many years, was undermined by the plague. The three orders of feudalism—clergy, nobility, and peasantry—had been challenged for more than a century by the rise of the urban bourgeoisie, and by the enormous, slow changes in productivity and in the cultivation of arable land. But the plague, ravaging the weakened feudal system from so many diverse and unpredictable quarters, tore it apart.

By far the greatest change in Western civilization that the plague helped hasten was a change of mind. Once the immediate traumata of death, terror, and flight had passed through a stricken town, the common lingering emotion was that of fear of God. The subsequent surge of religious fervor in art was in many ways nightmarish. Though medieval religion had dealt with death and dying, and naturally with sin and retribution, it was only after the Black Death that painters so wholeheartedly gave themselves over to pictures brimming with rotting corpses; corpses being consumed by snakes and toads; swooping birds of prey appearing with terrible suddenness; cripples gazing on the figure of Death with longing for deliverance; open graves filled with blackened, worm-eaten bodies; devils slashing the faces and bodies of the damned.

As the Black Death waned in Europe, the power of religion waned with it, leaving behind a population that was gradually but certainly turning its attention to the physical realm in which it lived, to materialism and worldliness, to the terrible power of the world itself, and to the wonder of how it works.

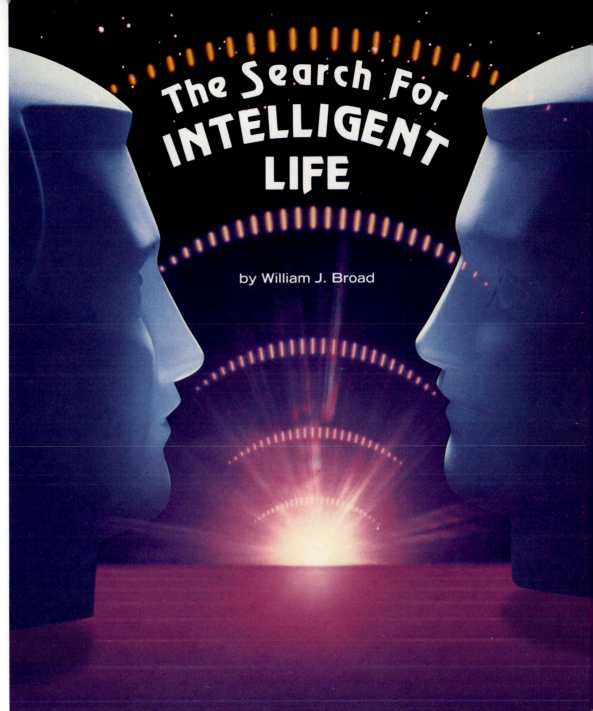

The Search For INTELLIGENT LIFE

by William J. Broad

After decades of dreaming, and sometimes scheming, a small band of earthlings has received federal money to embark on a $100 million, ten-year project to find aliens in space.

Their goal is to scan the sky systematically with dish-shaped antennas, listening for faint signals from advanced civilizations they say may dot the galaxy. Their continuing search for extraterrestrial intelligence is known by its acronym, SETI.

For years, about 20 scientists, many of them working for the National Aeronautics and Space Administration (NASA), have lived on shoestring budgets, thinking about, rather than doing, major searches.

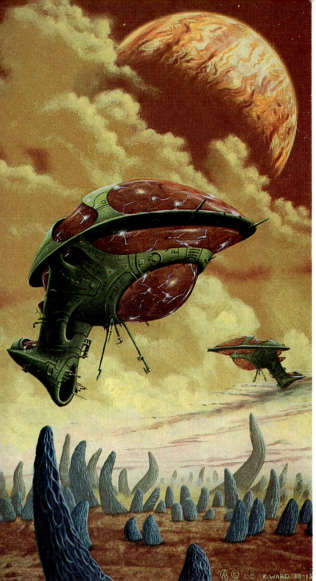

Scanning the universe for signs of intelligent life might turn up unexpected results. Life forms that evolved on planets with conditions dissimilar to those on Earth would likely appear bizarre to an earthling's eyes.

But in the 1990 fiscal year, for the first time, they received NASA funds to start building hardware for the world's first major examination of the heavens for signs of alien civilizations.

"This is the big step," says Dr. Frank D. Drake, a founder of the modern field of extraterrestrial searches, who is a professor of astronomy and astrophysics at the University of California at Santa Cruz. "It will allow us to do very sophisticated searching."

Advocates say the project is important because it will start to address with scientific rigor one of the ultimate questions that faces mankind: Are we alone?

There are many skeptics: from those who dismiss the whole effort as a waste of money, to those who doubt the existence of extraterrestrial life, but say the search is worthwhile. Hard-core skeptics say there is no evidence whatever that beings are out there waiting to be discovered; if there were such beings, they say, Earth would have heard from them by now.

Cosmic Haystack

Despite the skepticism, enthusiasts have conducted about 50 minor searches over the years. The first modern one occurred in 1960, when Dr. Drake used the 85-foot (26-meter) antenna at the National Radio Astronomy Observatory in Green Bank, West Virginia, to listen for possible transmissions from distant civilizations. Other searches followed, usually by individuals or small teams borrowing time on large radio telescopes primarily used in astrophysical studies. Both Harvard and Ohio State University now have programs that scan the heavens for extraterrestrial clues.

The challenge for enthusiasts today is not getting more radio telescopes to join the search, but finding powerful new ways to distinguish a genuine alien transmission from earthly static and cosmic background noise. The metaphor often used is finding a needle in a cosmic haystack.

The main federal project is to build a highly advanced radio receiver that will simultaneously scan 14 million channels of radio waves from existing radio telescopes around the world, searching for signs of intelligence in the universe. Scheduled to be switched on in October 1992, the 500th anniversary of Columbus' discovery of America, the receiving apparatus is to run until the turn of the century.

"In the first minute, we'll accomplish more than all the other projects combined," says Dr. Peter R. Backus, a principal investigator at the SETI Institute, a private, nonprofit group based in Mountain View, California, that works on the NASA project.

Getting to this point was no easy task. The scientists were attacked, ignored, ridiculed, and threatened with professional excommunication by traditional astronomers and astrophysicists. But they and their field survived, growing increasingly respectable. To flourish, as they are starting to do, the scientists also had to grow sophisticated in the ways of Washington.

"It's Been a Long Battle"

"It's been a long, arduous task to convince all the appropriate parties that this is the right thing to do," says Thomas Pierson, director of the SETI Institute.

Dr. Bernard M. Oliver, a senior SETI official at the NASA Ames Research Center in California, which heads the space agency's effort, says the short-term perspective that prevails in Washington was a real impediment.

"It's hard to elevate the consciousness of congressmen from mundane to heavenly matters," he says. "We've had to do a lot of advocacy. It's been a long battle."

At the start the field was shaken by setbacks. In 1978 William Proxmire, then a senator from Wisconsin, gave SETI's scientists a "Golden Fleece" award for what he said was their waste of federal money. Then, largely at Mr. Proxmire's initiative, the 1982 federal budget dropped funds for SETI altogether, nearly killing the fledgling effort.

An array of eminent scientists soon mobilized, writing Mr. Proxmire and other congressmen on behalf of the search. In 1982 a panel of the National Academy of Sciences (NAS) called the effort "a legitimate part of astronomy," and added that it was hard to imagine a discovery "that would have greater impact on human perceptions than the detection of extraterrestrial intelligence." In the 1983 budget, funds were restored.

Threat of Being "Flaky"

NASA's annual budget for SETI throughout the 1980s has ranged between $1.5 million and $2 million. The scientists have worked mainly on devising strategies and testing equipment for the big search, as well as polishing their image.

"SETI is always burdened with the threat of being declared flaky, or fringe, or pseudoscience," says Dr. Drake. "So you have to be very careful that very qualified, right-thinking people are involved."

Last year their political efforts faltered again as congressional budget cutters again proposed killing the effort, just as NASA wanted to start building receivers for the big search. "Tripling the size of its budget, in light of the severe budget constraints facing NASA, is unjustified," wrote a Senate budget panel.

The construction program was saved by Senator Jake Garn, a Utah Republican who rode the space shuttle into orbit and became convinced of the possibility of extraterrestrial be-

© Kevin A. Ward

Scientists have already determined that life as we know it does not exist on the other known planets in our solar system. People from planets beyond the solar system would need to travel many light-years to reach the Earth.

ings. Mr. Garn succeeded in restoring the funds, although they were slightly less than hoped. NASA had asked for $6.8 million for SETI, but received $4 million, a doubling of its budget.

Construction and Celebration

It was enough money, however, to start construction and celebrations at NASA Ames in California. "It's an exciting time for us," says Dr. Jill C. Tarter, a scientist there. "We'll go ahead and make the instruments we hope to turn on in 1992." Of course, there is no guarantee that the project will receive financing for its full

Popular culture has helped mold our perception of how the inhabitants of other planets would appear. The television show "Alien Nation" depicts beings from outer space as having a distinctly human demeanor (below). Even those films that strive to portray extraterrestrial beings in a truly alien light still cling to an essentially human body structure (left).

ten-year duration. The NASA scientists note that next year's proposed tripling of the budget is likely to meet resistance in Congress.

The apparatus now taking shape is to fill four refrigerator-sized electronic racks, Dr. Backus of the SETI Institute says. It will first divide incoming signals into 14 million channels, and then analyze each one for extraterrestrial clues, scanning for either continuous waves (similar to the carrier waves used on Earth broadcasts) or for pulsed signals (similar to rotating lighthouse beacons).

The scientists plan to use many radio telescopes, including the world's largest, the 1,000-foot (300-meter) one at Arecibo, Puerto Rico.

The search will have two distinct parts. One will examine specific targets, about 770 Sun-like stars up to 80 light-years, or 470 trillion miles (750 trillion kilometers), from Earth. Scientists believe such stars might have Earth-like planets, and possibly civilizations. The other part will survey the entire sky, including the Milky Way and its 400 billion stars.

"The typical guess is that there are 10,000 to 100,000 advanced civilizations in the Milky Way alone," says Dr. Drake.

The scientists think their equipment will be sensitive enough to pick up beacon signals sent out by advanced civilizations to broadcast their presence, and perhaps some routine transmissions that leak out as well.

"This first search will not be eavesdropping, unless their signals are very powerful," says Dr. Backus. If nothing is found in the seven-year search, the researchers hope to build more sensitive equipment and bigger antennas.

Both photos above: Photofest; facing page: © Kevin A. Ward

But they warn that time may be running out. Radio interference on the ground and from orbiting satellites is increasing so fast that this decade might be the last opportunity for Earth-based searches unimpeded by extraneous noise. "Beyond the 1990s, it's uncertain," says Dr. Backus. "The rate of increase for the number of transmitters is tremendous."

Interest in such searches is far from universal among astronomers and astrophysicists.

Skeptics Abound

The main criticism is that there is no firm evidence of any life elsewhere in the universe, despite the enormous number of stars and the possibility that there may be other habitable planets. "Maybe we're a fluke," says Dr. Robert T. Rood, an astronomer at the University of Virginia who describes himself as mildly skeptical of the NASA plan for a big search.

Another criticism centers on extraterrestrial logic. SETI scientists assume that advanced civilizations would not want to travel from star to star across the vast, empty reaches of space, and would opt instead for powerful radio communication. But critics say this assumption may be wrong: an advanced civilization could choose to build a large, self-sustaining colony, with many generations of citizens living on the spacecraft as it traveled across space. Thus, critics say, there would be no need for giant transmitters.

Further, if such feats of interstellar travel are possible, where are the aliens? The universe is so old—about 15 billion years—and the Milky Way is so vast, that many aliens should have visited us by now if biological evolution is as universal as the advocates believe.

Even if there are alien civilizations in the universe, says Dr. Rood, "I don't see them making the investment to communicate with primitive ones like us."

The advocates challenge such arguments. For one thing, they say, their equipment does not depend totally on beacons in distant solar systems; it is sensitive enough to do some eavesdropping. Further, they stress that their program is very conservative, making no assumptions about interstellar travel or other exotic possibilities. Instead, they say, the field deals solely in known quantities like planets, radio transmissions, biological evolution, and what has been proved technologically possible on Earth.

In the end, they say, the only way to try to resolve the riddle is to scan the sky.

"You can theorize forever," says Dr. Drake. "But intelligent life is so complicated in its activities and philosophies, as we know from ourselves, that it's quite impossible to psych out the extraterrestrials and deduce by logic how they might behave. The only way we can really learn the truth is to search."

The universe could contain thousands of planets with all the conditions necessary to give rise to humanlike life.

Portholes to the Past

by Jeffrey H. Hacker

A sponge diver off the southern coast of Turkey reports seeing strange "metal biscuits with ears" lying on the seabed at a depth of 150 feet (45 meters). Underwater archaeologists explore the site and find a Canaanite merchant ship dating to the 14th century B.C. The "metal biscuits with ears" are heavy copper ingots (metal cast into flat shapes for convenient storage) with handles. The vessel is laden with a treasure trove of other artifacts—pottery, glass, jewelry, weapons, and tools—from seven cultures of the Late Bronze Age.

Ocean salvagers discover a 17th-century Spanish galleon in 1,500 feet (460 meters) of water off the southwest coast of Florida. The low temperature and high pressure appear to have preserved much of the wooden vessel and its cargo. Dozens of amphorae (large jars), gold coins, silver ingots, and other relics are strewn about the wreck.

Deep-sea researchers locate the fabled luxury liner *Titanic* off the coast of Newfoundland.

Technological advances have enabled today's shipwreck archaeologists to make deeper, longer, and more detailed underwater explorations. Robert Ballard (above) used submersibles to discover the remains of the Titanic *(top) and the World War II battleship* Bismarck *(facing page).*

At a depth of 13,000 feet (4,000 meters), the ship is mangled and dismembered—but largely intact—more than 70 years after its calamitous demise. The absence of sunlight, heat, and parasitic life have preserved its contents in an eerie underwater museum: unopened wine bottles, china plates, a pair of boots, a doll's head.

Few developments in modern science capture the public imagination like the discovery of a sunken ship. Unseen for decades, centuries, or millennia, their shadowy hulls lurk in the depths carrying ancient artifacts and—who knows?—a safe or treasure chest filled with long-lost riches. Indeed, technological advances in underwater exploration have enabled a new generation of bounty hunters to turn up some multimillion-dollar caches.

For the underwater archaeologist, however, the real treasure of a sunken ship lies in its evidence of how people lived at a particular time in history. An ancient wreck is a time capsule of the society that produced it. The artifacts found on board are like pieces in a jigsaw puzzle, contributing to an overall picture of life in that civilization. The vessels themselves are invaluable tools for the student of history, since ancient ships typically represented the best a society could muster in technology, materials, and craftsmanship. More broadly, the accumulation of data from multiple wrecks sheds light on patterns of trade, exchange of technology, changing sea levels, and migration of peoples.

Plumbing the Depths

Underwater archaeology—the discovery, study, and preservation of historical artifacts from the oceans, lakes, and rivers of the world—is barely a half-century old. Only with the invention of scuba (*s*elf-*c*ontained *u*nderwater *b*reathing *a*pparatus) by the French explorer Jacques Cousteau and the engineer Emile Gagnan in the mid-1940s did the seas become accessible to hands-on scientific study. More recently the development of sophisticated new technologies has led to a host of major archaeological discoveries and salvage operations. Deep-sea submersibles, remote-detection devices, low-light video, and other technological innovations have allowed shipwreck archaeologists to locate, analyze, and preserve underwater finds with the same level of scientific rigor as they employ on land.

The most spectacular deep-sea discoveries of recent years have been made by manned and remote-controlled submersible vehicles that not even Jules Verne could have imagined. Most prominent among them is the DSV (Deep Submergence Vehicle) *Alvin*, which carried three men to the *Titanic* site in 1986. As of 1989, the 25th anniversary of its launching, *Alvin* had made more than 1,800 dives. Over the years, improvements in design and equipment have enabled the *Alvin* submersible to make deeper, longer, and more detailed explorations of the world's oceans.

Photos, below and facing page: © Woods Hole Oceanographic Institution

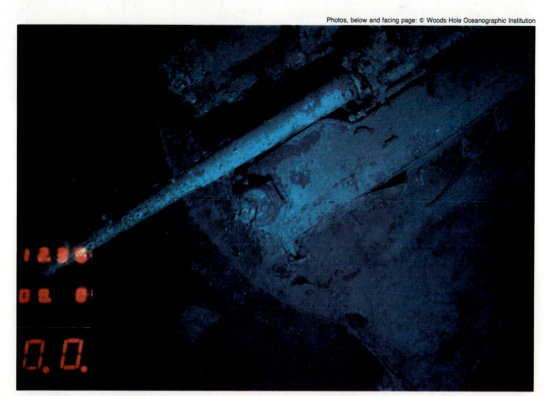

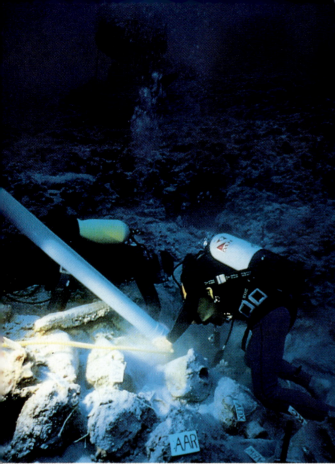

Using an underwater vacuum cleaner called an airlift (left), divers blow away sediment to uncover ship parts and other artifacts of long-lost vessels. Divers record their find by taking three-dimensional photographs using special hand-held equipment (above).

Above: Donald A. Frey/Institute of Nautical Archaeology; right: © Don Kincaid

The actual discovery of the *Titanic* was made during 1985 sea trials of a remote-controlled photographic "sled" called *Argo*, which could descend to depths of 20,000 feet (6,100 meters). In the 1986 return visit, close prowling of the *Titanic* was made possible by a small robotic camera, called *Jason Jr.*, tethered to *Alvin*. In 1989 the legendary Nazi battleship *Bismarck*, lost in the Atlantic in May 1941, was found and photographed by *Argo* in 15,000 feet (4,600 meters) of water.

Less expensive and more commonly used technologies in underwater archaeology include sonar imaging devices and other remote-sensing instruments. Side-scan sonars and sub-bottom profilers are battery-powered acoustic devices, normally towed by a small boat, that use sound waves to generate a topographic image of the seabed. A side-scan sonar transmits high-frequency sound waves that bounce off the ocean floor; the pattern of returning pulses produces a detailed image of the seabed surface. A sub-bottom profiler transmits low-frequency sound waves that penetrate the surface; returning pulses thereby produce a graphic record of underlying features. Recent discoveries made with sonar imaging devices include the late-18th-century privateer *De Braak* off the coast of Delaware.

Alone or in conjunction with sonar, scientists also use a remote-sensing instrument called a proton magnetometer. By measuring the strength of local magnetic fields, a magnetometer can detect high concentrations of buried or exposed iron (as from an anchor, cannon, or ship's fittings). The magnetometer has been a primary tool in the search for 17th-century Spanish galleons in the New World, where most wrecks are buried or broken up in sand or coral. On the outer shore of Cape Cod, the 18th-century pirate ship *Whidah* was located by magnetometer.

Other remote-sensing technologies used in locating wrecks include aerial and satellite imaging—by high-altitude photography, multispectral scanning (detection of electromagnetic radiation), or laser-beam pulsing. A hand-held acoustic "gun" is replacing the measuring tape for plotting site dimensions. Computers are used to store data on sites and artifacts. Actual excavation is done with an airlift, a sort of underwater vacuum cleaner that draws up sediment with compressed air. And underwater cameras or low-light video provide a visual record.

Yet, with all the high-tech systems finding their way into modern marine archaeology, many important discoveries have resulted from plumb luck—recreational diving, talking to local sponge divers, or dint of nature. The subsequent on-site work, normally done in scuba gear, is as painstaking as on any land dig.

Old World Wrecks

Since the advent of modern marine archaeology, no body of water has borne a richer bounty than the Mediterranean Sea. The formative cultures of Western civilization—Roman, Greek, Egyptian, and others along the Mediterranean—were involved in seafaring, fishing, and maritime commerce from the earliest times. The strong currents, sudden storms, and treacherous coasts of the Mediterranean spelled disaster for many of their sailing ships. Today nearly 1,000 wrecks from the classical and medieval periods have been found in the Mediterranean. From 50 to 100 new sites are discovered every year, each providing a unique opportunity to learn more about the ancient world.

The oldest known shipwreck is the 14th-century B.C. Canaanite (Bronze Age Phoenician) merchant vessel found off the southern coast of Turkey. Discovered by a young sponge diver near a promontory called Ulu Burun in 1982, the 50-foot (15-meter) wooden vessel has been excavated by George F. Bass (a pioneering figure in marine archaeology) and his team from the Institute of Nautical Archaeology at Texas A&M University. The thousands of artifacts recovered from the ship—including copper and tin ingots, cobalt-blue glass, a gold chalice, amphorae filled with glass beads and a yellow resin used as incense, an elephant tusk, a hippopotamus tooth, jewelry, weapons, and tools—were traced to seven cultures of the eastern Mediterranean (Canaanite, Mycenaean, Egyptian, Cypriot, Assyrian, Nubian, and Kassite). Their presence on a single Canaanite vessel suggested strong commercial ties among Bronze Age kingdoms and disputed a widely held theory that the Mycenaeans had a monopoly on sea trade. Analysis of individual items enabled Bass and his team to chart the ship's course, identify the origin of materials, and gain myriad new insights into life, industry, and commerce at the time of Tutankhamen and the fall of Troy.

In 1987, with the Ulu Burun wreck about half-excavated, Bass and his colleagues decided to visit the site of a previous Bronze Age dig off Cape Gelidonya (also in southern Turkey). Discovered in 1958, the Gelidonya wreck was the subject of the first methodological underwater excavation carried out to completion. Bass concluded that the vessel dated to the 13th century B.C., but some scholars believed it dated to the 12th century B.C. Upon returning in 1987 and 1988, Bass's group found artifacts they had previously overlooked, including two clay jars that seemed to confirm the earlier origin.

Another recent discovery in the "cradle of civilization" resulted from drought conditions in Israel during January 1986. As the waters of the Sea of Galilee receded, the outlines of a hull began to appear in its banks. Archaeologists excavated a 27-foot (8-meter) boat dating to between 100 B.C. and A.D. 70. The only ancient vessel ever found in the Sea of Galilee, the five-oar boat was of a class described in the New Testament (Gospels) and in the writings of the 1st-century Jewish historian Josephus.

Prior to excavation, all shipwreck sites are carefully mapped out. Divers use specially designed writing utensils to record their measurements underwater.

© Bill Curtsinger/Photo Researchers, Inc.

Resurrecting the Past

The most important artifact of all at a shipwreck site is the ship itself, a fact overlooked or ignored in the early practice of underwater archaeology. J. Richard Steffy, a top authority on the reconstruction of ancient ships, likens that approach to "wandering atop the Acropolis searching for ancient coins, but not noticing the Parthenon." In recent decades, however, nautical archaeologists have succeeded in raising, preserving, and reconstructing the remains of sunken ships spanning more than 2,000 years of history.

The oldest shipwreck to be fully retrieved and reassembled is a 4th-century B.C. amphora-carrier discovered near the town of Kyrenia, Cyprus, in the mid-1960s. Described as the best-preserved hull of the classical Greek period, the 45-foot (14-meter) wooden vessel was raised piece by piece over five years, treated with a water-soluble wax (polyethylene glycol) to prevent deterioration, reassembled with stainless-steel wires, and placed on exhibit in Kyrenia. Similar salvage and preservation methods have been used on five 11th-century Viking

Metal detectors (above) help divers find ancient coins and other metallic objects hidden beneath the sediment. Archaeologists use such items to date the shipwreck. Larger artifacts are lifted to the surface by means of a balloon (right).

sailing ships excavated from Denmark's Roskilde Fjord; the *Bremen Cog*, a 14th-century merchant vessel uncovered on West Germany's Weser estuary; and the ornate Swedish warship *Wasa*, which sank in Stockholm harbor on its maiden voyage in 1628.

As demonstrated in 1982, however, the raising of an ancient wreck need not be done piecemeal. That October the surviving hull of the *Mary Rose*, the flagship of King Henry VIII's war fleet, was recovered intact from the muddy waters off Portsmouth, England. Sunk in 1545 en route to battle with French invasion forces, the *Mary Rose* was lifted from the seabed by wires attached to a steel frame, and positioned on a protective cradle. The cradle was then lifted onto a barge and towed ashore.

Whether by reconstructing a wreck timber by timber or by examining a hull intact, marine archaeologists have learned a great deal about ancient shipbuilding and the evolution of technology. For example, Steffy points out, reconstruction projects have led scientists to conclude that classical Mediterranean ships were built shell-first; that is, the strength of the vessels derived from the planking, joined edge to edge, rather than from internal frames. Only by the 11th century were hulls built frame-first, with internal beams providing the basic support. Moreover, says Steffy, close examination of wood fragments and construction details may reveal facts about the craftsmen themselves. "You can learn a lot about the shipwright from the smallest of clues—and from them you may be able to extrapolate to the attitudes of a whole society. From construction features we have been able to gain insights into religious influences, superstitions, customs, even hygiene."

Following in the Wake

To test the seaworthiness of early craft and to answer questions of basic design, archaeologists have also joined with naval architects and classics scholars in building and sailing modern replicas of ancient ships. Culminating a 20-year preservation and research project on the 4th-century B.C. wreck at Kyrenia, modern Greek shipbuilders used ancient techniques in constructing a full-scale working model, dubbed *Kyrenia II*. By late 1987 *Kyrenia II* had made several voyages in the Mediterranean.

The Kyrenia wreck was also one of the few pieces of concrete evidence used in resolving a centuries-old debate among scholars and naval engineers—the design of ancient Greek triremes (warships propelled by three banks of oarsmen). Fundamental to Greek naval strength for centuries, triremes helped preserve the political and economic conditions in which ancient Athens made its great contributions to Western civilization. According to classical literature, use of the fast, easily maneuvered vessels enabled Athens to defeat a more powerful Persian sea force at the historic Battle of Salamis in 480 B.C.

Triremes were known to have been about 118 feet (36 meters) long and 18 feet (5.5 meters) wide, carry 170 oarsmen, and reach speeds of up to 12 knots. But because no wreck or plans of a trireme had ever been found, key questions remained unanswered: What did it look like? How could so many oarsmen be crammed into so small a hull? How could they row in unison?

In 1981 a group headed by Professor John Morrison of Cambridge University and John Coates, retired chief naval architect of the

A sunken ship is a time capsule of the civilization that produced it. The jewelry, religious items, glassware (below), and other artifacts found aboard the Conde de Tolosa, *a Spanish vessel that sank in 1724, help contribute to a picture of life in Spanish society of that era.*

Facing page, top: © Robert Holland; bottom: Donald A. Frey/Institute of Nautical Archaeology. Right: © Jonathan Blair/Woodfin Camp & Associates

The ancient Greeks owed their naval strength to fast, highly maneuverable warships called triremes. Relying on evidence from shipwrecks, ancient literature, and even old coins, archaeologists, classical scholars, and naval architects joined forces to build Olympia, *a full-scale replica of a trireme. With a volunteer crew,* Olympia *successfully achieved a top speed of 7 knots in a 1987 trial run on the Mediterranean Sea.*

British Ministry of Defense, set out to solve the riddle by designing and building a full-scale replica. Drawing on ancient naval accounts, limestone carvings, 40 years of underwater archaeological discoveries (including the Kyrenia wreck), and their own ingenuity, they finally succeeded—after six years—in constructing a fast, maneuverable trireme of authentic dimensions. They christened it *Olympia*.

The key design feature was a V-shaped hull, with the bottom row of oarsmen sitting inboard near the keel, the second row farther outboard, and the upper row on a scaffolding that protruded from the top. The secret to smooth rowing, they discovered, was precise coordination among vertical triads of oarsmen as well as equal strokes among rowers on the same level. With a crew of volunteer oarsmen, *Olympia* took to the sea in the summer of 1987 and achieved a top speed of 7 knots. With more practice and refinements in construction, the speed of the ancient Greeks could be equaled—and another mystery solved.

Treasure Hunting

By virtue of technological advances, improvements in methodology, and sheer accumulation of data, underwater archaeologists have reached a new level of skill and sophistication. As work in the Mediterranean continues to expand our knowledge of ancient life, explorations of the deep-ocean floor have unlocked mysteries of the more recent past. The discovery of the *Titanic* by Robert Ballard and his team in September 1985 settled a long-standing debate as to whether the ship sank in one piece or broke apart as it went down; the bow and stern were found more than 600 yards (500 meters) apart. Discovery of the Nazi warship *Bismarck* in June 1989, also by Ballard and his crew, suggested that it was scuttled by the Germans rather than sunk by the British navy.

Unfortunately, the very advances that have yielded the new wealth of information have also posed a new threat to underwater archaeology. With deep-sea vehicles making it possible to bring up artifacts from thousands of feet below the surface, private salvage companies have entered the business of high-tech treasure hunting—with some dramatic results. In 1985 salvagers off the southwest coast of Florida discovered the Spanish galleon *Atocha*, which sank in a storm en route to Spain in 1622; the gold and other treasures found on board were worth an estimated $400 million. In June 1989, another early-17th-century Spanish galleon was found by a private business concern in the same area as the *Atocha*, at a depth of 1,500 feet (460 meters). Videotapes of the site revealed gold coins, silver ingots, and other evidence of a vast fortune. More important, the wreck appeared to be intact and well-preserved, rare for a Spanish galleon. Thus, the search for sunken treasure ships by profit-minded corporate enterprises has both excited and alarmed underwater archaeologists: they are eager for new discoveries, but fear wholesale plundering.

Much as on land, however, the greatest single threat to underwater archaeology today is rampant looting by clandestine salvagers. Pottery, works of art, and precious metals have been stolen from literally hundreds of offshore sites in the Mediterranean. A thriving black market has sent prices soaring, leading to more looting and violence among rival gangs. Boats are burned, divers killed. In the south of France, they call it the Amphora War.

With all the progress in locating, salvaging, and reconstructing ancient shipwrecks in recent decades, no aspect of preservation has presented a more intractable problem than protecting sites from looters. Ironically, guarding wrecks from modern-day pirates poses the most fundamental challenge to underwater archaeology in the years to come. The spoils of the Amphora War are the riches of our cultural heritage. For archaeologists and historians, the knowledge gained from ancient shipwrecks and their artifacts is worth more than any treasure of silver and gold.

Shipwreck archaeologists must spend much of their time above water to properly analyze and catalogue their finds.

© Don Kincaid

WORDS OF LOST WORLDS

by Helle Bering-Jensen

It does not look like much when you hold it in your hand. A reddish brown brick, a couple of inches square, covered with tiny scribbles on both sides, cracked along the top edge and carefully glued. It feels slightly warm, as if still holding some of the heat of the desert sand from which it emerged after four millennia of oblivion.

Still, there is nothing like it in the whole world. This small clay tablet contains the closest Mesopotamian parallel version of the Old Testament Flood, complete with an ark and a Noah look-alike named Ziusudra.

Fortunately, nothing much will happen if you drop the tablet on the floor. Having been baked to perfection in modern high-temperature ovens, clay tablets are virtually indestructible. The tongue is the long-vanished Sumerian language, and the exquisitely neat and orderly symbols belong to the cuneiform (or wedge-shaped) script—humanity's first real system for written communication—pressed with a reed stylus onto flat clay tablets. Cuneiform is a legacy of the Sumerians, who also gave us the wheel, the 360-degree circle, and 60-minute hours.

Though the prize specimen, the flood tablet is only one in the tablet collection of the University Museum at the University of Pennsylvania in Philadelphia. The collection, which totals almost 30,000, is one of the largest in the United States, and derives from the excavations undertaken between 1888 and 1900 in the Sumerian cities of Nippur and Ur, in modern Iraq.

A Lexicographic First

The University Museum is one of the primary centers for the study of the ancient Near East, and it is there that the world's first comprehensive dictionary of Sumerian is in production. Ake Sjoberg, editor and professor of Assyriology, is proudly displaying a selection of his favorite texts, lodged in flat, cotton-lined draw-

Scholars reconstructing the languages of the ancient Near East (below) hope to gain a new understanding of the long-lost civilizations that created them (facing page).

ers in a section of the museum protected by the most perfunctory of security measures, a large, rickety wire screen equipped with a cowbell and dire warnings in Sumerian.

The Pennsylvania Sumerian dictionary is just one of the great comprehensive dictionaries of ancient Near Eastern languages in progress at American universities. Supported by grants from the National Endowment for the Humanities, the dictionaries will provide the groundwork for decades of research in fields spanning biblical history, philosophy, theology, archaeology, linguistics, and, of course, Egyptology and Assyriology. In addition to the Sumerian dictionary, work has been going on for decades at the Oriental Institute at the University of Chicago on dictionaries of Assyrian and Hittite. A comprehensive dictionary of Aramaic was begun about three years ago by scholars at Johns Hopkins University, Hebrew Union College in Cincinnati, and Catholic University of America in Washington, D.C.

Sjoberg wrote his first Sumerian index card in 1949 and his collection of more than 500,000 neatly typed entries forms the core of the enterprise. The first volume of the Sumerian dictionary published was the letter *B* in 1984. The letter *A*—a more difficult and voluminous letter—should be out next year. The team expects to see completion of the alphabet in another 20 years or so.

In terms of the study of the Bible lands, the great *Chicago Assyrian Dictionary*, with 21 published volumes, is the progenitor of them all. Begun in 1921, the dictionary is heading down the homestretch; the two final volumes are in page proof. (The name is slightly confusing. Actually, the dictionary is of the Akkadian language, dialects of which were spoken in Babylonia and Assyria.)

Encyclopedic Dictionaries

David Owen, professor of Near Eastern studies at Cornell University, says the Assyrian dictionary provides an example of the way a dictionary can change a field. For the study of Babylonian and Assyrian empires, "it is hard even to imagine the field without the dictionary," he says. (As a tool, it may be indispensable, but it isn't cheap: the price for a full set is $700.)

Like the other dictionaries now under way, the Assyrian work is almost more like an encyclopedia, citing variations of form and meaning, including the texts in which a given word can be found. "It is such an enormously rich resource for evaluating documents. You can trace the history of a particular occupation, object, idea. In fact, whenever I get a new volume, I sit down and read it through from cover to cover simply for the entertainment value," says Owen.

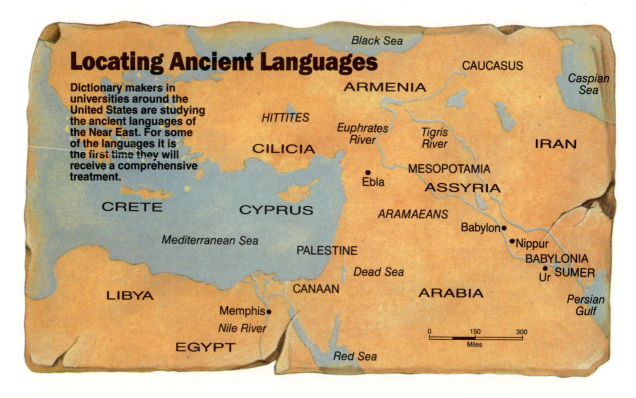

Locating Ancient Languages

Dictionary makers in universities around the United States are studying the ancient languages of the Near East. For some of the languages it is the first time they will receive a comprehensive treatment.

meticulous schooling of these scribes that we have been able to reconstruct the language at all.

While we tend to think of dictionary making as connected with names like Samuel Johnson and Noah Webster, the field is as old as writing itself. Says Sjoberg: "The Sumerians started collecting words about 3000 B.C."

At Johns Hopkins, Professor Delbert R. Hillers of the Department of Near Eastern studies is in charge of the Comprehensive Aramaic Lexicon, a project that will occupy scholars for the next 15 years or so.

Unlike a number of the other Near Eastern languages that are being studied, Aramaic was never actually lost. It is a close relative of Hebrew and a somewhat more distant cousin of Arabic. It remains spoken in some parts of Syria, and functions elsewhere as a religious language. Most people know it as the language Jesus spoke, but it is also the language of the Jewish Talmud, and parts of the Old Testament are written in Aramaic, including the story of Belshazzar's feast in the Book of Daniel, which remains somewhat of a puzzle.

Aramaic was also the language of commerce and official record for the ancient Assyrian and Persian empires. During its golden period before the coming of Islam, Aramaic was in use from Egypt to India. Indeed, the script now known as Hebrew is based on Aramaic letters. Nonetheless, there has never been a comprehensive dictionary of the language.

"Because it was a language that was preserved as a language of religion and study, it was treated very much in bits and pieces," says Stephen A. Kaufman of Hebrew Union College, who will take over as project leader next year. "Dictionaries would be made up of the various religious- or literary-based dialects. In addition, there have been discoveries of Aramaic literature dating from ancient times. Our job is to put all this together and come up with a picture of the language as a whole." The dictionary covers 900 B.C. to A.D. 1400, after which the language declined in importance.

Ancient lexicography is painstaking work. Ake Sjoberg (above, holding a cuneiform tablet) has worked since 1949 on a dictionary of the ancient Sumerian language; only the volume on the letter B has been published thus far. Only two out of 21 volumes of the Chicago Assyrian Dictionary (left) remain to be published; scholars have worked on the set since 1921.

"Before this, you spent your whole life looking up texts and making notes until you got in your 60s, and then you could start to write something general," says Erle Leichty, professor of Assyriology at the University of Pennsylvania, who works with Sjoberg on the Sumerian dictionary.

Among the good fortunes for students of the ancient Near East is the fact that many nations in the region were heavily multilingual; texts often appear in two or three parallel translations. Sumerian ceased to be spoken around 2000 B.C., but it was to Akkadian what Latin is to the modern world, a classical language that every scribe had to learn. It is thanks to the

Computer-aided Research

With the wealth of material available, the advent of computers has been a real blessing. Hillers estimates that the production of the actual volumes of the Aramaic dictionary will come within the last two years of the project. Right now, texts are being loaded into the system in great numbers, and he displays proudly the power of the computer to analyze each word, citing references and occurrences. "Aramaic has been a weak sister when it comes to the Semitic languages, even though it was a very important element of classical civilization. We certainly hope that once we have a single source, that we will be able to improve the status of Aramaic," says Kaufman.

Despite the fact that these languages go back to the dawn of civilization, studies of them are relatively young disciplines. "Nothing is more mysterious or more intriguing than the discovery of a hitherto-unknown language in an unreadable script, and few advances in scholarship are more dramatic than the decipherment of such a script and the subsequent translation of the language in which it was written," British archaeologist Colin Renfrew has noted. Lost for thousands of years, many have been recovered only in the 19th and 20th centuries.

Breaking the cuneiform script itself, of course, was the first step. Even though the writing is often amazingly distinct, the characters, which are a combination of stylized pictures and representations of syllables, did not lend themselves to easy decoding. That was the result of the determination and peculiar talents of 19th-century colonial officials, a combination of education and mathematical and linguistic abilities.

"Everybody in those days knew Greek, Latin, and Hebrew," says Cornell's Owen.

Cuneiform script, like that found on the statue's apron, combines stylized pictures and symbols of syllables. It proved impossible to translate for centuries.

The Metropolitan Museum of Art

"That was primary education even for military people, and many of them were posted to India or Iran or Iraq, so they knew Arabic and the Indian dialects. These soldier-scholars had an intellectual breadth few people have today."

Linguistic Connection to the West

Thus, the first person to suggest a connection between the languages of India and those of Europe (today known as the Indo-European languages, including also the Slavic and Persian languages) was Sir William Jones, a colonial judge in Calcutta, who in 1786 demonstrated the similarity between the classical Indian language of Sanskrit on the one hand and Greek and Latin on the other.

Within cuneiform studies the biggest breakthrough was achieved by Sir Henry Rawlinson, an officer of the British East India Company, who between 1835 and 1844 uncovered the inscription of Darius I on a cliff at Behistun, in what is now Iran. By focusing on names, Rawlinson managed to unlock the Old Persian cuneiform alphabet.

Two other languages in parallel translation on the rock were consequently deciphered. One was Akkadian, which belongs to the Semitic family of languages, which meant that much of the literature of the ancient Near East became accessible to scholars. Rawlinson's feat compares with the deciphering of the Rosetta stone by the French archaeologist Jean-François Champollion in the 1820s, which revealed the secrets of Egyptian hieroglyphics.

The exploration of these ancient languages continues. One of the latest arrivals is the language of Ebla, which has been recovered after excavation in Turkey in the 1970s. In fact, excavations in Syria, Iraq, and Turkey turn up clay tablets at the rate of about 25,000 a year.

While the potential for original research is great, with the huge number of unpublished texts and with the arrival of the dictionaries, many scholars are rather gloomy about the future of the field. The number of graduate students coming into ancient Near Eastern studies is minute, a function of slim career prospects.

"I have had a postdoctoral fellowship open for three years, and not one application," says Sjoberg of the University of Pennsylvania. The few existing graduate students are mentioned in terms normally reserved for the discussion of hatchlings of rare and near-extinct breeds of exotic birds. "We are so few workers," says Hillers, "with so much to do."

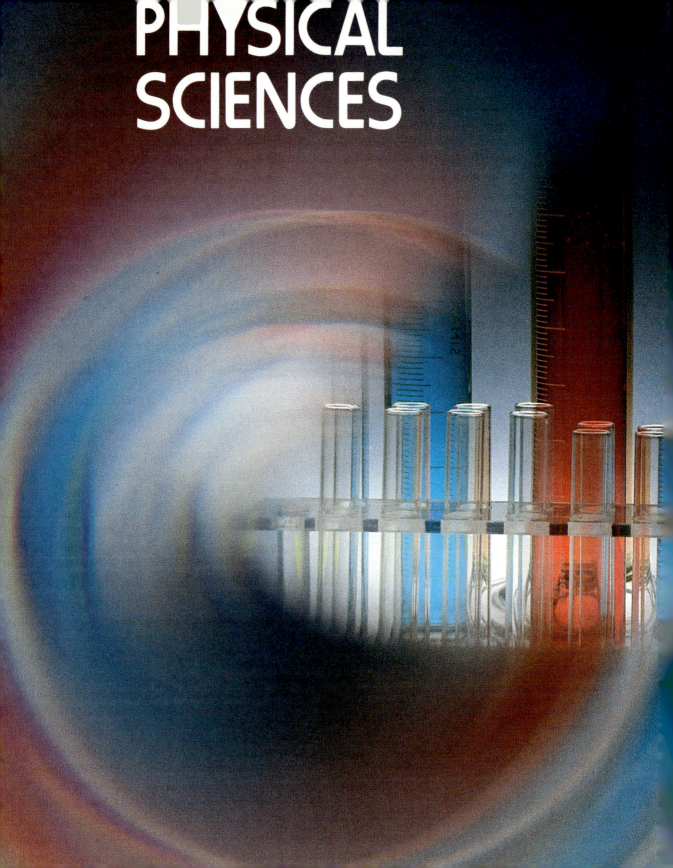

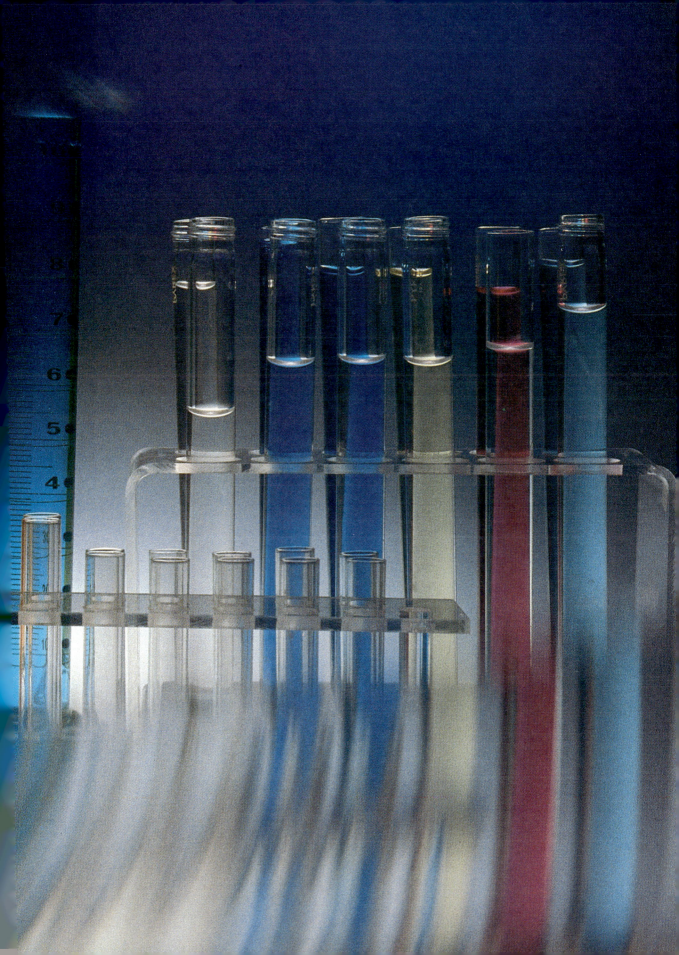

Review of the Year

PHYSICAL SCIENCES

Scientists learned more about superconductivity, explored the nature of subatomic particles, and developed ways to create synthetic enzymes. Improvements were made in FM transmission and steel production.

by Marc Kusinitz

Jonathan Player/NYT Pictures

Dan Taylor used a German formula discovered in World War II archives to develop a type of steel in which nickel is replaced by manganese and nitrogen.

Materials Science

Researchers at Sandia National Laboratories in Albuquerque, New Mexico, determined that corrosive chemicals add elements to the crack surface of glass, leaching out components of the glass and giving rise to a chemically distinct "alteration" layer. This work could lead to production of glass with increased resistance to cracking.

The managing director of a British foundry, looking for a low-cost substitute for nickel in the production of steel parts and equipment for the auto industry, developed an alloy called manganese-enhanced, austenitic (i.e., nonmagnetic) nitrogen steel, or "mean" steel. Mean steel uses nitrogen, held in place by manganese atoms, as a superglue to bind together molecules of the ingredients of the steel, producing a high-strength, corrosion- and heat-resistant, nickel-free steel. If the process can be applied to stainless-steel production, it could have a major technological and economic effect.

Superconductivity

The news was mixed from scientists studying superconductivity—a property of certain materials under specified conditions (usually including extreme cold) in which electrical current can flow unimpeded by resistance. Researchers at AT&T Bell Laboratories in New Jersey found that compounds of bismuth and thallium, which are superconductive at relatively high temperatures, lose their superconductivity when exposed to a magnetic field. Such a property, unless modified, would make bismuth/thallium compounds useless for applications in highly efficient electrical generators and magnetically levitated trains.

The good news came out of the U.S. Department of Energy's Argonne National Laboratory in Illinois, where scientists created a quarter-inch-thick superconducting ribbon capable of carrying at least 10 times the normal level of household current. They devised the ribbon by bonding a brittle ceramic material made from yttrium, barium, and copper oxide to a base of silver foil.

Catalytic Chemistry

Chemists improved the efficiency of abzymes, antibodies that work like enzymes to accelerate chemical reactions. They further demonstrated that some abzymes can work in organic solvents, a property likely to enhance their usefulness. Once perfected to work as fast as natural enzymes, abzymes could find wide use as antiviral drugs, drug purifiers, flavor enhancers, and protein synthesizers.

Harvard University scientists announced they had made artificial enzymes, called "chemzymes," to accelerate reactions in a way that virtually eliminates contaminating by-products. This achievement could herald a new era in the production of pure drugs with fewer side effects, and may open the door to production methods for other synthetic products.

Australian chemists using molecular chains as "rods" and copper atoms as "linkers" constructed large, hollow molecules containing spacious, interconnected cavities. These cavities could serve as catalytic sites where specific reactants enter and specific products leave. Such three-dimensional structures have potentially unlimited variations, and may lead to extremely light solid materials.

At the University of California at Los Angeles (UCLA), another team of chemists, led by Nobel laureate Dr. Donald J. Cram, succeeded in modifying large, artificially synthesized chemical "prisons" called carcerands so as to render them soluble—a vital step needed to purify crystals of the compounds and study their structures in detail. These "carceplexes" may ultimately be used to produce new forms of catalysts for promoting chemical reactions.

New Tools and Techniques

Physicists at the Argonne National Laboratory in Illinois demonstrated the wakefield effect for the first time. In this phenomenon a cluster of charged particles creates an undulating wake that boosts the energy of particles riding on that wake. The Argonne team boosted the energy level of a moving electron from 15 million to 23 million volts over a distance of less than 10 inches (25 centimeters). If developed successfully, the wake-field effect could be the basis for accelerators only 6 miles (9.6 kilometers) long, therefore eliminating the need for the enormously more expensive conventional accelerators with circumferences of up to 80 miles (130 kilometers).

Also at Argonne, physicists perfected a technique for making direct images of atoms within a molecule using a principle called a Coulomb explosion. By accelerating molecules of methane to one-fiftieth the speed of light, then smashing them into a thin film target 30 atoms thick, the molecules' electrons are stripped away, leaving the heavier nuclei to pass through the film.

Physicists at Stanford University created an atomic "fountain" by using laser pulses to push sodium atoms up and gravity to pull them down. During the descent, the scientists probed the falling atoms with microwaves to obtain measurements of transitions from one energy state to another. The results were accurate to within 2 hertz, a significant improvement over the 26-hertz precision of the present U.S. time standard.

FMX, a new form of FM radio transmission being adopted by some stations, caused a debate between its commercial developers, Broadcasting Technology Partners of Greenwich, Connecticut, and a critic at Massachusetts Institute of Technology (MIT), who claimed the new technology was inferior to standard FM. FMX supporters, however, claim that MIT used deficient equipment in its tests.

FMX works by adding a third channel to the two-channel FM transmission, an innovation claimed to improve reception quality and increase the station's range. Such advantages could produce increased advertising revenues for those stations that adopt FMX. Consumer-electronics firms have already geared themselves up to produce state-of-the-art FMX receivers.

Emil Torick developed FMX, a three-channel form of FM radio transmission claimed to improve reception quality and increase station range.

Families of Subatomic Particles

Teams of physicists in the United States and Europe, working separately, produced Z-zero particles. These massive subatomic particles were created by colliding high-energy electrons with positrons in particle accelerators. The research further bolstered scientific support for the so-called "Standard Model," a generally accepted explanation of the nature of matter that limits to three the number of families of fundamental subatomic particles. The families are represented by the electron neutrino, muon neutrino, and tau neutrino.

Physicists at the University of Washington in Seattle used inexpensive "tabletop" electromagnetic traps to determine that the radius of an electron must be less than 10^{-20} centimeters. This figure is less than one-thousandth the value of the previously accepted upper limit, as determined by studies using atom smashers. Researchers at the University of Colorado in Boulder used the trap to test various aspects of the Standard Model, including the influence of the weak nuclear force on electrons within an atom.

Physicists at the University of Washington and the National Institutes of Standards and Technology used electric fields oscillating at radio frequencies to trap individual ions enclosed in a near vacuum. The Washington group observed the ions "blink," i.e., emit visible photons during shifts to lower energy states. The Boulder physicists trapped single mercury ions and cooled them to nearly absolute zero, the temperature at which atomic vibrations cease altogether. Such work could lead to a better understanding of atomic behavior.

Halon's Threat to the Atmosphere

The highly effective fire-fighting chemicals known as halons, which appear to be even more destructive to the earth's ozone layer than are the more widely publicized chlorofluorocarbons (CFCs), have undergone an initial phaseout of use as a result of pressure from the EPA and environmental groups.

Bromine-containing halon compounds are less likely to harm electrical and electronic equipment in burning areas, and are required for use on board airliners. Their ability to be stored compactly, and their relative safety if inhaled by humans during a fire, are expected to make halons very difficult to replace.

The EPA has ordered halon fire-fighting chemicals to be phased out gradually.

Cosmic Time Travel

by David H. Freedman

Kip Thorne didn't set out to design a time machine. He set out to answer a question about space travel posed by a fiction-writing friend. After thinking about the problem for a while, he assigned it to one of his physics graduate students at Caltech. "A good professor leads his students into problems that he knows are solvable," says Michael Morris, now a physicist and cosmologist at the University of Wisconsin in Milwaukee. "I could clearly tell he knew what the answer would be. It had probably taken him a few hours to figure it out, and it took me a few weeks."

The problem involved wormholes. Not the terrestrial, mucus-lined variety, of course, but cosmic wormholes: rips in the fabric of the universe that—theoretically, at least—might tunnel from one region of space to another.

And Thorne's fiction-writing friend was none other than Carl Sagan, who, in the summer of 1985, had sent Thorne a draft of his novel *Contact*. Sagan wanted his characters to slither quickly across the galaxy through a wormhole. Being an astronomer himself, he was interested in keeping the physics as realistic as possible, and was hoping Thorne—a well-known authority on black holes—could tell him what such a wormhole might look like. Sagan had no idea that a wormhole might turn out to be a timehole as well.

Black holes and wormholes, as it happens, are cousins: both are progeny of general relativity, Einstein's theory of gravity. In that theory, gravity exists because massive objects curve the space around them; things fall, not because they are pulled by some mysterious force, but simply because they are following the shortest path through curved space. Wormholes and black holes have long been recognized as possible solutions to Einstein's equations. Both types of holes are severe space warps—that is, intense gravitational fields.

But there are two important differences between wormholes and black holes. For one thing, a black hole has a bottom—a "singularity" at which its gravitational field becomes infinitely strong. A traveler unlucky enough to fall into a black hole would never come out. He would be ripped apart by tidal forces (the grav-itational pull on his feet, assuming he were falling feetfirst, would be far more powerful than the pull on his head), fried by the intense radiation inside the hole, and finally crushed to infinite density as he approached the singularity.

Cosmic Shortcut

A wormhole, in contrast, does not have a bottom; it has two "mouths" connected by a "throat." The mouths may be very far apart in space even though the throat is short. Thus, a wormhole might in principle serve as a kind of cosmic shortcut, like the one a worm takes when it burrows through an apple rather than crawling the long way around the surface. A traveler sucked into one mouth of a wormhole and down the throat might emerge from the second mouth only a few moments later, but halfway across the cosmos. That's why science fiction writers, even before Sagan, have long been attracted to wormholes.

Their hopes have always been dashed, though, by the second important difference between wormholes and black holes: black holes exist. Or at least astrophysicists are convinced they must exist, even though they have never actually seen one. When a very massive star collapses under its own weight (as all stars do when they exhaust their nuclear fuel), it must, according to both general relativity and Newton's law of gravitation, form a black hole, simply because no known force could stop the collapse.

On the other hand, no one has ever figured out how a macroscopic wormhole might be formed and maintained in nature. The hypothetical wormholes that have been known to general relativists for decades have a troublesome feature: if one were ever created through some unknown process, its throat would pinch shut almost instantly. The two mouths of such a wormhole would be indistinguishable from black holes. A traveler, even one moving at the speed of light, would be unable to penetrate the wormhole before it shut, and the traveler would then be torn apart. That's why physicists, unlike science fiction writers, had pretty much given up on wormhole travel by the 1960s.

For the next 20 years or so, the idea was more or less shelved. Then Thorne got his letter from Sagan, asking him to design a better wormhole.

Thorne thought about the problem and realized, to his own surprise, that it was not unsolvable. In fact, there seemed to be a rather simple solution to Einstein's equations—actually, a

A traveler unlucky enough to fall into a black hole would be crushed to death by the extraordinary gravity. An entirely different experience awaits a traveler who falls into a wormhole—a hypothetical space tunnel that some physicists believe could hold the key to time travel.

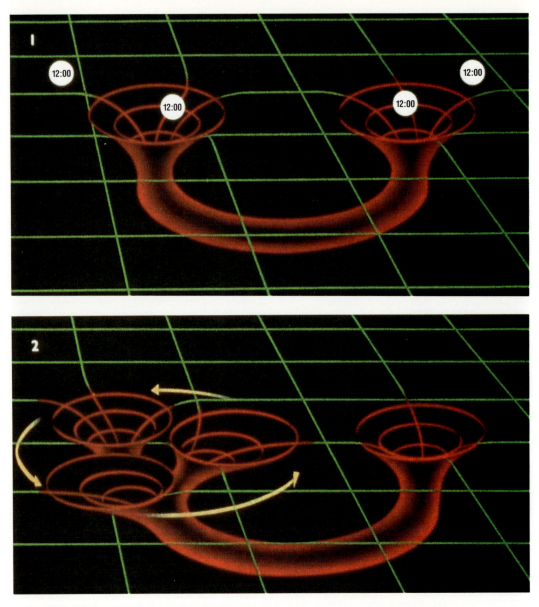

A wormhole might be envisioned as a black hole that has two "mouths" connected by a "throat." If such structures exist, the best place to test one, according to physicists, would be where the two mouths lie only one hour apart by spaceship (top). Assuming that wormholes can be manipulated, one of the "mouths" would have to be accelerated to near the speed of light for one hour (starting at noon, for argument's sake) before undertaking time travel (above).

whole new class of solutions—that had somehow been overlooked by previous theorists. Thorne asked Morris to work out the solutions.

Exotic Bricks and Mortar

The key to building a safely traversable wormhole, Morris found, was to prop the throat open with what he and Thorne call "exotic matter." "Exotic" is an understatement: to keep the throat from collapsing in on itself, the exotic matter would have to sustain a radial (outward) tension comparable to the pressure at the center of a neutron star. Indeed, the tension would be greater than the total density of mass-energy of the wormhole itself. For complicated reasons, this implies that the wormhole might appear to have negative mass and energy—a conclusion that, as Thorne and Morris put it, would gnaw at the foundations of "deeply cherished theorems of general relativity," not to mention common sense. (Once you have negative mass, for instance, you have antigravity.)

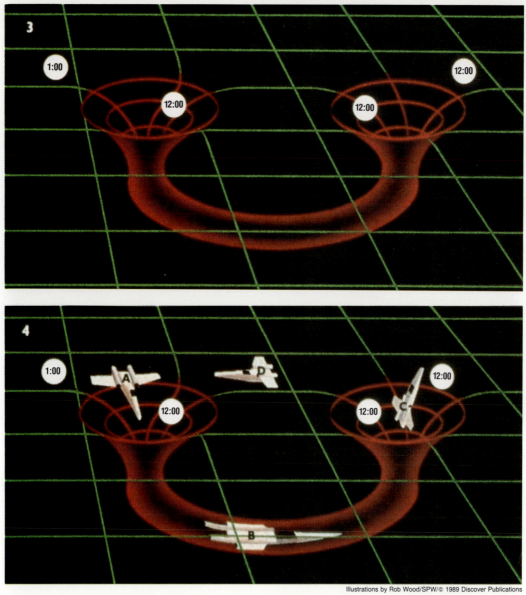

A clock outside the wormhole now reads 1:00 P.M., while clocks inside the wormhole and outside the stationary mouth have barely budged off noon (top). The wormhole is now ripe for time travel. If a spaceship (above) plunges into the mouth that moved (A), it can speed through the wormhole in an instant (B), emerge from the other mouth just seconds after noon (C), and return to the other mouth the long way around (D), in time to see itself going in.

What's more, keeping the wormhole open was not enough; to satisfy Sagan, human travelers had to be able to navigate the enormous tension in the throat without getting shredded into their constituent atoms. Thorne and Morris found two ways around this problem. In *Contact*, Sagan adopted the first, and probably less realistic, one: he stuck a vacuum tube down the center of the wormhole, thereby insulating the travelers from the exotic matter, which was confined to the walls of the tunnel. The other solution would be to assume that exotic matter, like the ephemeral particles known as neutrinos, almost never interacts with ordinary matter. That would enable the travelers to pass through the exotic matter in the wormhole (and it through them) without their feeling a thing—even though the stuff would be unfathomably dense.

It all sounds fantastic and a little ad hoc. There's still no known way such a wormhole could arise naturally, and Thorne and Morris

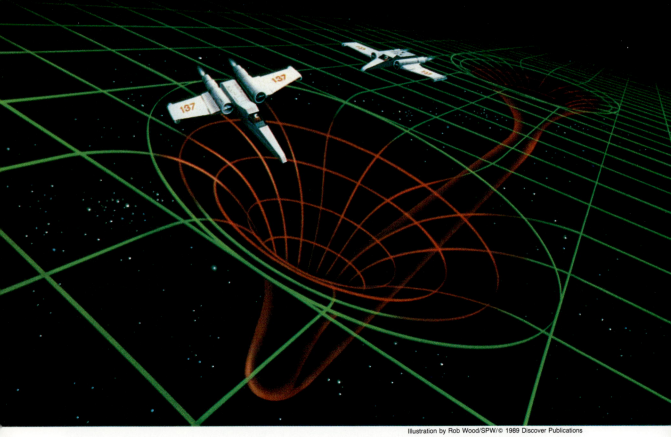

The idea of time travel raises a number of sticky questions, among them the concept of "causality violation": can something go back to the past and affect its future in such a way as to prevent its original journey through time?

have no idea how to build one or even how to create the exotic bricks and mortar. But that's not the point; those details can be left to the engineering wizardry of an "arbitrarily advanced civilization," as they put it.

The point is that Thorne and Morris had found nothing in the laws of physics that would prohibit a traversable wormhole. Our understanding of those laws, they said, does not yet rule out the possibility that someday somebody might somehow do what science fiction characters do now: flash through hyperspace to a distant galactic outpost, in effect traveling faster than light—faster, that is, than light that takes the long route through ordinary space.

Wormhole or Timehole?

But there was even more. At a conference in December 1986, Thomas Roman, a cosmologist from Central Connecticut State University, approached Morris. Traversable wormholes, Roman suggested, might have some disturbing properties with regard to time.

"At first I dismissed the idea," recounts Morris, "because I didn't see any way that wormholes could cause that sort of problem." But Roman, who was aware of this sort of thing from previous explorations of black-hole theory, finally persuaded Morris to take a closer look. This time, Morris found a way that an object entering two wormholes in quick succession might end up traveling, not through vast distances in space, but backward in time. "I showed it to Kip," he says, "and he came up with a way you could run into trouble by moving one mouth of a single wormhole."

Morris and Thorne had come up with a blueprint for a time machine: a timehole.

To understand how it works, you have to turn from Einstein's general relativity to his theory of special relativity. According to that theory, time slows down for a moving object, as measured by an observer considered to be stationary. The theory predicts, for example, that if one member of a pair of identical twins travels at near light speed to a distant star and back, he

288 PHYSICAL SCIENCES

will return to find that his clock has been running slow and his homebody brother has aged more than he has. This "time-dilation" effect has been confirmed by experiment: atomic clocks taken on long jet rides have been found to lag a few billionths of a second behind clocks that remained on earth.

Now imagine that a traversable wormhole has been constructed somewhere in the cosmos. Imagine further that it is possible to move one mouth of the wormhole (perhaps by using a heavy asteroid as a kind of gravitational tugboat). If the mouth is accelerated to a high enough speed and then returned to its original position, it will be just like the traveling twin. A clock inside the mouth will record an earlier time than one that has remained at rest just outside the mouth.

But—and here's the key—it will record the same time as a clock inside the other end of the wormhole. The two mouths haven't moved relative to one another, because they aren't connected by ordinary space. Nor has the stationary mouth of the wormhole moved with respect to the universe outside it. Thus, a clock just outside the stationary mouth will show the same time as the clock inside—and the same time as the clock inside the mouth that moved.

Again, all three clocks—the two inside the wormhole and the one outside the stationary mouth—will show the same time. And that time will be earlier than the time shown by the clock outside the mouth that moved. In other words, the two mouths of the wormhole are connected to parts of the universe that lie at different times. If a spaceship flies into the mouth that moved, it will emerge from the stationary mouth at an earlier time.

How much earlier? That depends on how fast the mouth is moved, and for how long. If it is moved at 100 miles (160 kilometers) per hour for two hours, for example, the time lag will be less than a billionth of a second. If the mouth is moved at 670 million miles (1.07 billion kilometers) per hour (about 99.9 percent of the speed of light) for 10 years, the time difference will be nine years and 10 months. By moving the mouth fast enough and long enough, the timehole's reach could be extended by years.

There is a limit, though, on how far back a timehole can take a traveler: it can't take you back any earlier than the creation of the timehole. The time-dilation effect only slows down time; it doesn't make time march backward. At very best, if the mouth is moved at the speed of light, it will be frozen in time at the moment it starts moving. As Tulane cosmologist Frank Tipler puts it: "Unless some really advanced beings have already made one of these things, we're not going back to visit the dinosaurs."

But we, or whoever builds the timehole, could still do some interesting things with it—and a lot of mischief. The latter is what interests Thorne and Morris. It casts a cloud over timeholes, and possibly over all of physics.

Causality Violation

Trying to speak to Kip Thorne about timeholes is a bit like trying to see the Wizard of Oz. ("Are you calling about Dr. Thorne's work on time machines?" his assistant asks politely. "Because if you are, Dr. Thorne has nothing to say about the subject.") A self-professed wallflower, Thorne is somewhat put off by the attention timeholes have attracted outside the rarefied world of his fellow cosmologists. He agrees to interviews only under duress.

In a way, that's ironic, considering it was a science fiction novel that led him to study time travel. What's more, the issue he's now focusing on is the one that would most tickle the fancy of the science-fiction-hungry public he finds so bothersome, and that is: if time travel is possible, how do you avoid violating causality? Or, what would happen if someone went back in time and killed his own grandmother?

Physicists aren't necessarily uncomfortable with the general concept of going back in time. (After all, Richard Feynman once showed that positrons, the antimatter counterparts of electrons, could be regarded as electrons that are moving backward in time.) Things get thorny, however, when something that has gone back in time is in a position to affect its own past. Consider a timehole with a one-hour time difference, and assume the two mouths are located near each other in space. A spaceship flies into the timehole's first mouth at 1:00 P.M., emerges from the second mouth at noon, and rockets back to the first mouth the long way around.

Suppose the spaceship arrives there at 12:55 P.M., just in time for its passengers to see their own ship getting ready to enter the timehole mouth. If they wave through their windows at themselves, then why don't they remember seeing a spaceship just like theirs, filled with people who looked exactly like them, waving to *them* when they first flew into the timehole? Far worse, what if they decide to head off the ship of fools, preventing it from entering the timehole at

Tim Hilderbrandt/© Discover Publications

Science fiction writers have developed several scenarios about what would happen if time travelers tampered with the past or the future. Although the theories hold little scientific appeal, wormhole physicists have for the most part declined to offer credible alternatives.

Metaphysical Safety Valves

Science fiction writers have proposed a number of solutions to the apparent paradox. Three are particularly popular: you are somehow prevented from doing anything in the past that would affect your future in a way that can't be reconciled with the time travel. Or, you can affect your future, but you and perhaps the whole universe are destroyed in the attempt. Or, you can affect your future, but you end up in a parallel universe whose future is different from the one you came from.

Not surprisingly, physicists aren't impressed with this list, but they haven't been able to improve on it substantially. "One possibility is that even if wormholes allow violating causality, human beings simply won't choose, for whatever reason, to do it," says Robert Wald, a theoretical physicist at the University of Chicago. In other words, causality violation isn't a problem as long as no one actually violates causality; and no one will do so.

Thorne and Morris are not yet ready to resort to this sort of metaphysical safety valve. Their hope is to come up with a description of time travel that allows a traveler to interfere with the past, but only in a way that doesn't muck up his future and create contradictions.

Thus, Thorne and Caltech graduate student Gunnar Klinkhammer are working on the billiard-ball version of the grandmother problem. They have already come up with the outline of a solution. In their scenario the billiard ball is thrown into a timehole in such a way that it flies out the other end at just the right time and angle to hit itself on the way in. If the ball is prevented from going in the timehole, of course, then there is a contradiction. But what if the incoming ball is merely deflected and enters the timehole anyway, at just the right angle to end up emerging and delivering itself a glancing blow? Voilà: consistent physics. Now the question is whether Thorne and Klinkhammer can prove there is always such a "glancing-blow" solution—and if so, whether it might apply to spaceships as well.

The whole issue may sound a little silly, but it is not to be taken lightly. If physicists cannot resolve the causality-violation paradox, they will be forced to choose among three unattractive conclusions: causality is not inviolable in our universe; general relativity is wrong, and thus the timeholes it describes are irrelevant; or, timeholes can't exist, even though no one yet knows a reason why they can't.

all? If the ship doesn't enter the timehole, then how could it have gone back in time to prevent itself from entering?

For some reason, most people who think about this problem tend to phrase it in terms of the person who goes back in time to kill his grandmother. (If the grandmother is killed, then the time traveler will never be born, in which case, who killed the grandmother?) Thorne, for his part, has been ruminating over a billiard ball that is hurled through a timehole in such a way that it flies out the other mouth and smacks into itself heading in. The problem is basically the same: how can something go back to the past and affect its future in such a way as to prevent its backward journey through time? This is the essence of causality violation.

290 PHYSICAL SCIENCES

The price of embracing causality violation would be high: physics as we know it might go out the window. "What physics does for you," explains John Friedman, an astrophysicist at the University of Wisconsin at Milwaukee who has been working with Morris on timeholes, "is let you specify a set of initial conditions, then use your equations to tell what things will look like at some later time. But things won't always work that way if there is causality violation." The problem, of course, is that after the initial conditions are specified, something might come back in time and change them.

Nevertheless, the ban on causality violation is not sacrosanct. "The statement that there can't be causality violation in the universe is very much a matter of opinion rather than fact," says Wald. "General relativity allows one to talk meaningfully about situations where there is causality violation." Einstein himself was vexed by this, notes Tipler: "He was aware of the possibility of causality violation right at the very beginning. He refused to rule it out a priori as nonsense, but hoped it would be ruled out on physical grounds."

Quantum Gravity

Might general relativity itself be wrong? The theory has withstood test after test over the past 70 years, and, needless to say, physicists are in no hurry to toss it out. Yet they know its limitations. It is not good at describing events that take place at extremely high energies (at the instant of the Big Bang, say) or over very short distances (inside an atom, for instance, or at the center of a black hole). In these situations, a new theory is needed, one that modifies general relativity by uniting it with quantum mechanics, the other great theory of modern physics. So far that unified theory, called quantum gravity, has not been found.

If and when a quantum theory of gravity is developed, though, it may only strengthen the case for timeholes. At the microscale of the quantum world, ephemeral regions of negative energy (one of the exotic properties of exotic matter) are commonplace, and space-time may have a foamy structure in which supertiny wormholes pop in and out of existence. Thorne and Morris have suggested that in the distant future, it might be possible to yank one of these wormholes out of the quantum foam and expand it to a useful size. "There will always be regions in which you can get the effect of exotic matter," says Morris. "The question is whether those regions will be large enough to get a traversable wormhole out of them."

In any case, says Morris, it would not be reasonable at this point to dismiss the possibility of timeholes out of hand. "You can't just say that the idea of causality violation is so terrible that there can't be wormholes," he states. "If this exotic matter can exist, then general relativity says you ought to be able to build a wormhole. And if you can build a wormhole, then it also seems you can turn them into time machines. And that's what bothers people."

Kip Thorne is not optimistic about the development of timehole technology. The chances of achieving any sort of time travel within the next thousand years, he says, are nil. To him, the payoff of a timehole theory would be a better understanding of the laws of physics. If timeholes prove to be impossible, physicists will have learned a great deal in finding out why.

Some mainstream physicists believe that wormholes, if somehow proven to exist, would hold a much greater likelihood of providing a means of time travel than have any of the other time-travel methods put forth to date.

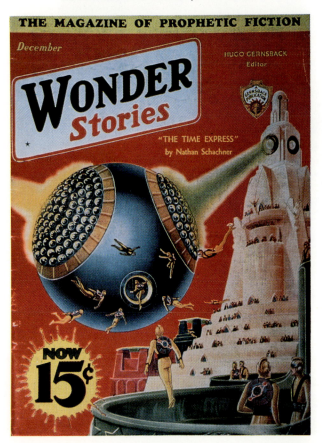

Farewell to the WOODEN BAT?
by Peter Gammons

A hot, sunny Cape Cod Sunday was giving way to the softness of evening. The visiting Cotuit Kettleers began their batting practice as a few fans settled into the stands at McKeon Field in Hyannis, Massachusetts, for the season opener of the Cape Cod League, an amateur summer league for leading pro prospects. Rich Juday, a Kettleers outfielder who plays for Michigan State, took his first round of eight swings. He hit a couple of line drives and a few grounders and sent one ball to deep left.

Juday finished his turn and walked behind the wire batting cage. "The ball just doesn't seem to go anywhere," he said to teammate Greg Haeger, a University of Michigan first baseman. "It still feels as if the bat's dead."

Haeger shrugged and said, "We'll get used to these things. That's what we're here for."

"These things" were the bats—wooden baseball bats—and on this night, for the first time in either of their lives, Juday and Haeger would step to the plate in an official game with a wood bat in hand. Both had played Little League, youth league, high school, American Legion, summer league, and Big Ten baseball, and, until now, as 20-year-olds entering their junior years in college, they had always hit with aluminum bats. "We've been working out for four or five days, trying to get used to wood," said Juday. "There have been a lot of broken bats, sore hands, and blisters. The coaches are telling us, 'You have to hit with the label *this* way—either up or down.' Most of us didn't know what the label on a wooden bat meant."

For more than 20 years, Major League Baseball has helped fund eight amateur summer leagues, including the Cape Cod League. This season, baseball gave the league $74,000 to offset the costs of balls, umpires, scorekeepers, and—most costly of all—wooden bats. Because it is one of only three of those amateur leagues to use wood, the Cape Cod League attracts some of the nation's best college players. "It's like a professional internship," says Central Michigan second baseman Darin Dreasky. "You know scouts are watching because they want to see how you make the adjustment from aluminum to wood. We all want to play pro ball, and pro ball is a wood game."

Even in the major leagues, the crack of the wooden bat will inevitably be replaced by the ping of aluminum. Explanations of such a dire prediction may vary, but they all boil down to one simple economic fact: wood bats split and must be replaced; metal bats don't break.

For now it is, but it won't be for long. Pressed by economic forces, the low minor leagues are likely to begin playing with aluminum bats within two years. By the turn of the century, even the majors will probably have put down the lumber and picked up the metal. Like it or not, the *crack* of the bat is inevitably being replaced by a *ping*.

An Endangered Species

Some 70 miles (112 kilometers) up the road from Hyannis, the day after the Kettleers' season debut, Boston Red Sox outfielder Mike Greenwell was sorting through his bats. "I ordered mine the second week of spring training, and finally got the shipment the first week of June," said Greenwell. "I've used Dwight Evans's bats, Ellis Burks's bats, Jody Reed's. I guess it's just not as simple as saying 'I need two dozen bats,' and having them a week later."

"Greenwell is a .326 lifetime hitter, the [1988] MVP runner-up," says Toronto Blue Jays assistant general manager Gord Ash. "And he's having trouble getting bats. Imagine what it's like for guys in the A leagues and kids in rookie ball, who are using wood for the first time and breaking a bat a day. The problems are acute in the minors, in terms of cost, availability, and quality."

Cost is the biggest reason the wooden bat is an endangered species. Wood bats break; metal bats don't. It's as simple as that. The college game came to grips with this harsh reality in 1974, when the NCAA approved the aluminum bat. "I wish we were all subsidized by the major leagues and could have wood throughout the game," says North Carolina coach Mike Roberts. "But it's economically unfeasible. In college, each player gets an aluminum bat [cost $70], and most of the players use that one bat the entire season. Over the course of the fall and spring season, we'd probably use more than 50 dozen wooden bats." At $14 per bat, that's a difference of about $7,500 a year, a figure no college athletic director can afford to ignore.

Los Angeles Dodgers hitting coach Ben Hines is concerned about the availability of wood for bats. "If good wood is getting harder and harder to find for the top major league hitters, then you see what minor-league players have to use. College, high school, and Little League teams would be getting sawdust glued together if they still used wood. You're going to see aluminum creep farther and farther up the professional pyramid."

© V. J. Lovero/Sports Illustrated

Manufacturers handcraft wooden bats (left) to meet the individual specifications of each major-league player. Makers of aluminum bats use more technologically advanced equipment to produce their goods (below). By its very nature, an aluminum bat has a larger area of high performance, or sweet spot, than does a wooden bat (chart, facing page).

"We've been told to prepare for a severe wood shortage over the next few years," says Bill Murray, director of operations for Major League Baseball, and chairman of the Rules Committee. "We may have to start thinking about an alternative to the wood bat."

"I certainly see a time in the not-too-distant future when everyone will be using some alternative bat—aluminum, graphite, or some composite," says Jack Hillerich, the third-generation president of Hillerich & Bradsby, which, because of its Louisville Sluggers, has been synonymous with baseball bats for more than 100 years. "A wood bat is a financially obsolete deal. If we were selling them for $40 apiece instead of $14 or $16.50 [the company's prices for minor league and major league bats], then we'd be making a sensible profit. But we aren't. We can't charge that much. The time will come when even the majors will use aluminum or graphite."

An Inefficient Proposition

Hillerich, whose company's bat production is now more than 50 percent aluminum, says that the availability of wood isn't the primary cause for concern—at least for H & B, which grows its own timber. But other bat makers do experience shortages. And all of them, Hillerich says, have found wood bats to be an increasingly inefficient proposition.

"While once we were making 7 million wood bats a year for all levels of baseball, now we're making a million and a half, 185,000 of which go to the major leagues," says Hillerich. "Major leaguers want specific orders, so we make three orders [one dozen bats per order] for one player, then shut down the operation. Then we make three more for another player, and shut it down. That's impractical, and it's highly expensive."

All of the bat companies (H & B, Rawlings-Adirondack, Worth and Cooper) have had trouble filling wood-bat orders this season. "I'm having to stop taking orders," says H & B salesman Paul Shaughnessy, who services several major-league teams.

"No one even wants the major-league business anymore," adds Chuck Schupp, H & B director of professional bat sales. "We do it, but partly because of the 100-year relationship we have had with baseball. When we make a bat, we use 40 percent of the wood, at most. If we sell the billets to other industries, nearly 100 percent is used."

The bottom line on the wooden bat is that the bat companies don't want to make them, and the pros are having trouble getting good ones. The alternative is clear. Says Ash of the Blue Jays: "They should use the aluminum bat in a rookie or low A league next season and see what happens." Adds Major League Baseball's Murray: "It's a subject baseball has to take a long, hard look at in the next year."

Crack versus Ping

Anyone who's thirtysomething or older and grew up playing baseball remembers the Louisville Sluggers and Adirondacks that were the staples of the game. Every young ballplayer imagined swinging the bat hard enough to duplicate the gunfire crack that Mickey Mantle and Frank Howard produced when they smacked the ball just right, the crack you could hear on the radio above the din of 40,000 fans. Today, though, on sandlots across America, that sound has given way to the metallic ping reminiscent of a golf ball ricocheting off a flagpole, an aesthetically inferior experience, to be sure.

In a sport as steeped in romance as baseball, that ping alone is enough to make purists resist the aluminum bat. More important, however, wooden bats and aluminum bats are very different beasts, and the evolution from one to the other is changing the game in significant ways.

How different are the two bats? Seattle Mariners rookie Ken Griffey, Jr., was recently asked to name the biggest difference between high-school baseball and the majors. "The wooden bat," he quickly responded. Says Dodger Kirk Gibson: "Wood bats drove me crazy at first. I must have busted hundreds of them." Boston's Greenwell says, "It took me two years to adjust to wood. My hands ached. I had a bone bruise the shape of a horseshoe on my palm that I couldn't get rid of."

"If you hit a ball right with a wooden bat, it'll go about the same distance as a ball hit with aluminum," says Seattle catcher Scott Bradley, who wielded aluminum during his college days at North Carolina. "But with wood, you have to hit it right. You have to use your hands, get the bat head out, and hit it on the sweet spot. If you get jammed or hit it off the end of the bat—unless you're Bo Jackson or José Canseco or Kevin Mitchell—the ball doesn't go anywhere, you often break the bat, and your hands hurt.

"With aluminum, you can make contact most anywhere on the bat and get the ball through or over the infield. Watch a college game and see how many hitters get jammed and

Chart by Nigel Holmes/Sports Illustrated

HOW SWEET IT IS

The sweet spot (deepest purple) is technically a single point. Studies show that a ball struck by a big-league-quality wood bat in that spot will go farther than one struck by an aluminum bat in the corresponding spot. But an aluminum bat has a larger area of high performance (dimensions in circle). It also weighs less than a wood bat of the same length, so a hitter can generate more bat speed. A four-ounce reduction means a batter can eye a pitch for five more feet before swinging.

ALUMINUM 4"–5"

WOOD 2"–3"

still hit flares into the opposite field. The center of balance is different, and when you've been swinging one kind for 20 years, there's a big change."

When Texas Rangers general manager Tom Grieve was managing a rookie league team in 1982, he used to hand all his players a wooden bat and take them to a nearby telephone pole. "I had each kid tap the bat against the pole until he found the spot on the bat where he couldn't feel any vibrations coming down to his hands," says Grieve. "That's the sweet spot, and on an aluminum bat, it's larger than on a wooden one. By finding the spot on a wood bat where he has to hit the ball, a kid can see what he's dealing with before he starts the adjustment process. With an aluminum bat, most kids can take the same swing at every pitch. When they see that they have to hit the ball on a certain spot on a wooden bat, they find out they have to swing differently, according to the pitch."

"What you see with the aluminum bats are a lot of long, sweeping swings, because you can hit the ball most anywhere on the bat and drive it anywhere in the ballpark," says Oakland A's player-development director Karl Kuehl.

Aluminum bats have still another advantage over their wood counterparts. "The weight difference between aluminum and wood bats makes it a lot easier to generate bat speed with the aluminum, and bat speed is what determines power," says Oakland hitting coach Merv Rettenmund. "Most 34-inch aluminum bats weigh 29 or 30 ounces [while a standard 34-inch wood bat is 2 to 4 ounces heavier]. That's a big difference."

Photos, below and facing page: © Anthony Neste/Sports Illustrated; chart, facing page: Nigel Holmes/Sports Illustrated

Today's Little Leaguers (left), raised on aluminum, may grow to adulthood knowing only the ping of the bat. Even the College World Series (below) has become exclusively a heavy-metal show.

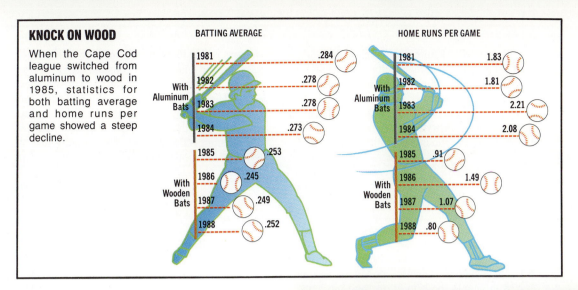

KNOCK ON WOOD

When the Cape Cod league switched from aluminum to wood in 1985, statistics for both batting average and home runs per game showed a steep decline.

Altered Pitching Style

In their debuts with wood on Cape Cod, Juday and Haeger each went hitless as Cotuit lost the opener to Hyannis 3-1. "The scores are pretty low for the first few weeks," says Cotuit manager Pete Varney. "Most of the kids make the adjustment in four or five weeks, but anyone who doesn't think there's a difference should look at the before and after stats."

In 1984, the last year of the aluminum bat in the Cape Cod League, the league batting average was .273 and the ERA 4.86. Games averaged 11.74 runs scored and 2.08 homers. Last summer's figures: batting average .252, ERA 3.68, runs scored 8.98, homers 0.8.

When hitting is altered, pitching is sure to follow, and the aluminum bat has dramatically affected the styles of the pitchers who face it. "Because of aluminum bats, pitching has become an outside game," says Toronto scout Tim Wilken. "Pitchers don't want to pitch inside because, even if they jam a batter [using an aluminum bat], he can dump it to the opposite field for a hit."

"I try to get my pitchers to work inside and forget the opposite-field, jam-shot singles," says Long Beach State coach Dave Snow. "But it's tough to get them to do it." Says A's right-hander Mike Moore, who was the first pick in the 1981 draft out of Oral Roberts: "It took me years to learn to pitch inside."

"Hitters get conditioned to diving out over the plate to hit the ball," says Wilken. "And because the majority of players in the big leagues today have grown up with that outside style, you see so many hitters get hit or nearly hit by pitches that are very close to being strikes. When major-league pitchers use the inside part of the plate, a lot of hitters collapse on the ground as if Bob Gibson had just thrown at them."

Boston right-hander Mike Boddicker says, "The strike zone has been pushed out so that most umpires won't call strikes on the inner few inches of the plate. It's practically impossible to get a strike on the inside corner today."

Amateur baseball leagues that still use wooden bats attract major-league prospects. Once there, the players adjust to the feel of lumber by swinging at tennis balls (above). Even after mastering wooden bats, hitters can rarely match their aluminum stats (chart).

Even more disturbing to some: the aluminum bat practically demands that a pitcher throw breaking balls. "With the aluminum bat, the idea is to make the hitter swing and miss," says Grieve. Many in professional baseball believe that college coaches frequently force their pitchers to throw too much breaking stuff.

Says Mets vice president Joe McIlvaine: "When you get 19- and 20-year-old kids throwing a ton of breaking balls, not only do they often hurt their arms, but they don't build up their arms, and therefore their fastballs, at a time when they are developing physically. I cringe when I see some of these college games, and coaches are calling 50 and 60 percent breaking balls. The aluminum bat has done a lot of damage to the game, both to hitters and pitchers."

Livelier Bats?
At last month's College World Series, in Omaha, Nebraska, 98 percent of the bats used were made by the Easton Company of Los Angeles, a firm that went from manufacturing archery products to becoming the largest producer of aluminum bats in America. Easton makes no wood bats, so, not surprisingly, Jim Easton, the company's president and a son of its founder, downplays the differences between wood and metal. "There's more myth than fact to the complaints about the aluminum bat."

Easton cites a 1976 Stanford study that shows that aluminum bats do not drive a baseball farther than wooden bats do, another claim sometimes made about aluminum, and that more variation can be found among baseballs from the same box than between the two kinds of bats. However, the Stanford study was done before advances in aluminum-bat technology began to soup up the bats. "In the late 1970s, guys in college all knew that the bats suddenly got livelier," says the Mariners' Bradley.

"Technology changed radically at that time," says Hillerich. "There's a lot you can do with aluminum, and we all learned to do it. The kids wanted it. We offered wood bats free to one local league, and the league was enthusiastic about accepting them. After a few games, the kids all wanted to go back to the aluminum."

Composite Bats
Easton, H & B, and Worth have all experimented with a ceramic-graphite composition bat, which produces a sound more like that of wood and—except near the handle—performs more like wood than aluminum does. "The problem is that those bats involve two elements, and they don't hold up to constant usage," says Ottman, whose company does not make such bats. "They're not cost-efficient, and I can't see the day when they'll be perfected. Take batting practice for a couple of hours with one, and it's not the same. The graphite bat is really just to please those concerned with the aesthetics of how they sound."

Says Easton, whose firm does manufacture a graphite composition bat: "A composition bat won't yet outperform aluminum, but it has the potential to be a lot better. Our tests show that our composition bat holds up very well. Graphite is an option, not a replacement for wood."

Such debates will only grow more heated as nonwood bats creep closer to the big leagues. For example, if aluminum bats are more potent, are pitchers endangered by shots back through the box? "If some of these major leaguers used aluminum, someone would get killed," says California Angels hitting instructor Deron Johnson. Easton spokesman Jim Darby counters by saying, "Incaviglia, Cory Snyder, and all those guys used aluminum bats in college, and no one got killed."

What about a metallic assault on the game's sacred record books? "If aluminum gets to the majors," says Boston's Wade Boggs, "I'm certain there will be another .400 hitter." Says Easton: "Baseball is obsessed with records, and pro baseball people keep telling us that with the aluminum bat, all records would be ruined. But what about the way records are changed by the differences in ballparks? How can you compare records of someone playing in Fenway Park and the Astrodome?"

Finally, will aluminum breed a new generation of artificial hitters, juiced up by a sweeter sweet spot? The Dodgers' Hines believes, to the contrary, that the aluminum bat has made for better hitters. "The first thing in hitting is having the confidence that you can hit," he says. "The aluminum bat gives you that. Kids now don't grow up with the fear of hurting their hands. It's made a tremendous difference, especially to kids living in the North. With a wooden bat, it's nearly impossible to hit on cold days until mid-May, and by then most college and high school seasons are over. With an aluminum bat, you can hit no matter what the temperature. Between the cost factor and the weather factor, aluminum bats are allowing a lot more kids to play. Isn't that good for the game?"

Ping. Get used to it.

Photos: Courtesy Underwriters Laboratories Inc.

Behind the Laboratory Door

by Bernie Ward

At Underwriters Laboratories, abuse is the order of the day.

Without it, the little "UL" symbol unobtrusively attached to a host of new wares wouldn't carry nearly the weight it does now. As it is, the mark represents sort of a *Good Housekeeping* Seal attesting to the fact that the blow-dryer you just brought home has been suf-

Underwriters Laboratories tests thousands of products each year for potential safety risks that might endanger life or property. The organization is particularly well-known for setting flammability standards.

PHYSICAL SCIENCES 299

Burglary-resistant glazing material has to withstand high-energy impact testing before it receives UL approval.

ficiently battered and beaten, pummeled and pounded to convince the tough critics at UL that it's safe enough to plug in.

After 95 years of abusing products to the breaking point, those UL testers have become experts in spotting potential safety hazards in hundreds of thousands of devices and materials that have passed through their hands en route to market. Consequently, many American consumers have come to look for that UL tag when they shop, even if they're not exactly sure what it means or where it came from.

An Influential Imprimatur

The UL tag is not ubiquitous. It just seems that way. The mark now appears on more than 6 billion new products each year, representing some 12,500 different types of goods. Last year alone, UL's engineering division conducted more than 81,000 investigations for its 40,000 clients to uncover any potential safety hazards. According to UL, those products ranged from "... toasters to computers; from wood paneling to aerial ladders; from air conditioners to smoke detectors; and from teddy bears to magnetic resonance imagers."

Occasionally, however, the products UL is asked to evaluate jump right off the bizarre scale—the electrically heated canary perch, or the 120-volt worm harvester.

"Those were probably the best of the worst," laughs Bill Schallhammer, a UL senior managing engineer. "There are products like that which we will not investigate, or ones that would cause injury when used in their intended manner—a hand grenade, for example, or a gadget that would shock someone trying to steal your car."

Products do not have to carry the UL imprimatur, but the influence of the mark, the reputation it has achieved over nearly a century of testing, is enough to make a manufacturer think twice about going to market without it.

Burgeoning Testing Industry

While UL may be the best-known lab in town, it certainly is not the only one. Pretested products are no longer niceties but necessities in our technology-driven, quality-conscious, litigious-minded economy. Thus, testing labs have grown in number and sophistication to keep pace with a skeptical society no longer willing or able to accept its goods at face value.

Joseph O'Neil is the executive director of the American Council of Independent Laboratories (ACIL), a trade association of engineering and scientific firms that provide testing, analyt-

A simulated component failure can help determine a product's potential to catch fire under adverse conditions.

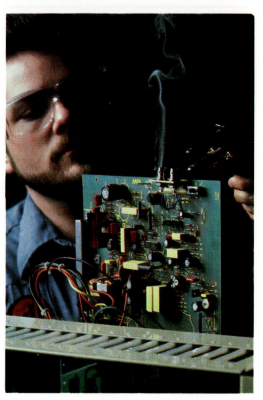

Specially designed equipment determines the temperature and pressure at which a gas or vapor is likely to explode.

ical, and research and development service to industry and government. More than 300 companies representing some 1,000 testing facilities make up the council.

O'Neil says the industry as a whole has expanded significantly over the past decade, while the segment devoted to testing on environmental issues " . . . has grown dramatically from practically nothing.

"The other driving force that has increased the demand for testing is a much greater focus on quality—quality of products in particular, but quality of life overall," O'Neil adds.

Nearly every conceivable product or material undergoes rigorous testing before it is released to market today. And to handle the massive demands for safety and quality assurance, an enormous range of scientific and engineering disciplines find their own niches in the testing labs. They include: analytical chemistry; agronomy; environmental sciences; food and drug technology; fuels and energy evaluation; research and development; chemical, metallurgical, and nondestructive testing; electrical, nuclear, petroleum, geotechnical, and materials engineering.

However, for all their expertise, the labs that are members of ACIL do not set standards for the products they test, O'Neil says. Developing standards is one activity; testing to those standards is another.

At this point the ACIL and UL part company. Almost from its founding, UL has been a leading developer of product-safety standards in the United States. The fact that UL has published more than 500 safety standards (80 percent of which are recognized by the American National Standards Institute) explains why millions of American consumers have grasped at that UL sticker as a lifeboat in a sea of uncertainty.

Silent Chaperone of Safety

Underwriters Laboratories both models and mirrors America's progress through the Age of Technology. It began with the great Chicago Columbian Exposition in 1893, when a new force—electricity—emblazoned the exposition's midway and illuminated the night sky in the first awesome display of this power that changed the world.

Equally astonishing was the way it repeatedly set fire to the spectacular, but highly flammable, Palace of Electricity. The Chicago Board of Fire Underwriters, worried that electricity would consume itself along with all the exposition visitors, invited William Henry Merrill, a crack investigator from Boston, to come to the fair and point out hazards.

Merrill found the palace was crisscrossed by new and untried electrical hookups and jerry-built connections that did indeed light the vast structure, but also made it a potential time bomb.

Merrill sensed that if the world knew little about electricity's incredible new power, it knew even less about the unknown safety hazards hidden within this Promethean gift. Merrill remained in Chicago, founded Underwriters' Electrical Bureau, and began offering his services as a fire-prevention expert focusing on the burgeoning electrical industry.

By 1901 Merrill's tiny company was caught up in the sweeping business-industrial surge of the 20th century. The company was renamed Underwriters Laboratories that year, and expanded its expertise to areas other than those purely electrical.

During succeeding decades, UL kept pace with the industrial and technological growth of America. Both world wars, for example, involved the lab in testing a host of war-related materials and devices. In the Roaring Twenties, when America was discovering flight, UL was the first to begin a program of pilot certification. When the federal government later took over that function, it had the UL standards as a convenient point of departure.

All these many decades, UL has been a silent chaperone in America's love affair with the automobile. We might not have grown so fond of our cars had it not been for UL's role in the development of safety glass, headlights, and bumpers.

The leap into space-age technology in the 1960s presented UL engineers with an unimaginable array of new challenges. As noted in a company report, " . . . spectacular changes in everything from leisure goods to atomic technology were changing the face of much UL testing." Some products were downright weird.

" 'Tracer' golf balls that trailed a red smoke screen were one bizarre example of a new testing problem. Prosaic things like manhole heaters were also tested. And exotic items like radioisotope containers and atomic-energy air filters."

Gearing Up for the '90s

But all that was mere child's play compared to the technological complexities and innovations that have characterized the 1980s and that have yet to transform the '90s into an era beyond conjecture. From investigating the fire hazard of a crude piece of wire at the Columbian Exposition, for example, UL tests have advanced to determining the flammability of adhesives used to hold miniature electronic chips inside consumer products and appliances.

Today UL's safety investigations are concentrated in six different departments. The largest is the Electrical department, which evaluates hundreds of different electrical and electronic products used in homes, businesses, schools, and factories. The primary concern of the Fire Protection department is classifying fire-

No safe can claim to be completely burglar-proof until it stands up to a UL technician's blowtorch.

UL's Marine department tests life jackets against its own standards as well as those of the U.S. Coast Guard.

resistant materials and testing fire extinguishers, upholstered furniture, modular homes, solid-fuel-burning appliances, and fire-fighting equipment.

The Heating, Air Conditioning and Refrigeration department performs tests on a variety of heating and cooling appliances from oil-burning furnaces to solar-powered pool heaters. Investigators in the Casualty and Chemical Hazards division look for safety defects in such products as cleaning fluids, flame-resistant fabrics, step-ladders, floor coverings, and automobile accessories; while the Marine department evaluates recreational boats, emergency-rescue equipment, lifesaving equipment, and marine sanitation devices.

Schallhammer says that while the form of UL's mission has remained the same over the years, the content of its investigations changes constantly with the introduction of new materials, new technologies, and, consequently, new types of hazards.

One example has been the evolution of television sets from tube to solid-state. Thirty years ago UL's safety tests focused mainly on input, temperature, and maybe a few fault tests. Today, however, with the circuitry concentrated on a single board, the test program has expanded to the point where it must also evaluate plastics and other new materials never seen before to determine how combustible they might be under high voltage.

"Today we have a similar situation involving lasers and compact discs," says Schallhammer. "We have to start worrying about the possibility of eye damage from the lasers that operate the compact-disc players, so we're constantly reorienting our approaches to hazards that might result from new products."

On the other hand, Schallhammer cautions about overestimating the meaning of the UL mark. Too many consumers believe that the mark is UL's endorsement of a particular product over a competitor's product. Not true.

"We don't test to see if one toaster does a better job of toasting than another," he says. "We just make sure it meets the proper level of safety.

"Some consumers believe the UL mark means that the product is absolutely safe, but that's not true, either. It means that the product has met certain safety standards. However, if consumers don't use the product in the way the manufacturer intended, then they could still be exposed to hazards," Schallhammer concludes.

PHYSICAL SCIENCES

THE NEW SCIENCE OF SOUND

by Gino Del Guercio

In the year 1898, a young Harvard University physics professor named Wallace Sabine used an acoustical formula he had recently discovered to design a new concert hall—the most difficult of all acoustical spaces. The result was Boston's Symphony Hall, considered by many musicians to be the best concert hall of its size in the world.

Today Sabine is known as the father of architectural acoustics. Yet for 60 years after he discovered his formula, which predicts how long a loud sound will reverberate inside a room, the designing of concert halls remained more art than science. During that time, several acoustical lemons were built, which ruined careers and left an industry in tatters.

Now, it seems, architectural acoustics is finally yielding to science. The acoustics of simple concert halls can be predicted with confidence, and acoustical engineers struggling with experimental designs can use an arsenal of techniques to predict how a new hall will sound. This knowledge is improving the acoustical environment in more humble settings as well, such as movie theaters, restaurants, home stereo rooms, and condominiums.

Acoustics for Unamplified Music

Concert halls are one of the last bastions of live, unamplified music. Harry Ellis Dickson, a violinist with the Boston Symphony Orchestra for 49 years, also does a great deal of conducting, including the orchestra's youth concerts. He tells his young audiences that "real" music is practically extinct.

"People listen to recordings, to stereo, to radio," he says. "And if you go to a rock concert, you don't hear anything about the original sound. It's all amplified."

For 40 years, Leo L. Beranek, one of the world's leading acoustical engineers, has traveled to concert halls from Argentina to Zurich to discern the characteristics that separate extraordinary halls from ordinary ones. He has listened to, and studied the plans of, more than 60 halls. Recently I visited the Boston office of Mr. Beranek, a dapper man with white hair.

Architects have developed a number of new techniques to accurately predict the acoustics of a proposed concert hall no matter how radical its design. For the Evangeline Atwood Concert Hall (left) in Anchorage, Alaska, architects first constructed a scale model (inset) to test its revolutionary "electronic architecture"—a system that alters the hall's acoustics to best suit the needs of a symphonic concert, an opera, or a Broadway musical.

Facing page, top: © Christopher Little/Courtesy Hardy Holzman Pfeiffer Associates; inset: Hardy Holzman Pfeiffer Associates

To judge a concert hall, Beranek says, one must listen to a full orchestra playing something from Beethoven, Brahms, Mahler, or something else from the golden age of the symphony orchestra. Beranek buys two tickets. He starts by taking a seat in the middle of the main floor, then at intermission he moves to another spot, usually in the balcony and to the side. He rarely sits where the music sounds best, in the first or second balcony straight down the center. Those seats are not particularly useful for evaluating halls.

Beranek's first criterion for a great hall is that it feel intimate, that he feel "connected with the orchestra." The music must also sound alive; the hall must be filled with music that reverberates throughout the enclosure. Balance is important: the basses, cellos, and violas should be as loud as the violins. And there are two things he should not hear: noises coming from outside, and echoes coming from inside. (An echo is a sound reflection that comes long enough after the original sound to be perceived as a separate sound and is loud enough to be distracting.)

One characteristic, which Beranek cannot easily judge from a seat in the audience, Dickson the violinist knows well. It is called ensemble. The musicians must be able to hear one another playing. "If we can hear each other and we feel a rapport with each instrument," says Dickson, "then we think it's going out well and it sounds well with the audience."

Optimizing Reverberations

In much the same way that Beranek studies concert halls, Sabine studied lecture halls decades earlier. Three years before he helped design Boston's Symphony Hall, he surveyed a group of lecture halls at Harvard and found that good acoustics depend on a hall's reverberation time. This is the length of time a loud noise will bounce around in a room (without becoming a separate echo) before it dies away.

To build better halls, Sabine needed a way to predict reverberation time, so he and two assistants carried pillows (he had observed that furnishings such as pillows shorten reverberation time) into different halls on campus. He sounded a loud organ pipe and timed how long it took the reverberation to die away. For this purpose he needed the quiet of night. So night after night for three years, pillows went back and forth, until his activities became something of a campus joke. Eventually he arrived at a simple

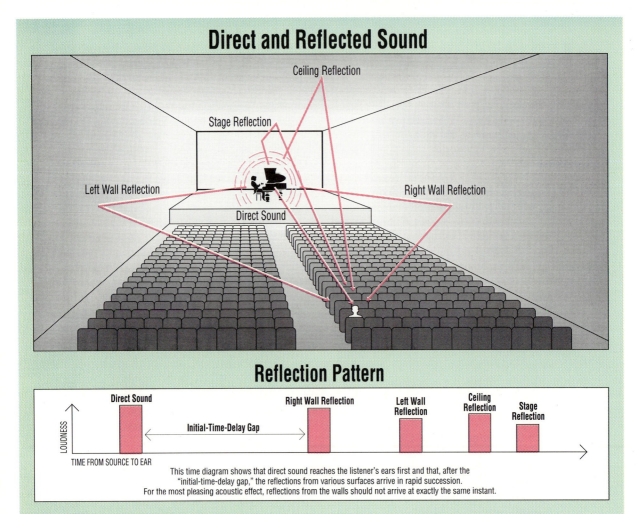

formula that states: The larger the room, the longer the reverberation time; and the more people or furniture in the room, the shorter the reverberation time.

Short reverberation times were found to enhance clarity of sound; longer times enhanced "liveness."

A room's ideal reverberation time depends on its purpose. Lecture halls should have a reverberation time of about 1.4 seconds, so the audience can hear the speaker clearly. Opera houses should have a reverberation time of about 1.6 seconds, so the audience can hear the singer clearly, but the music sounds alive. Concert halls should have a reverberation time of about 2 seconds, so the music sounds alive. Halls with reverberation times much longer than 2 seconds begin to sound like caves. Composers such as Beethoven wrote music with the acoustics of specific concert halls in mind.

For Symphony Hall, Sabine used his formula to select among proven designs. He recommended against enlarging the design of the Neues Gewandhaus in Leipzig, Germany, because it would have had too long a reverberation time, and instead suggested duplicating Boston's Old Music Hall. Symphony Hall's reverberation time is 1.8 seconds, shorter than the 2 seconds he had predicted. It might be even better acoustically if Sabine had gotten the reverberation time right. In fact, Sabine's formula was usually wrong by a small, but significant, amount.

Why? It wasn't until the 1950s that Leo Beranek discovered the answer. A concert hall's main source of absorption of sound is its audience. Sabine was figuring the sound absorption of the audience by counting the number of people. Beranek found that it wasn't the *number* of people that matters, but the *area* they cover.

306 PHYSICAL SCIENCES

Avery Fisher Fiasco

By the late 1950s, Beranek was so eminent that he was asked to consult on a hall to be built for the New York Philharmonic Orchestra. He proposed a rectangular design with proportions similar to those of Symphony Hall.

"That hall, we think, would have had quite beautiful acoustics," says Beranek. "It would have been at least as good as Symphony Hall—in fact, better."

When the proposal was published in *The New York Times,* many people complained there were not enough seats.

"They crabbed so much," says Beranek, now in his mid-70s, "that the building committee said to the architect, 'We're not going to go along with what we agreed to; we're going to do more seats.'"

With little time left, the architect patched the design by spreading the walls and creating steep, plunging balconies instead of flatter ones. Cost overruns forced him to put in smooth walls, leaving out the bumps and niches that would have softened the sound as it bounced off them.

When Philharmonic Hall was finished in 1962, the orchestra sounded dry and lifeless. The bass was weak, and the musicians could not hear one another. Harold Schonberg, music critic for *The New York Times,* likened the hall to a stereo with the treble turned way up and the bass way down.

Dickson remembers the first time he played there with the Boston Symphony Orchestra: "It sounded as if we were playing in an open field," he says. "There was just no resonance. It was a catastrophe."

Philharmonic Hall, renamed Avery Fisher Hall in 1973, was eventually gutted. A new auditorium similar to Beranek's original design was built inside, but most acoustical engineers believe Avery Fisher Hall is still not a great concert hall.

"We got a terribly bad name out of that hall," says Beranek, who never designed another. And the field of acoustical engineering, which had been dominated by his consulting company, Bolt Beranek & Newman, splintered into many small acoustical-design companies scattered around the world.

Source-Area Acoustics

Ironically, at about the time of the Philharmonic Hall disaster, several major discoveries were bringing acoustical design into the modern age. Again Beranek was a pioneer—with his discovery of the importance of the gap between when the listener first hears the direct sound and when he hears the first reflection.

"The initial-time-delay gap turns out to be one of the most powerful measures of the acoustical quality of halls for music," Beranek wrote in his book *Music, Acoustics & Architecture,* which was published the same year Philharmonic Hall was completed. According to his survey, the best halls had the shortest initial-time-delay gaps, usually between 8 thousandths and 21 thousandths of a second. He concluded that a short time-delay gap is critical to a hall's feeling of intimacy, and thus far more important than reverberation time, which determines the hall's liveness.

The area in which the orchestra plays, called the source area by acousticians, is also important. In some halls, such as Symphony Hall, the orchestra is behind a proscenium in its own smaller three-walled space, called the stagehouse. However, other great halls, such as the Grosser Musikvereinsaal in Vienna, have no proscenium, with the orchestra right in the front of one large room. And modern halls, such as the Berlin Philharmonie, have the orchestra out in the audience so there are people sitting all around.

"The information from the source has got to be good," says Christopher Jaffe, founder of the Connecticut consulting firm Jaffe Acoustics. "It's the concept that if you have the best hi-fi set in the world and have a scratchy record, you've got a bad sound. So source-area acoustics are critical. They provide balance and transparency."

Classical acousticians such as Cyril Harris, who redesigned New York's Avery Fisher Hall, and Beranek like to stick with the shoe-box design, which they understand and know will work. However, much younger acoustical engineers see the merits of more modern and revolutionary shapes and designs.

"The problem with the shoe box isn't so much acoustical," says Jaffe. "It's that you're sitting at the end of a bowling alley, and there's no real contact with an audience."

Scale Models

Modern architects, in their desire to build halls that are grand and unique, like radical designs, but no one knows how they will sound. So acoustical engineers have taken to building scale models. These are giant dollhouses, 15 to 20

SYMPHONIC VARIATIONS

The Grosser Musikvereinsaal in Vienna (left), generally regarded as the best concert hall in the world, typifies the so-called "shoe-box" design, in which the orchestra sits at one end of a long rectangular room. Symphony Hall in Boston has a similar design. The Berlin Philharmonie (above) surrounds the orchestra with the audience.

THE GREATEST CONCERT HALLS

First a word of caution. Acoustical engineers believe that, although concert halls definitely vary in quality, it is in much the same way that works of art do.

When asked what he considers the 10 best concert halls, Christopher Jaffe, founder of Jaffe Acoustics, replies with the rhetorical question: "What are the 10 best sculptures in the world? I don't think you can really do that. It's a matter of taste. It's a matter of sound. It's a dynamic thing."

That said, I asked him and others to list their *personal* favorites.

The Grosser Musikvereinsaal in Vienna, Austria, is generally considered to be the best concert hall in the world. It is smaller than Boston's Symphony Hall, which leaves room for the latter as the best in the next category up in size, although this is not so universally accepted.

Acoustical engineer Leo Beranek loves the Concertgebouw in Amsterdam, the Berlin Philharmonie, and the Stadt-Casino in Basel, Switzerland. For a new concert hall, he likes Segistrom Hall in Orange County, California.

After Symphony Hall, Boston musician Harry Ellis Dickson likes Carnegie Hall in New York City, although it was recently changed, and the sound quality, many say, isn't as good as it originally was. The Grosser Musikvereinsaal "of course is marvelous," he says. And he likes the Berlin Philharmonie, although he says that when he first played there, it seemed strange to be surrounded by the audience.

Christopher Jaffe puts the Concertgebouw in Amsterdam at the top of his list, although he prefers halls with more transparency, which are usually the surround halls, such as the Berlin Philharmonie and a new concert hall in Mexico City. Of the shoe-box concert halls, Jaffe agrees with the others that Vienna's Musikvereinsaal can't be beat.

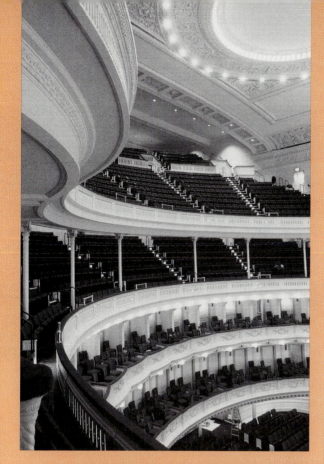

Japan's Suntory Hall (below) places musicians slightly off-center. The hall's reverberation time was found to be 2.2 seconds, slightly longer than the ideal of 2.0 seconds. New York's Philharmonic Hall (above) received much harsher reviews when it opened in 1962. Subsequent acoustical improvements include a concrete floor below the stage to block out subway noise.

feet (4.5 to 6 meters) long. Miniature people are placed in miniature upholstered seats, and music is piped in through two tiny speakers placed on the stage. Because the model is one-tenth the size of the proposed hall, the music has to be 10 times higher in pitch in order for the sound waves to bounce around properly. So the music is played at high speed, like a 33-rpm record played at 45 rpm, only more so.

Air absorbs high frequencies more readily than it does low frequencies, so, to prevent distortion of results in the model, the air is replaced with a gas like nitrogen that absorbs all frequencies at the same rate. The engineers must wear gas masks to work with the model. Finally, two little microphones are stuck like ears in a tiny Styrofoam head, and this miniature listener is placed in the audience. The high-pitched sound from the two microphones is recorded and played back at a slower speed, so designers can hear what the hall would sound like at full size. The Japanese, who are building many Western-style halls with their newfound wealth, love this concept, and copy the halls down to the finest detail.

There are two major problems with this use of models, according to Beranek.

"You cannot get the feeling of a distributed orchestra," he says, because the sound is coming from just two speakers. This makes predicting balance impossible. In addition, the models cost $100,000 and take months to build, so changing designs is expensive and time-consuming.

Beranek recently traveled to Japan to listen to some of the halls. He found them to be good, but not among the world's best. One was Suntory Hall, a large, somewhat round structure with a high ceiling and a small number of seats behind the orchestra. Beranek found the reverberation time to be too long: 2.2 seconds.

"Japan has a reputation for having halls that have too short a reverberation time, so they said we aren't going to have any more of those," he explains.

"I think they overshot it a little."

Everyday Acoustics

One day this spring, I joined a couple of acoustical engineers during their lunch break, and we searched for a quiet restaurant. The first, a little Italian place with the best food of the lot, was so noisy we could hardly hear each other. The floors and much of the walls were covered with ceramic tile, which resulted in a long reverberation time and sharp reflections. The acousticians, who worked just above the restaurant and ate there often, had been trying to convince the owners to take a little acoustical advice, perhaps in exchange for a few meals. But the restaurant was doing well, and the owners didn't want to change a good thing.

The restaurant we finally selected was acoustically designed by the engineers. The job was fairly simple: some thick carpet on the floor, and acoustic tiles on the ceiling. The walls were covered with paintings and knickknacks that gave a varied surface from which sound could reflect. The food wasn't as good, but the sound quality made for a pleasant meal and a fruitful interview.

These small projects are the bread and butter of most acoustical engineers. And problems can sometimes lead to solutions that can later be applied to major projects.

"Designing concert halls doesn't allow you to learn much new," says Beranek. "The best learning bases for these kinds of problems are apartment houses and public buildings, and then you just use that knowledge in the concert halls, because there is no money available to experiment."

Electronic Architecture

Of all the new concert halls, probably the most futuristic is in Anchorage, Alaska. The Evangeline Atwood Concert Hall, which was completed last year, is based on a traditional European opera-house design, with multiple balconies sweeping around the side walls to the stage, and a magnificent ceiling that looks vaguely like an aurora borealis. The Atwood Hall has a short reverberation time of only 1.6 seconds, which is designed to sound good for light opera and amplified Broadway.

With the flip of a switch, however, the reverberation time suddenly jumps to 2 seconds for symphonic music. The trick is accomplished with a system called "electronic architecture." A tiny microphone hangs down over the orchestra. This picks up the music, makes up to 48 multiple copies, and pumps them out through speakers in the walls and ceiling. These provide the additional reflections and increase the reverberation time needed for symphonic music.

"Electronic architecture" was first used on a limited basis in the 1970s, but it was not until the development of sophisticated digital electronics that realistic-sounding systems could be created. The modern electronic system never increases the volume of the music or changes its sound. In fact, the audience is never aware sound is coming from the speakers, because of a trick the human ear plays that tells you the source of noise is where it is loudest.

"It's very important that people understand how subtle it is," says one of the new breed of acoustical engineers, Mark Holden, who designed the system used in Atwood Hall. "We're providing additional reflections and additional reverberation, enhancing it by 10 or 15 percent to change the overall perception of the room's quality. We're not tampering with the direct sound that comes out."

A survey of the ballet companies, opera companies, and other users of the hall since its opening gave its acoustics an A+ rating.

"It's just simply a tool. It's something that we've used where there is no other choice," says Holden. "It's a good thing, and not a negative thing."

For all the praise by critics, Holden sounds slightly defensive when he talks about the system, but that might be expected, considering acoustics' turbulent past and the aversion of classical musicians to anything electronic.

Could "electronic architecture" have saved Avery Fisher Hall? Holden says he thinks not. For the system to work well, the hall must be acoustically sound in the first place. The system cannot eliminate sounds, such as unwanted echoes; it can only add additional reflections and reverberation. Thus, it may be able to improve the best halls.

"If you take a very, very good hall, a great hall, with this system, you can add in more reflections than could occur naturally," says Holden. "At some point, you go, 'Whoa! That's bigger than reality.' There's this temptation to say that's so big and beautiful that it's too good."

So far, no one is willing to say the advances in acoustics have resulted in dramatically better halls. However, there are several major advances just on the horizon. For instance, computer programs are now being written that accurately predict the sound of a proposed hall without ever having to build a model.

Some acoustical engineers, like Holden, are confident that in the future, concert halls will sound better than ever. Others, like Beranek, are more cautious. He says simply that the advances will result in more consistently good halls. And that in itself would be no small feat.

The 1989 Nobel Prizes

PHYSICS AND CHEMISTRY

by Glenn Alan Cheney

A revolutionary discovery in molecular biology and two significant advances in physics brought the 1989 Nobel Prizes in Chemistry and Physics to three Americans and two former East Germans.

Thomas R. Cech of the University of Colorado and Sidney Altman of Yale University shared the prize in chemistry. In separate research efforts, the two scientists discovered that RNA plays not only genetic but also catalytic roles in cellular activity. The latter function was previously thought to be performed only by protein-based enzymes.

The prize in physics went to Norman F. Ramsey for his work in developing the atomic clock and the hydrogen maser, and to Hans G. Dehmelt of the University of Washington and Wolfgang Paul of the University of Bonn for their work isolating ions and subatomic particles.

The work by the laureate physicists gave scientists powerful new tools for analyzing the nature of atomic structure.

The Prize in Physics

The 1989 Nobel Prize in Physics went to two parties that have given science a new vision of basic physical laws, including those involving time and space. Dr. Norman F. Ramsey of Harvard University received half of the award money for his work, which led to the invention of the atomic clock and the hydrogen maser.

The other half of the prize money was split between Hans G. Dehmelt of the University of Washington and Wolfgang Paul of the University of Bonn. The two physicists, both born in what is now East Germany, developed ion traps that allow scientists to analyze isolated atoms and subatomic particles.

Dr. Ramsey began his work shortly after World War II. He first developed a method of studying atomic structure by shooting an electronic beam of atoms through two oscillating electromagnetic fields, inducing shifts in energy levels.

The procedure led to the invention of the cesium atomic clock, which defines a second as 9,192,631,770 oscillations of the cesium atom. Such precision allows measurement of continental drift and astronomical observations.

Dr. Ramsey's research also made possible the development of the hydrogen

Norman Ramsey shared the Physics Nobel Prize for work that led to the development of the atomic clock. The clock works by detecting subtle movements of the poles of the cesium nucleus, which oscillate in a circular motion much like a top. The oscillations affect the electron circulating around the nucleus. When a cesium atom is beamed between electromagnetic fields, the electron emits detectable energy at intervals precise enough to be used as a timekeeping medium.

The New York Times

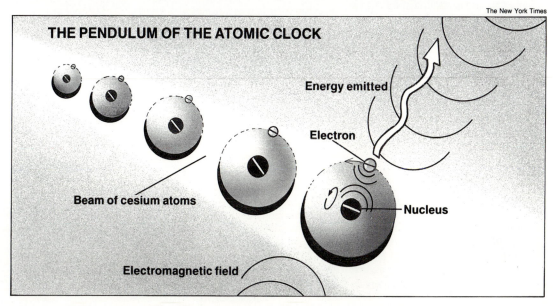

Hans Dehmelt (left) and Wolfgang Paul (above) shared half of the physics prize for developing the ion trap. The other half went to Norman Ramsey (right), whose work led to the atomic clock.

Photos: AP/Wide World Photos

maser. This laserlike instrument focuses excited hydrogen atoms so they emit a specified frequency of microradiation. More stable than the cesium clock for short time spans, the hydrogen maser is the most accurate timepiece available for taking highly precise frequency measurements.

Wolfgang Paul and Hans G. Dehmelt share their half of the Nobel Prize for their work with ion traps. In the 1950s Dr. Paul built an electromagnetic trap—now known as the Paul trap, or radio-frequency trap—that could capture and retain small numbers of ions for periods long enough to allow study. It became a standard instrument of modern spectroscopy.

Dr. Hans Dehmelt of the University of Washington in Seattle is best known for refining radio-frequency traps toward an ideal machine that would suspend a single particle in an environment where it would not interact with anything.

He devised the "Penning trap," which uses a strong magnetic field and a weak electric field to isolate single electrons.

By cooling an isolated electron, Dehmelt was able to analyze the "g-factor," the ratio of magnetic and angular momenta of the electron, to within four parts per trillion. Such accuracy makes it possible to measure important electron properties and observe the quantum jump of a single ion.

Norman F. Ramsey was born in Washington, D.C., on July 27, 1915. He made significant contributions to the development of radar during World War II. After the war he joined the faculty of Harvard. He was later president of the American Physical Society, and in 1989 co-chaired the federal committee that investigated the first reports of cold nuclear fusion.

Wolfgang Paul was 76 years old when he won the Nobel Prize. He was born in Lorenzkirch, which is now part of East Germany. Currently he is a professor at the University of Bonn in West Germany.

Hans G. Dehmelt was born on September 9, 1922, in the city of Görlitz in what is now East Germany. In the early 1950s, he came to the United States to teach at Duke and eventually at the University of Washington. He won the Davisson-Germer Prize of the American Physical Society in 1970, the Humboldt Prize in 1974, and in 1978 was elected to the National Academy of Sciences.

An atomic clock defines a second as 9,192,631,770 oscillations of a cesium atom.

Royal Greenwich Observatory/Science Photo Library/Photo Researchers, Inc.

The Prize in Chemistry

Thomas Cech (pronounced *Check*) and Sidney Altman did not work together on their research into previously unsuspected enzymatic characteristics of ribonucleic acid, or RNA, but their independent projects inspired each other. Nobel committee member Bertril Andersson said that development of their concurrent discoveries went back and forth like a Ping-Pong ball as each learned from the other and went on to make further advances.

Enzymes accelerate the biochemical reactions that make life possible. Until the discovery of the enzymatic properties of RNA, scientists assumed that all enzymes were proteins. Enzymes in the liver, for example, convert alcohol to simpler substances. Enzymes in plants make photosynthesis possible at a rate fast enough to sustain life.

The chain of discoveries began when Cech and his colleagues at the University of Colorado at Boulder were investigating how the RNA transcript of DNA is changed into the messenger RNA that controls the synthesis of protein. To splice together this specialized transcript of RNA, certain sections of the RNA strand that are not involved in protein production must be removed.

It was always supposed by scientists that these sections, known as introns, were selected by specialized enzymes that reacted with them and thus sliced them from the RNA.

To search for the enzyme that controlled this editing process, Dr. Cech and his associates isolated RNA before it produced the cells where the enzymatic reactions take place. They then added enzymes from a cell nucleus, expecting them to remove introns one by one. The last enzyme to react would be the one that controlled the process.

But the experimenters observed, to their surprise, that the process needed no enzymes at all. The basic RNA transcript was capable of acting as its own enzyme, editing out the unneeded introns and splicing itself back together.

Further experiments showed that under certain conditions, the intron segments would twist themselves into complex knots, then break free and splice the remaining RNA together again.

The discovery opened the way for Dr. Altman's experiment with ribonuclease P, a unique enzyme composed of both a protein and a piece of RNA. Before his experiments, it was assumed that the protein and the RNA worked together as a catalyst. But Altman determined that the RNA molecule alone was playing the role of an enzyme. From that determination he was able to establish certain catalytic RNA segments that became known as ribozymes.

The discovery seems to indicate that life on earth probably started as RNA material that was able to function before protein or DNA evolved. A more contemporary and immediately useful result of Cech's and Altman's work is the new field of enzymology and an array of new biotechnologies. Future applications of the technology might apply to every disease from the common cold to terminal cancer.

Thomas R. Cech, born December 18, 1947, in Chicago, Illinois, earned a doctorate at the University of California at Berkeley in 1975, worked briefly at MIT, and then joined the faculty at the University of Colorado.

Sidney Altman was born in Montreal, Quebec, in 1939. Now a citizen of both Canada and the United States, he studied at MIT before pursuing his doctorate at the University of Colorado. He has been with Yale University since 1971, serving as dean of Yale College from 1985 until 1989.

Sidney Altman (left) and Thomas Cech shared the Chemistry Nobel Prize for separate research in which they discovered that RNA (ribonucleic acid) plays not only genetic but also catalytic roles in cellular activity.

Photos: AP/Wide World Photos

TECHNOLOGY

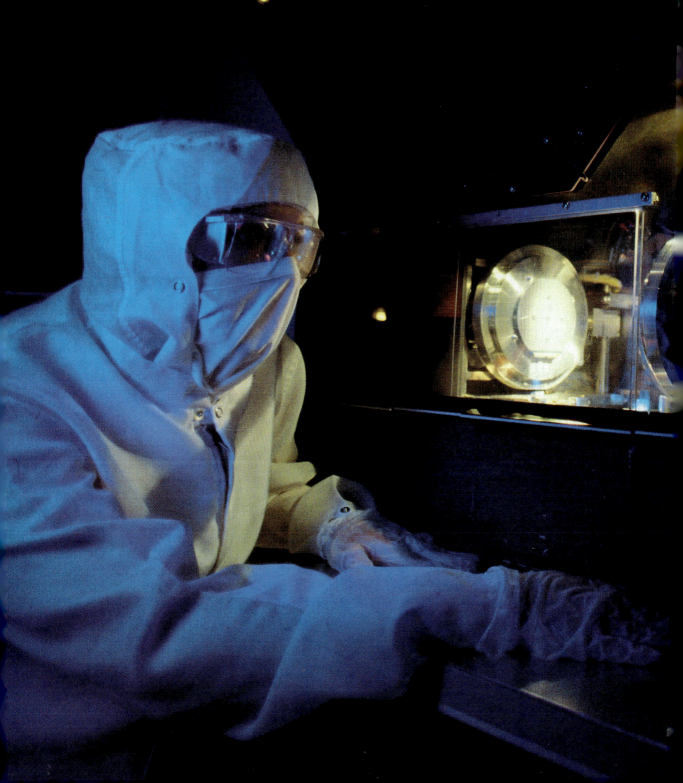

Review of the Year

TECHNOLOGY

New technologies and improved technologies—created and refined in response to societal needs, consumer wishes, and business mandates—continue to reveal the myriad directions of cultural evolution.

by Donald Cunningham

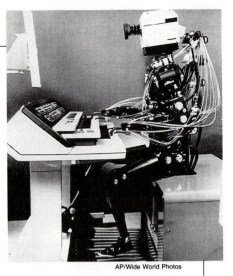

Wabot II, a Japanese musical robot recently exhibited at the Chicago Museum of Science and Industry, can play a keyboard and create original music.

Consumer Technology

Patrons of certain Arby's restaurants can now key in their orders, thereby reducing the amount of "talk time" between customer and counter worker.

Fast food getting faster? Arby's restaurant introduced access terminals for customers to key in their orders, thereby cutting down on the amount of "talk time" between customer and counter worker. The Blodgett Corporation created, for restaurants, a computerized pizza machine that makes and cooks a pizza in three minutes. Pizza Hut installed ovens in student cafeterias, shopping malls, and hospitals, reducing the time between ordering and picking up a pizza to about 30 seconds. For home consumption, hamburgers, hot dogs, fish sandwiches, soups, cakes, and cheese dip that can be prepared in seconds in a microwave oven became available.

A California bicycle manufacturer and a division of Du Pont teamed up to create a revolutionary three-spoked bicycle wheel. The 2.6-pound (1,200-gram) wheel, though slightly heavier than a conventional bicycle wheel, is more aerodynamic. Made of a carbon-fiber composite, the three wide spokes—separated by 120-degree angles—produce less drag than do the 36 spokes of a normal wheel and are less susceptible to the effects of crosswinds.

A couple in Des Moines, Iowa, invented a two-passenger kayak that can be dismantled and reassembled in minutes without tools. The kayak's frame, made of plywood struts and brass connectors, is assembled in two pieces; the pieces are then slid into either end of a waterproof skin made of synthetic rubber sheeting and a blend of canvas and acrylic. The kayak's 26 parts fit into two containers the size of standard luggage.

An inventor in New Mexico devised a transparent two-passenger submarine that can withstand depths down to 100 feet (30 meters). The egg-shaped acrylic hull houses 10 104-ampere batteries used to power six electric motors. In a novel arrangement, ballast tanks attached to the ends of struts are extended at each side and filled with water in order to submerge the craft. To surface, the driver swings the tanks underneath the submarine and empties them of water, so that they push the craft upward.

Safety air bags for drivers became a standard feature in millions of new models of American automobiles. The Big Three American carmakers employed a single technology: a canister of sodium azide and an inflatable sac installed in the center of the steering wheel. In a frontal collision, sensors initiate a reaction of the azide, producing nitrogen gas, which inflates the bag in less than one-twentieth of a second. The automobile must collide at a speed of at least 14 miles (22.5 kilometers) per hour to initiate the reaction. The air-bag system is effective in preventing serious head injuries in front-end collisions.

An Iowa couple invented a two-passenger kayak that in minutes disassembles into 26 parts that fit into two containers the size of standard luggage.

Materials

A Michigan firm created a degradable plastic lawn bag based on a technology developed by a Canadian company. The plastic contains cornstarch and vegetable oil. When the bag is exposed to the environment, bacteria feed on the cornstarch, thereby exposing material to the oxidizing effect of the oil. After reacting with salts in the soil, the oil chemically attacks the polymer chains in the plastic, breaking it into pieces small enough to be eaten by microorganisms.

The unveiling of the *Starship* composite craft was perhaps the most dramatic display of new uses of composite materials (see page 318) but other advances were equally important. The new space shuttle *Endeavor*, for instance, will have over 100 parts made from a composite composed of carbon fibers fused in a carbon matrix. The composite retains its strength even at the extremely high temperatures the shuttle will likely encounter.

The serial process needed to produce composites—weaving, curing, soaking, reheating—has resisted efforts at automation. In making the B-2 "Stealth" bomber, however, Northrop Corporation reportedly achieved an automated process. Workers at North Carolina State University invented a machine for braiding composite fibers that can handle 9,000 fiber ends simultaneously. The Textron Company developed a process that produces continuous fibers of silicon carbide. They sandwiched sheets of the fibers between titanium foil and heated them under pressure. The resulting composite is stronger at 1,000° F (537° C) than is unreinforced titanium at room temperature.

Defense Technology

The B-2 Stealth bomber made a successful maiden flight, following a decade of development. The tailless, boomerang-shaped jet rose a modest 10,000 feet (3,050 meters) into the air above the California desert. The plane went through a number of turns at three speeds—all much lower than the speeds at which it is designed to fly ultimately. To help ground crews track the radar-evading plane during the flight, a special signal-emitting device was placed on board. Meanwhile, the military's current Stealth bomber, which the B-2 is meant to eventually supersede, went into action during the invasion of Panama.

The Pentagon conducted the first test of a particle-beam device in space, one of many technologies that make up the Strategic Defense Initiative (SDI, or "Star Wars"). The device, lofted 125 miles (200 kilometers) by a rocket, fired beams of neutral hydrogen atoms with energies of 1 million electron volts (MeV) in a variety of directions, while television cameras recorded their trajectories. In practice, the beam might be used to discriminate between enemy warheads and decoys. A beam about 100 times stronger could destroy enemy missiles.

In another SDI advance, scientists conducted a successful test of a powerful chemical laser called Alpha. The compact laser, which uses hydrogen and fluorine gases, produced a burst of laser light for one-fifth of a second at an undisclosed power that was something less than its intended goal of 2.2 million watts. Alpha is scheduled to be brought into space in 1994 in order to test pointing and tracking capabilities.

Added to the array of SDI technologies under development was a new idea called "Brilliant Pebbles." Conceived of by Lowell Wood of Lawrence Livermore Laboratory, "Brilliant Pebbles" would comprise a multitude of small, brainy rockets that would home in on enemy missiles and destroy them on impact. Because each rocket would contain a small supercomputer and a system for optical sensing, it would not require the aid of bulky and costly satellites or even ground controllers for guidance.

The Navy met with less success in its initial attempts to launch the Trident 2 missile from a submarine. The Trident 2, which is designed to carry nuclear warheads, had been successfully tested on land. After two of three submarine launchings resulted in failure, the Navy stated that designers may have misjudged the effects of pressure on the ballistic missile as it moves through the ocean before lifting into the air. The Trident 2 is much heavier and longer than its predecessor, the Trident 1. Spokesmen cited the nozzles on the missile's first-stage rocket as being too weak to withstand the turbulence created as the bigger missile moved through the water.

Both photos: Lockheed Corporation

The YF-22A advanced tactical fighter can engage hostile targets head-on using rapidly updated gunsight information. The cockpit's color screens provide flight and target data.

Communication Technology

The evolution of the telephone accelerated in 1989, thanks to further miniaturization and simplification of electronic components. Major manufacturers such as AT&T and Panasonic developed smaller, more mobile phones that produce better sound quality. Motorola unveiled a palm-sized cordless cellular telephone that weighs only 12.3 ounces (348 grams) and operates for up to 75 minutes with a standard battery.

American Telephone and Electronics Corporation of San Francisco, California, and the Hattori Seiko Company of Japan announced the joint development of a wristwatch paging system. The new device displays short messages on the face of a digital wristwatch. Its rapid FM-radio network can transmit the messages anywhere in the world.

The demand for cellular-telephone service reached such a high pitch in 1989 that urban systems began gearing up for a switch from analog to digital telephone technology. The digital technology will have three times the capacity of current technology. The switch will be gradual, since the equipment now in use requires analog technology.

Advances in speech-recognition systems brought researchers closer to creating translating telephones. Workers at Carnegie-Mellon University developed a system that recognizes continuous spoken English based on a vocabulary of 997 words. Meanwhile, a Washington firm unveiled a hand-held, computerized translator. The device recognizes an individual spoken word, translates it into a foreign language, and speaks the word in that language.

Facsimile transmission, or fax, continued its strong growth in 1989, including even the use of fax machines with cellular car phones. Fax amenities now include superfine resolution (up to 200 horizontal lines and 400 dots per inch along the vertical), halftone images, and devices to automatically switch from a voice mode to a data mode—for systems in which a telephone is used for transmission of voice and facsimile codes.

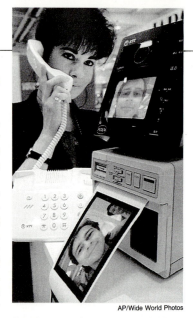

AP/Wide World Photos

Reach out and see someone? Visual telephones are quickly becoming a reality. Within ten seconds, the device above delivers a picture of the conversation partner over telephone lines.

Aircraft

The first commercial airplane made mostly of carbon fibers and other composite materials was sold in 1989. The 10-passenger *Starship* has better fuel efficiency, better resistance to structural fatigue, and more passenger headroom than small planes made of aluminum. The tough, lightweight, less brittle, manmade materials made the improvements possible. The *Starship* also features new design elements destined to find a place in larger aircraft of the future. These include small wings (or canards) on the nose of the plane, which stabilize the craft during takeoffs and landings; small vertical wings (or sails) at the ends of the conventional wings, which replace the tail rudder; and paired turboprop engines placed at the rear-center of the plane in order to reduce noise in the cabin and ensure a smoother flight should one engine fail.

Airlines made strides in safety thanks to technologies on the ground and in the air. A microwave landing system—20 years in the making—was installed at a small airport in New Hampshire, the first of what experts hope will eventually be worldwide implementation. The microwave system can spread out the pathways of incoming and outgoing airplanes by allowing for a wider variety of approaches; it should therefore reduce congestion in the air around airports, and consequently enhance safety. Airports with short runways, rough terrain, and poor weather will no doubt welcome the new system.

A retired astrophysics professor developed a computer program for tracking aircraft using satellites. The program combines a computerized map with information received from the continuous satellite monitoring of aircraft to pinpoint the geographic locations of the aircraft. Pilots are then automatically told the locations of nearby planes.

The ten-passenger Starship *is made almost entirely from synthetic materials.*

Beech Aircraft Corporation

Trains

The fastest commercial high-speed train began service between Paris and cities on the Atlantic coast of France in September 1989. Using mainly existing technologies but more powerful engines, the French *train à grande vitesse* (TGV) can travel over 200 miles (320 kilometers) per hour—much faster than the 130-mile (210-kilometer)-per-hour rate of Japan's bullet train. In Texas a group of investors began the process of creating a high-speed train system for their state.

Meanwhile, the first magnetically levitated train (mag-lev) to be operated in the United States is under construction in Las Vegas, Nevada. The German-designed "people mover" will initially travel a 1.1-mile (1.8-kilometer) route at a modest speed of about 25 miles (40 kilometers) per hour. Authorities plan to increase the range of the system to service points around the city, such as the airport, and will eventually test the train at speeds closer to 50 miles (80 kilometers) per hour. Japan and West Germany continue their development of high-speed mag-levs along separate technological avenues: the Japanese system uses superconducting magnets and is based on the repulsion of oppositely charged metal rails; the German system makes use of magnetic attraction.

Conventional trains aren't about to retire anytime soon. Production and worldwide sales of locomotives by General Electric (GE) and General Motors (GM) have been growing steadily. New innovations include computerized controls, refined engine-braking systems, and more powerful engines (4,000 horsepower). Under development are main engines that burn coal slurry rather than diesel fuel, and wheel-drive systems that use alternating electric current instead of direct current.

Mirroring the airline industry, American and Canadian railroad companies began to develop electronic, computerized dispatching systems that operate using information received from transponders placed on trains. Such systems will be more precise than the conventional system of electrical signals sent to dispatchers when trains periodically pass certain points along their routes.

Below: NYT Pictures; bottom: Joe Traver/NYT Pictures

Train technology has made significant advances in recent years. Two-hundred-ton locomotives (right) with computerized controls and refined braking systems have entered the final assembly stage. Computerized tracking and dispatching systems (above) should improve service and scheduling.

NEW LIGHT on the OLD MASTERS

by Jeffrey H. Hacker

If, as Michelangelo wrote, art is "the shadow of divine perfection," no work in the history of Western culture more nearly reflects the state of grace than his own frescoes in the Vatican's Sistine Chapel. In just over four years (1508–12), Michelangelo covered the entire ceiling and upper walls with majestic depictions of the Book of Genesis, Old Testament prophets, and ancestors of Christ. When his work was done, according to a contemporary, "It was such as to make everyone speechless with astonishment."

Nearly five centuries later, in the early 1980s, visitors to the Sistine Chapel expressed similar astonishment upon viewing the first newly restored sections of Michelangelo's masterpiece. Dirt and grime removed, some of the

Art scholars, historians, and scientists are applying the latest technology in an effort to conserve and restore some of the world's greatest art treasures. In the process, they have shed new light on techniques of Renaissance artists, uncovered previous restoration attempts, and even developed foolproof ways to detect art forgery.

Above: © Gianni Tortoli/Photo Researchers, Inc.; facing page, far right: © Wally McNamee/Woodfin Camp; near right: © Chuck O'Rear/Woodfin Camp

most cherished images in Western art had undergone a radical transformation. Known to generations for their deep hues and dark, timeless shadows, the frescoes now emerged as works of dazzling color. In addition to altering aesthetic perceptions of the frescoes themselves, lifting the veil of 500 years yielded important insights into the artistic techniques of a genius.

The ten-year, $4.2 million cleanup of the Sistine Chapel—completed early in 1990—was the most ambitious, most courageous, and indeed, the most controversial of many recent projects to restore and protect classic works of art. In an unlikely marriage of old-fashioned, painstaking handicraft and state-of-the-art analytic technology, art historians and scientists have joined forces in efforts to conserve such other treasures as the Ice Age wall paintings in the Lascaux Caves in France, Tintoretto's *Paradiso,* and Leonardo da Vinci's *Last Supper*. Digital image processing and other analytic aids—now including X-ray fluorescence, scanning electron microscopy, gas chromatography/mass spectrometry, and a host of others—are being used to analyze pigments and peek beneath the surface of paintings at the artist's underdrawings. Beyond helping restore great works and revealing artistic techniques, the new technology is also helping solve another age-old problem: forgery.

Michelangelo Meets the Computer

That the restoration of the Sistine Chapel took more than twice as long as the painting itself underscores the meticulous care and exacting procedures that the team of restorers brought to their task. The small team, headed by the chief of the Vatican's painting restoration laboratory, Gianluigi Colalucci, began with an exhaustive survey to determine the condition of the frescoes, the techniques used in the original renderings, and the composition of the later deposits. The analysis—which continued section by section throughout the restoration—was conducted with the help of a specially programmed computer-graphics workstation that provided the team with a detailed topographic map at the touch of a button.

Before the restoration even began, said Colalucci, researchers spent a year and a half in the Vatican's Secret Archives digging out every detail about previous cleanup efforts. Meanwhile, a Rome consulting firm began a photogrammetric survey of the chapel's barrel-shaped vault. The three-dimensional coordinates of every point on the ceiling were charted in the computer and passed through a stereo digitizer. Installed 65 feet (20 meters) above the chapel floor on scaffolding similar to that used by Michelangelo, the computer (with 70 million megabytes of disk memory) showed every

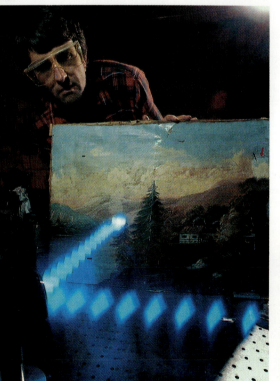

Art-restoration teams (below) use advanced microscopy to study the pigments used by the original artist. Such information helps determine which restoration method best lends itself to a particular work of art. Lasers (left) can help remove layers of surface grime from many paintings.

The restoration of the Vatican's Sistine Chapel produced one of the most radical transformations in the history of Western art. After years of painstaking labor (left), the familiar deep hues and dark, brooding shadows of Michelangelo's frescoes (above) emerged as works of surprisingly brilliant color (facing page).

Top: Photo courtesy of NTV Tokyo-Musei Vaticani/Scala/Art Resource; above and facing page: Art Resource

bump, sag, crack, and hollow in the surface of the plaster.

The library of information stored and displayed on the computer also derived from a series of highly sophisticated tests and procedures to get beneath the surface of the frescoes. These techniques—including ultraviolet fluorescence, sodium monochromatic light, infrared spectrometry, liquid chromatography, and atomic absorption spectrophotometry—were conducted on each section of the work before cleaning. Infrared light, for example, could penetrate overpainting and varnish to show original fine details. Ultraviolet light proved useful in identifying touch-ups and other conservation efforts of previous centuries. To analyze foreign matter on the surface, tiny pigment samples were suspended in polyester resin and allowed to harden; cross sections were then cut and magnified, clearly distinguishing dirt and other substances from original paint.

Through the magic of digital image processing, data covering every square foot of the Renaissance masterpiece could be displayed simultaneously on a computer screen: heavy green lines marked the beginning and end of each day's work by Michelangelo; diagonal green lines identified loose plaster; red lines indicated cracks; yellow crosses marked metal clamps inserted in the ceiling to anchor weakened plaster; and blue circles showed areas damaged by previous restorations. Thus, section by section for ten years, man and machine labored together in reilluminating the "shadows of divine perfection."

Agony and Ecstasy

Commissioned by Pope Julius II to paint the Sistine Chapel ceiling, Michelangelo began work in 1508 at the age of 33. Renowned as a sculptor

but with relatively little experience as a painter, Michelangelo resisted the massive undertaking and pleaded with Pope Julius to find someone else. The pontiff insisted, and Michelangelo had no choice but to accept.

Following the tradition of his Florentine training, Michelangelo is believed to have worked entirely *a fresco*—painting only while the plaster was still wet and adding no significant touches after it was dry. He began each section by spreading a smooth layer of thin, wet plaster *(intònaco)* atop a rougher layer of dry plaster. In most cases, the outlines of the desired image were transferred to the wet surface by means of a paper design, or "cartoon," tacked onto it. The transfer was accomplished in either of two ways: by running a stylus along the borders of the cartoon and digging grooves in the plaster, or by sprinkling charcoal dust through tiny holes in the cartoon drawing. (Some of the grooves and dots are still visible in the ceiling.) With the image thus outlined, he then applied water-based paint to the wet plaster. Each section was called a *giornata* (day) because it took one day to dry.

The work was agonizing. "I live here in great toil and great weariness of body," Michelangelo wrote, "and have no friends of any kind . . . and haven't the time to eat what I need." Day after day he stood atop the scaffolding, back bent, arms stretched to the ceiling. Paint dripped in his beard and eyes. His earliest images were spoiled by a hazy "mildew" resulting from too much water in the *intònaco*. He begged the pope for a reprieve, but none was forthcoming, and the work dragged on. According to the graphic record provided by modern technology, it took Michelangelo a grueling 29 days—even with helpers—to complete the 18-square-yard section depicting *The Flood*. His pace did pick up, however, for by the time he reached *The Creation of the Sun and Moon*, it took him only seven days to cover an equal area, laboring alone.

His patron, meanwhile, raged at him to work faster. The tyrannical Pope Julius even threatened to throw him off the scaffolding for not completing the job sooner. When finally the task was done, Michelangelo could write to his father: "The Pope is well satisfied."

Even if Michelangelo himself was more than just "satisfied," he might not have predicted that the Sistine Chapel would become the crowning jewel of all Renaissance art. The sublime, ecstatic quality with which the work has become associated was described by the 20th-century American art historian Bernard Berenson: "Michelangelo," he wrote, "joined an ideal of beauty and force, a vision of a glorious but possible humanity which has never had its like in modern times." The genius of the frescoes, many critics have felt, lay in its architectural composition and solid, sculptural renderings of the human form—not at all surprising from an artist who signed his letters, "Michelangelo, sculptor."

Wiping Away the Years

The agony of Michelangelo's labors could not have been lost on the team of professionals who undertook the restoration project. Inch by inch, they analyzed and treated the frescoes, careful not to damage or expunge any of the original work. Here and there they came upon bristles from the master's brush, small indentations where he had tested the setting plaster, altered images, and abandoned outlines. The stress and strain on the restorers derived not from a badgering patron but from their constant contact with genius and their fear of destroying it.

Exposing the original bottom layer meant somehow removing five centuries of accumulated dirt and other foreign substances. Archival records and high-tech analysis together identified a variety of extraneous materials. In addition to the dust and sooty grease of metropolitan Rome, the frescoes were yellowed by the burning of tallow candles on the floor below. (Electric lights were not installed in the Sistine Chapel until the 20th century.) At the beginning of the 18th century, wine was used as a cleaning solvent, but this, too, left a pallid residue. Most damaging, perhaps, were the varnishes made of animal glue used to brighten the surface. Applied repeatedly over the centuries, the glues had the desired effect for a while but gradually darkened the frescoes even further.

The key to the actual cleaning lay precisely in the original *a fresco* technique. As confirmed by subsurface testing, the water-based paint had chemically bonded with the drying plaster, lightening the colors and keeping them from fading. Within hours after Michelangelo painted, a hard layer of calcium carbonate had begun to develop, protecting his images from the dirt and grime of the ages. Restoring the frescoes, therefore, could be accomplished by removing the top layer of foreign matter.

The cleaning began by wiping a small section of the fresco with a natural sponge soaked in distilled, deionized water. Then, with a soft-bristle brush, a special cleaning solution was applied. The essential ingredient was a solvent called AB-57, specially developed some 20 years ago for use on frescoes. Made of bicarbonates of sodium and ammonium, the AB-57 was added to a fungicide and antibacterial agent and mixed in carboxymethylcellulose and water. The result was a gel that clung to the surface (without dripping) for about three minutes, then got wiped off with a sponge and water. With it came the dissolved surface substance and none of the original paint. Where the dirt was especially thick, the procedure was repeated 24 hours later.

The first years of the restoration (1980–84) focused on the upper lunettes—flat, semicircular panels around the top of the chapel windows, depicting Christ's ancestors. Then came the ceiling and spandrels, including the narrative of Genesis and the portraits of the prophets. (Another four years will be devoted to restoring *The Last Judgment*, painted 30 years after the ceiling on the chapel's western wall, above the altar.)

As each section of the restored masterpiece came to light, art historians and the ordinary viewing public—not to mention the restorers themselves—marveled at the brilliance and variety of colors. Gone were the somber tones and brooding darkness so long associated with the work. Tempered were the powerful architectural and sculptural elements so long regarded as Michelangelo's trademark. To the late-20th-century eye, the most startling feature of the frescoes was the range and brightness of hues— the soft, earthy greens; the shimmering Naples yellows; the wide palette of blues, reds, and oranges. Here, they seemed to say, is the "real" Michelangelo.

Science for Art's Sake

For art historians and museum curators, the traditional means of verifying an artwork's age, authenticity, and ownership history have been to examine the style and technique, then track down every available piece of written documentation. Although these methods have continued to play an important role, the process was enhanced in the early and mid-20th century by such innovations as X rays and radiocarbon dat-

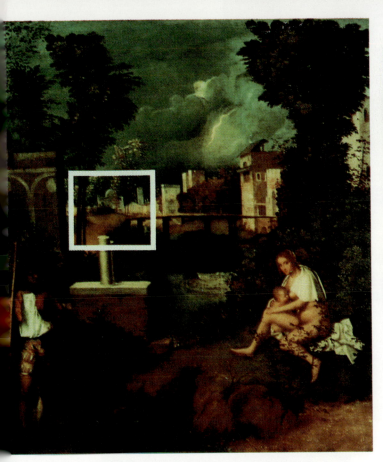
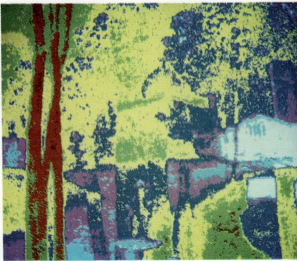

Art historians use a method called computerized infrared reflectoscopy to peer below the surface of a finished work and analyze the layers of paint it contains. Such an analysis of Giorgione's masterpiece *La Tempesta* (left) revealed details (inset, enlarged above) painted over by the artist nearly 500 years ago.

Photos: Courtesy Olivetti

ing. In recent years, however, largely due to the success of the Sistine Chapel project, the use of high technology in the study of classic art has taken a quantum leap. Computers linked to gamma-ray detectors, infrared sensors, and other sophisticated instruments are being employed in museums throughout the art world. Not only is the new technology facilitating the restoration of paintings, sculptures, pottery, and other relics, but it is also helping discover lost works, identify forgeries, and solve some of the oldest mysteries of art history.

New York's Metropolitan Museum of Art is just one of many to employ infrared reflectography for looking behind the scenes of old paintings at the artist's original handiwork—rough sketches, repaintings, and erasures; using a computer and video display screen, the artwork can be superimposed over the infrared image, clearly showing any differences. The Louvre in Paris has installed a particle accelerator that uses high-energy ion beams to analyze paint pigments, metals, ceramics, and various other materials. At the National Gallery of Art in Washington, D.C., scientists have borrowed techniques from the field of immunology and applied them to pigment analysis. The American physicist John Asmus has completed a detailed computerized reconstruction of Leonardo da Vinci's masterpiece *Mona Lisa,* using image-enhancement technology developed by the National Aeronautics and Space Administration (NASA) for processing satellite photos; the study was made with a scanning microdensitometer that measured (in three wavelengths) and digitized the light given off by 6 million points on the painting.

The results have been valuable—and often surprising—to art historians and curators. The studies at New York's Metropolitan Museum, for example, have yielded important new insights into the drawing style of northern Renaissance painters. Asmus' examination of the *Mona Lisa* revealed underlying "ghost" images *(pentimenti)* of a bead necklace and a mountain range; the finding supported scholars who believe that da Vinci began the work as a conventional Florentine portrait, then gave it a more enigmatic character in later years. Other discoveries:

• A long-lost painting by the 19th-century French artist Jean-François Millet, the critically

Technology used in the restoration of Giovanni Bellini's Feast of the Gods (left) revealed that the painting had been radically altered twice. Apparently, shortly after Bellini's death, a Venetian duke commissioned a journeyman court painter named Dosso Dossi to modify the painting for use in his private bedroom. The bright yellow and green foliage in the Dossi version (below) clashed alongside the room's other artwork, however, leading the duke to hire the Venetian artist Titian to cover up Dossi's contribution with an elaborate grove of trees (below left).

assailed *Captivity of the Jews in Babylon,* was found lurking on the canvas beneath another of his paintings, *The Young Shepherdess.*

• An examination of *Madonna and Child* by the 16th-century Venetian master Titian revealed a praying saint, never before seen, under the baby's legs.

• At the National Gallery of Art, the use of gas chromatography/mass spectrometry, infrared reflectography, and 220x magnification of paint specks by stereomicroscope settled the long-standing mystery of Giovanni Bellini's masterpiece of the Venetian Renaissance court, *Feast of the Gods* (1514). Long believed to have been retouched by Titian, the work was shown in the 1950s to have been largely *repainted* by the latter artist. The most recent tests, however, revealed the brushstrokes of yet a third artist, a court painter named Dosso Dossi.

• X-ray fluorescence testing of works by the 19th-century French Impressionist Paul Gauguin during his Tahiti period disclosed a high

Photos: National Gallery of Art, Washington; Widener Collection

level of barium sulfate in the lead-white pigments, indicating that the paint was of poor quality; either he was cheated by his Paris supplier or could not afford better paint.

With their remarkable powers of detection, today's advanced technologies are also finding new applications in another realm of the art world—testing for forgeries. Although X rays and radiocarbon dating have been used for decades to authenticate valuable artworks, more accurate subsurface examination and more sensitive chemical testing of materials have enabled experts to identify fakes in virtually every art medium. And new methods are constantly being explored. At the Smithsonian Institution's Conservation Analytic Laboratory, for example, chemists have used gas chromatography to study the way oils degrade over time, a technique that would allow more precise dating of paintings. Also at the Smithsonian, researchers are studying whether artificial aging—the process used to fake old documents—causes the same chemical changes as real aging.

Computers alone—without linkup to gamma-ray detectors or infrared cameras or thermographic detectors—provide a valuable tool to art historians and restorers. Nowhere is this more evident than in efforts to repair damaged sculptures. In Bologna, Italy, for example, a computer analysis of body positions showed restorers how to reassemble a 15th-century terra-cotta statue of Christ. In Rome a computer designed a flawless electronic "mold" of a bronze statue of Marcus Aurelius. That capability has raised the prospect of an "electronic archive" for all of Italy's art treasures, so that any damaged work could be accurately restored. Such an archive already has been created for cataloging the thousands of frescoes and mosaics among the ruins of Pompeii.

Other Perspectives

With all its valuable applications, the use of high technology in art research naturally has its limits. Although high-tech systems and processes have secured a permanent place in the art historian's toolbox, few in the art or scientific community would argue that they will ever replace the human eye or conscientious scholarship. "We're doing more science than we ever have before," said the chief of conservation at the National Gallery. "But if you want to understand why Raphael painted Madonnas, it's more profitable to look at the social environment in which he lived than to take X rays."

Reliance on technology in the particular area of art restoration raises thornier questions. Not surprisingly, perhaps, the issue came most prominently to the fore in regard to the most ambitious restoration project to date—that of the Sistine Chapel. During the course of the cleanup, a number of historians and artists expressed strong opposition to the entire endeavor. In 1987, 15 American artists—including Robert Rauschenberg, Robert Motherwell, and George Segal—signed a petition asking Pope John Paul II to halt work. Some critics contended that the restorers used corrosive chemicals and had no way of predicting their long-term effects. Others disputed the assumption that Michelangelo's technique was purely *a fresco,* and feared that the restorers had expunged vital elements added after the plaster had dried. In most cases the doubts stemmed at least in part from the unexpected brightness of the colors. "[The restoration] turned Michelangelo into an ordinary, second-rate painter," howled the Italian art critic Toti Scialoja. "They have destroyed the very quality and thinking of Michelangelo."

Defenders of the restoration—representing the majority of the art community—pointed first to the careful, conscientious, objective methods of the Vatican team. Materials were tested, assumptions verified, methods practiced to the point of expertise. As for the issue of color, scholars brought out several points: that Michelangelo would have used bright colors to compensate for the dim candlelight and to make the ceilings visible from 65 feet (20 meters) below; that the high tones had been part of his training and were close to those of another work, his *Doni Tondo* in Florence's Uffizi Gallery; and that resistance to the cleaned-up frescoes was but the reflection of long-held but mistaken historical judgments. "Art historians are always surprised when a work of art is revealed in its entirety," said one conservation expert. "They confuse grime with mystery."

On one point, neither side can disagree. It is impossible to restore any work of art to its original state. To whatever degree, no matter how it is executed, the piece begins to deteriorate the moment it is done. Like the fingers of Adam and God in the most famous of Sistine Chapel frescoes, the art of the restorer and the art of the painter will never quite meet. That fact alone should inspire ever more skill on the part of the restorer, ever more precision in technological analysis, and ever closer cooperation between the two.

McDonnell Douglas Helicopter Company

a new lift for HELICOPTERS

by Susan Katz Keating

The sight of the thing is enough to give even Chuck Yeager a case of the pre-flight jitters. Like a last-stage tadpole waiting to blossom into a frog, this helicopter has almost all the right parts: bubble cockpit, landing skates, main whirly-gear, and tail boom. But that's as far as it goes.

Instead of the familiar set of smaller blades at the tip of the boom, this chopper has—well, nothing, except two strangely angled little fins. The McDonnell Douglas Helicopter Company calls it the NOTAR, for "no tail rotor." Two versions have been unveiled and will be available within the next few years. The design looks about as safe as an airplane without a tail, but the aviation world is hailing it as the hottest thing since the turbine-powered helicopter debuted in 1951.

"It is extremely revolutionary," says Chris Colligan, editor of *Vertiflite* magazine.

"There is no question this is a major advance," says Mark Paris, communications director for the American Helicopter Society. "This is a milestone in aviation history."

Tackling Torque

The benchmark is not in the class of such aviation achievements as the development of the jet engine, but it does provide an alternative solution to a problem that long has concerned heli-

A new helicopter design replaces the traditional tail rotor with an air-circulation system (art below). Compared to conventional helicopters, the "no-tail-rotor," or NOTAR, craft is safer, more maneuverable, and, in military situations (facing page), less visible to radar.

copter designers: What do you do about torque? Physicists will explain that torque is the twisting effect exerted by a force acting at a distance on a body, or, in other words, the force that produces rotation. In helicopter language, that translates into an unpleasant result: when the main rotor whirls in one direction, the body rotates in the other direction.

"The spinning effect is nice. You can have 360-degree vision," jokes aviation pioneer Ralph P. Alex. "But it can be embarrassing to the passengers and pilot."

Previous attempts to solve the torque problem have included a tandem main rotor system, in which two counterrotating sets of blades operate on a single shaft, and a set of large blades, fore and aft, placed to overlap but not collide. The first solution may be the helicopter's wave of the future. The second, used in the Army's Chinook helicopter, operates much like a set of blender blades: the rotors overlap so closely they appear in danger of grinding together.

The most common answer to the torque problem has been a counterspinning device in the form of a tail rotor. But tail rotors bring special problems of their own. They increase both weight and drag on the aircraft, add noise to an already cacophonous machine, increase the chopper's signature on enemy radar, produce vibrations that can damage other devices on board, and are susceptible to damage from bushes, trees, bullets, even birds. A helicopter is only as sound as its tail rotor, for once that is gone, so is the aircraft. Tail rotors are also dangerous on the ground. They can maim or kill even when they appear to be slowing to a stop.

Rave Reviews

The new system controls torque with a fan-driven air-circulation system inside the tail boom. Air is vented through two horizontal slots along the side of the boom and through a variable thruster at the rear. The thruster consists of a series of holes on each side of the boom, opened to various degrees by a rotating can. The pilot uses foot pedals to shift the can within the boom, allowing air out in controlled amounts, thereby turning or straightening the craft.

The pressurized boom also allows downwash from the main rotor to circulate around the boom, providing lift. This concept does not work in other helicopters because their tails are not pressurized and because turbulence from the tail rotor interferes with the downwash flow.

The new system earned rave reviews for three engineers who worked on the concept demonstrator. The American Helicopter Society awarded its 1987 Howard R. Hughes Trophy to McDonnell Douglas engineers Andrew H. Logan and Evan P. Sampatacos and project chief

A New System Holds Sway

The "no tail rotor" system controls torque with a fan-driven air circulation system inside the tail boom. Air is vented through the two horizontal slots on the side of the boom and through the variable thruster at the rear. The thruster consists of several slots, on each side of the boom, opened in varying increments by a rotating cylinder.

- Air intake
- Main Downwash
- Circulation Control Boom
- (Hover and Right Sideward Flight)
- Air Jet
- Thruster Jet (Left Sideward Flight)

SOURCE: McDonnell Douglas Helicopter Co.

James R. Van Horn. The honor is reserved each year for an outstanding achievement in helicopter technology.

Watching the ship in action is much like witnessing a combination aerial circus and mechanical ballet. A video produced by McDonnell Douglas features the helicopter whipping, spinning, climbing, and diving with such grace that it is reminiscent of the "airs above the ground" performed by the famed Lippizaner stallions of Austria.

The company claims the craft can shift 360 degrees, while hovering, in less than four seconds, and can whirl a complete circle in four seconds while flying at 30 knots. It can whip along sideways, in seeming defiance of aerodynamic laws, at 40 knots. It can climb and twist at high angles of attack, and seems impervious to winds that would stall ordinary helicopters.

Tail in the Trees

The new vehicle's pièce de résistance, however, is its tail-in-the-trees routine. The pilot can back the flying craft into a clump of trees or bushes with impunity; a tail rotor would crack up immediately from such a move, sending chopper and pilot into a plunge.

The craft has become the talk of the industry, and pilots in particular are taken by its charm. "I've been flying helicopters for 30 years," says John Zugschwert of the American Helicopter Society. "The NOTAR is so easy to fly, compared to the helicopter. Just one flight on the NOTAR made me get lax on the pedals [in a regular helicopter]. If pilots get their training in a NOTAR, they won't be qualified to fly a [conventional] helicopter."

The difference, he says, is in the new craft's response to torque. An ordinary chopper requires the pilot to compensate for torque so frequently that the maneuvers become an automatic part of flying. "When I pick up on the collective to go up or down, I automatically compensate with my right pedal. As a pilot, I just know that—I have to compensate. But in the NOTAR, I had the nose pointed exactly at a building and took it straight up and down without using the pedal."

The result is to decrease the duties of the operator. "This is a much safer aircraft," says Zugschwert. "It reduces the work load."

Other pilots have commented on the ease of handling in wind conditions that would stall other aircraft, as well as increased maneuverability, thanks to the lack of air currents created by the motion of tail blades.

Helicopter History

Helicopter history is both lengthy and sluggish. The first known rotary-type device is a flying top created in China in A.D. 400. A helicopter existed on paper in 1325, when a Flemish manuscript depicted a string-powered device. Perhaps the most famous early design is that of Italian inventor Leonardo da Vinci, who sketched his idea for a spiral airscrew in the 1480s.

Tail-in-the-trees: the NOTAR helicopter's missing tail rotor affords the craft safe passage in tight spots.

McDonnell Douglas Helicopter Company

The wing-mounted tilt rotors of the XV-15 helicopter enable it to rise like a helicopter and fly like an airplane.

Other models and writings provided the groundwork for development of the helicopter, but it was not until 1907 that a full-size rotary-winged craft lifted a man and hovered with him for a sustained period. In 1923 Spaniard Juan de la Cierva flew his Autogiro. The contraption evolved into something akin to a cross between a helicopter and a propeller-driven airplane. It derived power from a central tractor-propeller engine mounted in the nose, and got its lift from a large main rotor. The craft also contained vertical stabilizers in the rear. A helicopter, by contrast, gets both its lift and power from the main rotor. Still, Cierva is credited with making great strides in the advancement of helicopters.

The airplane, which was ready for war only a few short years after its invention, developed rapidly. But helicopter evolution took a much slower route, and did not get under way seriously until 1937, when the German Focke-Achgelis FA-61 became the first in its genre to remain airborne for more than an hour.

Famed engineer Igor Sikorsky ushered in a flurry of developments in the 1940s. Among the techniques Sikorsky Company designers attempted was a pressurized tail boom that deflected air in an effort to eliminate the tail rotor. "We called it the Flettner rotor," says Alex, one of the project engineers. The craft was, essentially, a NOTAR. It was named for the German designer Anton Flettner. "It did work, in a way. But it wasn't considered practical."

Sikorsky abandoned the project after a few years. "If it takes a hell of a long while to make a big improvement, the development needs to be looked at carefully," says Alex. "If it takes more than three to five years, then you have some basic problems." The lengthy developmental process of the rotorless tail, which has been under way for eight to ten years, leaves Alex with a measured dose of skepticism. "The tail rotor has been a handicap for many years, but it has survived quite well."

"The tail rotor is vital to the performance of the helicopter," says Jean Ross Howard, executive director of the Whirly-Girls Association. But, she adds, "to do without it would be a wonderful thing."

McDonnell Douglas, meanwhile, is putting the finishing touches on its two production versions. The MD 520N, an extension of the existing MD 500, will be ready in 1991, while the MDX, a brand-new twin-engine helicopter, is slated for delivery in 1993. "This aircraft will change the way operators look at helicopters," says Mark Holland, program manager for the MD 520N version. "Once you've flown it, you become an immediate believer in NOTAR."

SPACE SPIES

by William E. Burrows

On Route 101 in Sunnyvale, California, at the northern fringe of Silicon Valley, stand four large dish antennas. Nearby is a pale-blue windowless six-story building dubbed the Big Blue Cube—one of the government's most highly classified facilities. Guards with submachine guns are stationed behind the front door. Workers in the cube learn only what they need to know, and go only where necessary for their jobs.

This is Onizuka Air Force Base, a site that has neither planes nor runways. The base is the federal government's Satellite Control Facility, which uses the large white antennas to operate U.S. reconnaissance satellites.

The recent end of a 2½-year hiatus in spy-satellite launchings as space-shuttle missions resumed, together with the debut of the Titan 4 rocket for unmanned launches, have enabled specialists within Sunnyvale's Big Blue Cube to use superior new spy satellites. Advanced technology aboard these satellites is thought to be powerful enough to capture images of grapefruit-size objects, or capable of detecting heat from the afterburners of military aircraft. There's also an important new type of spy satellite deployed that provides intelligence agencies with photolike images even when the Earth below is shrouded in thick clouds or darkness. By the mid-1990s, the number of large imaging satellites may triple to about 12.

Four types of satellites are involved in gathering intelligence: TV-camera-like sensors —something extraordinarily sensitive to infrared light—are coupled with powerful telescopes used aboard the newest "optical-imaging" satellites. These sensors detect missile launchings and capture detailed images of weapons and other activities. A "radar-imaging" satellite launched in December 1988, originally code-named the Lacrosse, requires huge solar panels to power its cloud-penetrating microwave transmitter. So-called "signals-intelligence" and "ferret" satellites are supersophisticated radio receivers that capture the radio and microwave transmissions from foreign countries. Finally, "relay" satellites transmit data from spy satellites to U.S. ground stations. High-speed computers on Earth play a key role in making torrents of data intelligible.

Several top-secret surveillance satellites orbited by the U.S. government can capture an unprecedented level of detail. Such "space spies" provide high-resolution reconnaissance pictures even at night or on cloudy days.

Having this all-weather, round-the-clock spying capability is a valuable new tool for detecting cheating on arms-control agreements. The task is much harder, though, as missiles become smaller. With federal drug-enforcement agencies stepping up the war on narcotics trafficking, spy satellites have still another new role (see the box on page 337).

Snapshots from Space

Spaceborne espionage started in the summer of 1960, when the Central Intelligence Agency (CIA) began photographing the U.S.S.R. with its Discover satellites. Discoverers, like the photographic intelligence (PHOTINT) satellites that followed, circled the Earth in roughly polar orbits so that the continents spun beneath their film cameras. The spacecraft were then rocketed out of orbit, parachuted northwest of Hawaii, and snatched in midair by U.S. Air Force transports. After film processing in Hawaii, couriers rushed the pictures to Washington, D.C., for thorough analysis.

Later, picture-taking spy satellites flew under a program code-named Keyhole. Authorities dropped the shorthand labels KH-8, KH-9, and KH-11 after they appeared in the book *Deep Black*—although most of the media still haven't. Most Keyhole satellites took either wide-area pictures of large land areas for routine surveillance and weapons targeting, or close-up pictures of special-interest objects from a low 80-mile (128-kilometer) altitude.

The 1971 KH-9 satellite, dubbed Big Bird because of its 30,000-pound (13,600-kilogram) Greyhound-bus size, developed film for wide-area images on board and beamed the pictures to earth as TV signals. A special Kodak camera snapped the area-surveillance images. For close-up photos, parachuted back to Earth in six film-return pods, a Perkin-Elmer-built mirror telescope served as Big Bird's telephoto optics. Despite Big Bird's sophistication, however, its limitations frustrated intelligence personnel. The amount of film and the number of pods the satellite carried limited the number of high-resolution close-up photos they could obtain. Also, poor weather and logistical problems often delayed intelligence for weeks. And while wide-area images transmitted to Earth could be obtained quickly, the TV camera and radio link limited their resolution. That made them nearly useless to photo interpreters who need to scrutinize radars, aircraft, ships, and other weapons in minute detail.

Intelligence experts believe that the latest KH-11 spy satellite has the advanced optical equipment needed to zoom in on and clearly photograph objects as small as grapefruits. Specialists using computer-enhancement techniques may even be able to read license plates and book titles (see photo sequence). The KH-11's flexible telescope mirrors compensate for atmospheric distortions. Infrared-sensitive cameras permit night photography. The newest KH-11 can also capture images in a variety of other wavelengths. Analysts compare images of objects recorded in different portions of the spectrum to determine the material from which they are made.

Solid-State Imaging

All that changed in 1976, when the first KH-11, code-named Kennan, began operating from a special orbit that keeps the Sun at a constant angle with respect to objects on Earth. The advantage: shadows recorded from space help detect movement and the size of objects. KH-11 was an important advance in espionage because it could provide detailed high-quality images almost as an event was occurring. Space-shuttle astronauts orbited an improved KH-11 in the series in August 1989.

An important technology advance for the KH-11 is the solid-state image sensor it uses, a charge-coupled device (CCD). The CCD (or similar semiconductor sensor) is now used in millions of consumer video camcorders and TV cameras.

Bell Labs' invention of the CCD in 1969 provided satellite engineers with a lightweight, low-power, extremely sensitive image sensor. The surface of a CCD is covered with a rectangular grid of microscopic light-sensitive elements. A postage-stamp-size CCD chip available in the early 1980s for scientific space telescopes, for example, had an array of 800-by-800 elements. Images it produced would have 640,000 picture elements, called pixels. (Chips can be clustered together to generate, say, four times as many pixels.) An image begins as light falls on individual pixel sensors, creating an electrical charge proportional to the light intensity.

CCDs aren't the only type of sensors used. Telescopes in another type of optical-imaging satellite, the Defense Support Program (DSP) series, are equipped with sensitive infrared detectors to spot the hot plumes of missiles. DSP satellites in fixed geosynchronous orbits are designed to provide an instant warning against missiles launched from either land or sea. Once missiles are detected, other satellites can zero in on the launch site to provide more-detailed views. Space-policy expert John Pike of the Federation of American Scientists believes the first Titan 4 booster put the latest DSP infrared satellite into orbit in mid-June 1989. The satellite maker, TRW, reports that this third-generation model weighs 5,200 pounds (2,350 kilograms), operates at 1,274 watts (double the power of the previous model), and has some 6,000 infrared detectors.

Although CCDs and infrared-sensitive sensors have become widely accessible for imaging and other electronic tasks, another photographic technology aboard KH-11s, adaptive optics, has not. Adaptive optics employs flexible mirrors that can be bent slightly under computer control to compensate for the distortion caused by Earth's atmosphere. Such advanced imaging technology aboard the newest KH-11 satellites is expected to provide views of some objects some 5 inches (12 centimeters) across.

Space Radar

Despite their eagle-eyed spies in orbit, intelligence agencies have long faced a serious restriction. "During late fall, winter, and early spring, the U.S.S.R. and much of Eastern Europe are covered by clouds up to 70 percent of the time," explains a former U.S. intelligence official.

In December 1988, when the space shuttle *Atlantis* thundered from the Kennedy Space Center, its payload bay held a solution to the problem of obscuring clouds. At a 275-mile (440-kilometer)-high orbit, Air Force Colonel Mike Mullane released the $500 million Lacrosse satellite from the bay and helped rocket it to a higher orbit. The main image sensor aboard Lacrosse, a radar transmitter and receiver, demands the high power provided by its large solar panels. While optical-imaging satellites passively capture light from the Earth below, Lacrosse is active: it beams microwave energy to the ground and reads the weak return signals reflected into space. This new breed of spy satellite can "see" objects on the Earth's surface through clouds and in darkness.

Photo sequence: © The Eames Office 1990. Excerpted from the Eames film "Powers of Ten" © 1977 Charles and Ray Eames

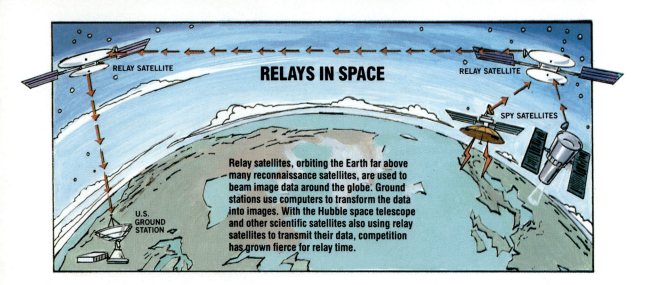

RELAYS IN SPACE

Relay satellites, orbiting the Earth far above many reconnaissance satellites, are used to beam image data around the globe. Ground stations use computers to transform the data into images. With the Hubble space telescope and other scientific satellites also using relay satellites to transmit their data, competition has grown fierce for relay time.

Lacrosse's rectangular antenna, reportedly some 48 feet (14.6 meters) long and 12 feet (3.6 meters) wide, differs from standard mechanically scanned radar antennas. The satellite antenna is covered with rows and columns of small transmitting and receiving elements. Radar signals can be aimed or swept back and forth in phased-array designs such as this by changing the electrical signals fed to each small antenna element.

Spy-satellite views may have triggered U.S. accusations that Libya is building a chemical-weapons plant near Tripoli. Libya denies the charges. The image below, with a resolution of 30 feet, led West Germany to the same conclusion. The plant may have burned down in 1990.

© 1990 CNES provided courtesy of SPOT Image Corporation, Reston, Virginia

There's another difference in the way satellites such as Lacrosse capture images. Most radars provide a constant stream of images, as echoes from targets are instantly converted on a display. Lacrosse, however, records a series of quick "snapshots" as it arcs over the Earth. The return signals from objects in each of these snapshots shift slightly because of the Doppler effect, which causes wavelength to shorten as a surface feature is approaching, and lengthen as it recedes. The time required for radar signals to make a round trip (range to targets) also changes. Signal strengths help determine the brightness levels of elements in the final radar image. To obtain these pictures, signals are sent to ground stations, where computers can blend the snapshot data into detailed images. This technology, called synthetic-aperture radar (SAR), provides image-resolving power that normally would be possible only with a much larger antenna, collecting entire radar images instantaneously.

How detailed are SAR images? John Pike believes the advanced Lacrosse can resolve objects about 3 feet (1 meter) across. That level of detail is necessary to identify military hardware such as Soviet mobile missiles for arms-control agreements. Pike thinks four Lacrosse-type satellites will be orbiting by the early 1990s.

"We are moving into a verifications business that requires 24-hour-a-day monitoring by remote sensing," says Michael Krepon, head of the verification project at the Carnegie Endowment for International Peace. "If we're going to be able to monitor the Intermediate Nuclear Treaty or other arms agreements," notes a former U.S. intelligence official, "we have to be able to see under clouds. It's that simple."

Although imaging satellites more readily capture the public's imagination, those that collect radio signals, the ferret and signals-intelligence (SIGINT) satellites, bear the brunt of U.S. space spying. Because the military and political organizations of the major powers could not function without almost continuous electronic communications, intercepting these signals often provides the first indication that something important is about to happen. Intercepted signals are analyzed, and decrypted if necessary, at the National Security Agency (NSA) in Fort Meade, Maryland. At times the information suggests that something is worth photographing from space.

SIGINT satellites with a variety of cryptic code names have at one time or another floated in geosynchronous orbit some 22,300 miles (35,880 kilometers) out in space. Huge umbrella-shaped antennas capture microwave telephone traffic and other radio signals from the U.S.S.R. and anywhere else the United States wants to eavesdrop.

Rhyolite, an early SIGINT satellite that operated throughout the 1970s, could simultaneously monitor 11,000 telephone calls and walkie-talkie traffic from orbit in addition to its primary task, snooping on telemetry signals from Soviet and mainland Chinese missiles during test flights.

Relaying Spy Data Home

Also playing an important role are the Tracking and Data Relay Satellite Systems (TDRSS) and other platforms, which relay data from satellites to Earth stations. *Discovery* launched the third TDRSS last March. Most observers believe the newest spy satellites are designed to send their data to Earth via relay satellites. Relaying data not only speeds it to the other side of Earth when necessary, but sending the data upward to higher-flying satellites complicates the job of the Soviet technicians who try to intercept it.

Relaying may also create problems, however. According to a March 1989 *Science* magazine article, collecting data from Lacrosse through relay satellites limits the new radar satellite's performance. The main problem is the inability of relay satellites to handle data fast enough. Robert Cooper, former head of the Defense Advanced Research Projects Agency (DARPA), says the new radar satellite puts out prodigious amounts of information. He says that a space radar with a resolution of about 1 foot (30 centimeters) can generate raw data at many billions of bits per second—far beyond the 300-

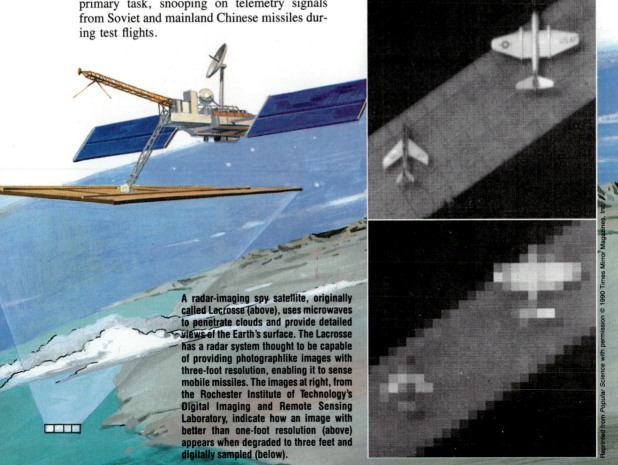

A radar-imaging spy satellite, originally called Lacrosse (above), uses microwaves to penetrate clouds and provide detailed views of the Earth's surface. The Lacrosse has a radar system thought to be capable of providing photographlike images with three-foot resolution, enabling it to sense mobile missiles. The images at right, from the Rochester Institute of Technology's Digital Imaging and Remote Sensing Laboratory, indicate how an image with better than one-foot resolution (above) appears when degraded to three feet and digitally sampled (below).

megabits (million bits)-per-second maximum capacity of one of TDRSS' two high-capacity channels. "The data rate limits everything: resolution, gray-scale accuracy, and field of view," notes a designer of the Earth Observation Satellite (EOS) at the National Aeronautics and Space Administration's (NASA's) Jet Propulsion Laboratory (JPL).

Transforming the raw data into images on board the satellite and sending them through TDRSS would require the radar satellite to carry one of the world's largest supercomputers, says Cooper. One way to avoid the data bottleneck, notes the *Science* article, is to store bursts of data on recorders and then transmit them intermittently at a slower rate.

Organizing Satellite Spies

Spying from space is so vital to national security that it is controlled by a special organization, the National Reconnaissance Office (NRO), headquartered in the Pentagon. The NRO is so secret that even its letterhead is classified. The NRO, a collaboration of the Air Force and the CIA, gets satellites built and operating under Sunnyvale's control. In addition to its Washington, D.C., headquarters, officially known as the Office of Space Systems, the NRO has offices in El Segundo, California, near several satellite manufacturers.

The NRO has the biggest budget of any organization in the U.S. intelligence community. Coming up with cutting-edge spy technology while remaining hidden from public scrutiny makes the NRO and its contracting companies "a playpen for engineers," says John Pike.

The bill for keeping track of what the rest of the world is up to is staggering. Considerably more than $5 billion is spent annually on designing, testing, building, launching, and operating reconnaissance satellites. Many additional millions are spent maintaining ground stations around the globe, and operating a variety of aircraft to snoop along the periphery of the Soviet Union and other areas of the world. Most likely, the net worth of the entire operation exceeds $50 billion and is increasing rapidly as new arms-

Reprinted from *Popular Science* with permission © 1990 Times Mirror Magazines, Inc.

Infrared detectors aboard Defense Support Program satellites sense missile launches as they occur. New models (right) are already undergoing tests. Although these satellites have relatively poor resolution, as the artist's concept suggests, their extraordinary sensitivity to certain infrared wavelengths can capture views of exhaust trails and warheads during missile tests or of jet aircraft flying with afterburners.

The government now uses spy satellites and data from commercial satellites in its war on drugs. Earthbound observers (above) apply computer-enhancement techniques to the images to reveal minute details (below).

SPACE WAR AGAINST DRUGS

The tiny specks of light green clustered along Peru's Huallaga Valley in the Landsat satellite photo are fields of crops. The surrounding dark areas are tree-cloaked mountains. A fraction of these crops are coca, from which cocaine is derived. But this image displays half of the world's annual cocaine crop, according to an October 14, 1989, *New York Times* story. Intelligence and drug-enforcement agencies, the article says, are now employing the nation's spy satellites—as well as commercial satellite data from Landsat and SPOT—in the drug war.

Computerized images, such as this one, can reveal remarkable details. "Suppressing foliage is a pretty straightforward technique," says Mark Labovitz, manager of remote sensing services at ST Systems Corporation in Lanham, Maryland. "You can tone down the foliage, penetrating through to get information on understory areas." Hidden roads, buildings, and runways used for drug production might be unveiled. Sequences of images over time are useful, too: "Let's say you took three images," says Labovitz. "One showed a clearing. Then a regulatory agency went in and shut things down. The next image showed vegetation had grown back. If a third photo showed the area has been cleared again, you'd ask, 'What's going on here?' "

Top: © Dan Helms/Compix; above: Earth Satellite Corporation

While analysts have perfected satellite reconnaissance of common crops from space, they're still learning how to pinpoint drug-related plants. "Up until now, coca hasn't been what you'd call a national cash crop," says retired Army intelligence expert Nelson Johnson. He stresses that human intelligence about production, not just technology, is crucial in the drug war.

JOHN FREE

control agreements require verification and more nations acquire the ability to make strategic weapons.

Costly computers have a crucial spy-satellite function. Because the data fed to ground stations are digital, those who interpret the imagery can use computers to manipulate the data, enhancing certain elements and reducing or eliminating others. If an object is hidden in shadow, for example, computers can lighten the shadow, but at the same time maintain the contrast relationship between adjacent picture elements so the object becomes discernible to the eye with its details intact.

Today's biggest technical hurdle, though, is programming high-speed computers to extract crucial intelligence information from the deluge of incoming data. "In the past the problems have been mostly connected with sensing the data. Now they are more in filtering and interpreting it," Thomas Rona, a former Pentagon official and now deputy director of the White House Science Office, told *Science*. "You have to filter it in terms of geography, but also in terms of targets that are interesting," he said. "Humans—photointerpreters—used to do this. Now there are attempts to automate it."

"Somewhere in that data there's a target," explained former Air Force Secretary and former head of the National Reconnaissance Office Edward (Pete) Aldridge, Jr. "Now, how do you find it . . . unless you take the population of the U.S. and make them photointerpreters?" asked Aldridge in *Science*. Experts believe that the answer lies in the development of automated image interpreters. In theory, high-speed computers could search spy-imagery data and recognize the signature of a Soviet weapon. But this task is formidable. "We're still five years away from the point where some data comes in and rings a bell and says, 'I've got a target X in location Y,'" said Aldridge.

The most valuable contribution of computer analysis, according to Rona, may be in matching up information from various sensors so that one instrument can correct the other's blind spots. While a KH-11 might be fooled by a detailed plastic decoy that looks like a tank, for example, the attempted deception might be unveiled by the radar on Lacrosse. Multispectral analysis, as this technique is called, allows a rough chemical analysis of an object's surface from space so that an object's composition can be correlated with its appearance. Devices that can monitor up to 224 frequency bands in the electromagnetic spectrum are being developed. "All sensors lie a little," Rona said. "The reason you coalesce information from all sorts of sensors is that you don't trust any of them."

Attempts to write computer software capable of comparing and evaluating data from many different sensors have run into problems, notes *Science*. Military sources estimate that working prototypes of these "data-fusion" systems will not be available for several years.

Spy Satellites

The coming decades will bring extraordinary improvement to surveillance satellites, not only in their observational capabilities, but in their survivability, too. The threat of antisatellite weapons and nuclear explosions in space in time of war requires hardening all military satellites to radiation and other weapons. Radiation-resistant integrated circuits will be used increasingly for vital components. Stealth technology—such as non-radar-reflective surfaces, epoxy finishes that absorb radar waves, and low-heat maneuvering thrusters—will be used.

There are also plans for huge radar satellites with antennas larger than football fields but thinner than cardboard. They would orbit at 1,000 miles (1,600 kilometers) or more. Other platforms, outgrowths of an experimental program called Teal Ruby, will use giant arrays of infrared sensors to stare at whole continents instead of scanning them as with current technology. These sensors could detect virtually anything important giving off heat, from cruise missiles, to aircraft, to missiles under test.

Weapons such as these are not only becoming smaller, but increasing numbers of them are mobile, and so, tougher to locate from space. Even worse, tactical and strategic weapons are beginning to proliferate in many parts of the world at a rate that officials in both Washington and Moscow find alarming. Alleged poison-gas facilities in Libya, Iraq, and elsewhere represent one new aspect of the surveillance challenge. Another is the potential for nations in Asia, the Middle East, and South America to produce their own ballistic missiles, not to mention the nuclear warheads for them.

In the minds of U.S. intelligence officials, this worldwide threat will eventually supersede that from the Soviet Union and strain the nation's space-surveillance capability to the limit. The game of hide-and-seek is about to become infinitely more complicated.

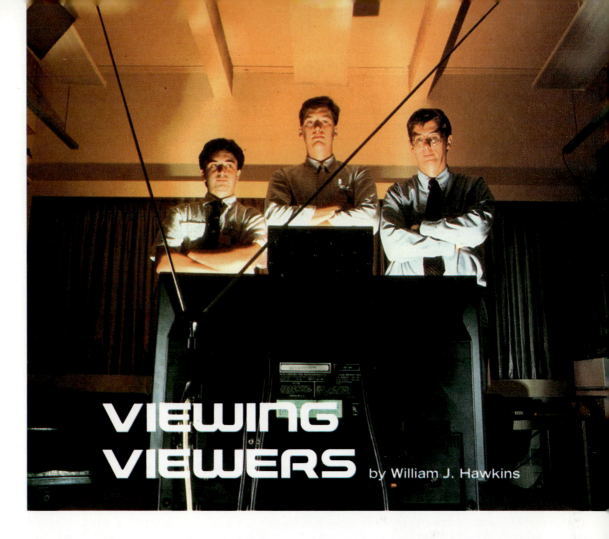

VIEWING VIEWERS
by William J. Hawkins

Dr. Curtis R. Carlson is comfortably seated in front of a video monitor while colleague Matt Clark begins to type simple commands on a nearby computer keyboard. In a moment a still snapshot of Carlson's face pops up on the monitor's screen, but the eyes, nose, and mouth are surrounded by colorful squares. Clark presses a few more keys. "Gotcha," he says in a low voice. "The system now knows who you are."

The *system* is part of an eight-year project that researchers at the David Sarnoff Research Center in Princeton, New Jersey, call smart sensing. "It's a face-recognition system," says Carlson, briefly leaning back on his chair to look at me. "It recognizes people."

Carlson, director of the information systems research laboratory, turns back to the video monitor once again. Now the image is live—moving as he moves. "Once the computer learns what you look like, it can recognize you later by looking at your facial characteristics," he says. I watch as the computer seems to lock onto his face, momentarily freezing the picture and drawing a colored square around his head. The name "Curtis" flashes on the screen.

Automatic Recognition

The smart-sensing system is being developed for Nielsen Media Research, the ratings company that helps networks determine what we watch on TV. Currently, 4,000 Nielsen participants—dedicated TV watchers—must press buttons to tally electronically the times and TV shows they watch. The system is fallible, argue some network executives, because children are lax at using it, and adults tire of it quickly—neglecting to "punch out" during a dash to the kitchen or bathroom.

Shades of Big Brother? To gather more accurate viewership data, Nielsen Media Research is developing a computerized system that recognizes viewers and combines that information with the time and selected TV channel.

Photo by John B. Carnett/Reprinted from *Popular Science* with permission © 1990 Times Mirror Magazines, Inc.

The smart-sensing technology will make future Nielsen home installations passive—capable of automatically recognizing family members and combining that information with the time and selected TV channel. And it will be more accurate: bury your head in a magazine or newspaper—or just look away from the TV—and you're instantly discounted as a viewer for that length of time.

In the past, image-recognition systems required massive—and expensive—computing power, which basically memorized every conceivable view of the subject. "They wouldn't work here," says Carlson. "Nielsen's passive people meter must be no bigger than a VCR to fit on top of the TV, relatively inexpensive, and capable of recognizing a face."

Comb your hair differently, smile, grimace, or frown—to a computer, your face is an inexhaustible source of varying data. Add to that a background of movable furniture, blowing curtains, and playing children that come and go in the room, and a simple picture-matching scheme is, well, out of the picture. To make smart sensing work, Carlson and his colleagues had to limit the amount of information the computer "sees," yet let it gather enough data for recognition. How? They did it by studying—and borrowing—tricks of the human eye.

A Face in the Crowd

"We began our research with two groups of people," says Carlson. "One group worked on how a computer could do recognition tasks, and the other studied how the human visual system works." Gathering picture information via computer was the easy part. The image seen by a conventional CCD video camera is converted to digital data and stored.

However, narrowing and interpreting a scene full of data to identify one or more specific individuals became the challenge, and it wasn't long before the two groups merged. "We began to appreciate just how really clever and wondrous the human vision system is," says Carlson. "We kept borrowing tricks from it in the way we do the processing."

Like our own vision system, smart sensing limits the amount of information by determining what's important in a scene. "We only see

The passive people meter first scans the room for people, then checks each person's face for identification characteristics stored in memory. When a match is found, the system adds in channel and time, and sends the data via phone lines to the Nielsen statistical computer.

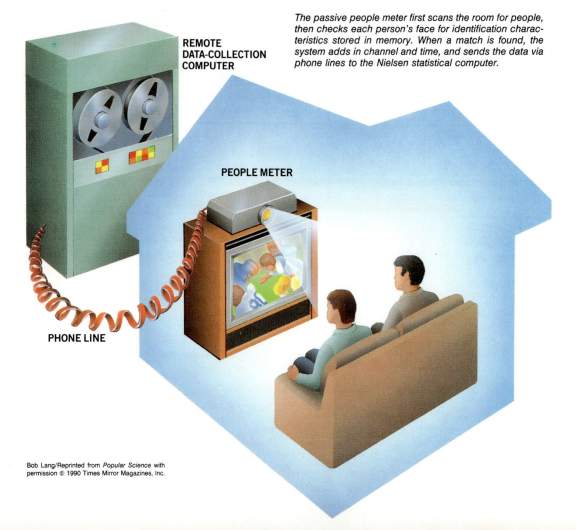

REMOTE DATA-COLLECTION COMPUTER

PEOPLE METER

PHONE LINE

details in the center of what we look at," says Carlson. "Periphery images are blurry." Being attentive to only the center of the image drastically reduces the processing required by about 1,000 to 1, according to Carlson. But the periphery of the scene isn't ignored entirely.

Carlson holds one hand out to his side and begins to wave it at me. "You're looking at me, but you still see my hand," he says. "If you want to, you can forget about me and pay more attention to my hand in your peripheral vision." The ability of the system to shift emphasis on what it sees lets it process only small pieces of an image in high resolution. That makes it fast and lowers the amount of memory needed to scrutinize a scene carefully.

In the Nielsen people meter, it means the clever box can switch between a broad low-resolution stare to a high-resolution observation. "When you enter the room, it will just track you as you walk," says Carlson. "Once you sit down in front of the TV, it will determine if you are a person." That eliminates pets like dogs and cats (and maybe some relatives). "Finally the system tries to recognize who you are."

The Nielsen box identifies you by matching the face it sees against the characteristics of the faces it has stored in memory. I learned that the squares I saw earlier superimposed over Carlson's nose, mouth, and eyes were the main facial areas the computer retains for comparisons. "It checks obvious things like the shape of your eyes," says Carlson, "but it also looks for more subtle characteristics like certain kinds of wrinkles on the side of your mouth."

Changing Variables

Of course, the match is seldom perfect—wrinkles come and go, and the angle at which the system sees your face can change drastically, depending on whether you're watching TV from the floor or the easy chair. For those situations, the system calculates what you *should* look like. "It uses the views it has stored to interpolate a view it doesn't know," says Carlson. "That's one of the keys to making this work."

Can it be fooled? "Sure," says Carlson. Then he smiles and takes off his glasses. "But so can you. I look quite different to you now." Major facial changes, like removing eyeglasses, are easily remedied: "We just teach the system what you look like with them on and off," adds Carlson. Minor changes, such as a slightly different hairstyle, will be acceptable to the system. "It's tolerant of normal facial changes,"

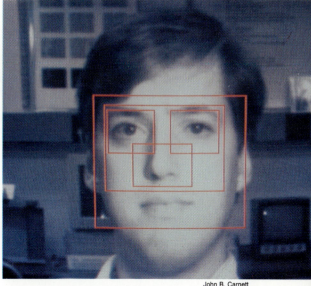

Squares superimposed over the video image indicate the areas that the system scans in order to "learn" a face.

says Carlson. "It should work as long as people aren't actively trying to deceive it."

Should work? While the demonstration I saw was impressive, Carlson and his associates still have another three years of work ahead of them before the Nielsen passive people meter actually hits a living room. The time will be spent shrinking components into a VCR-size box and determining what effects the variables of everyday living—from eating popcorn to changes in room lighting—will have on the system. (Carlson points out that those variables could be more of a challenge to overcome than what they have accomplished so far.)

But Carlson is already far beyond that, thinking of new applications. For instance, in the medical field, the tedium of reading pap smears could be replaced by a smart sensor that doesn't tire and make mistakes. In industry, manufactured components could be examined for quality, despite their position on the assembly line. In law enforcement the face of a suspect could be checked speedily against those on file. And in the home of the future, a smart sensor might recognize you at the front door to give you access.

An electronic butler? Carlson turns back to the glowing video monitor, and the computer again recognizes him. "Why not?" he says. "With smart sensing, it's all how you look at it."

Review of the Year

WILDLIFE

Habitat conservation as a means of preserving wildlife generated controversy in Florida, the Pacific Northwest, and elsewhere. Programs to reintroduce endangered species to the wild saw greater success in 1989.

by John Nielsen

The northern spotted owl found itself caught between loggers who covet its forest habitat and environmentalists who want the threatened bird preserved.

Owls Versus Loggers

With a bureaucratic flourish, the northern spotted owl was transformed from a small nocturnal raptor into an environmental kingpin. Like the Furbish lousewort and the snail darter before it, the forest-bound owl was rarely known until its designation as a threatened species.

Now, in the vague language of the Endangered Species Act, the government is required to ensure the survival of "minimum viable populations" of the spotted owl. To accomplish this, the habitat of the owl must now be preserved—a habitat coveted by loggers in the Pacific Northwest.

Since it grew teeth back in 1973, the Endangered Species Act has often been used to call attention to habitat issues. In 1979 environmentalists tried and failed to persuade Congress that protecting the snail darter's habitat was more important than building the Telico Dam project in Tennessee. In the early 1980's, the need to protect the lousewort plant helped foul plans to build a dam on the St. John River in Maine. In Alaska, using the same logic, some have argued that old-growth logging threatens the bald eagle. In the eastern part of Southern California, an endangered rodent has repeatedly been invoked as a means of slowing urban growth.

Politically, however, the spotted-owl fight dwarfed them all. For years the Reagan administration helped Northwestern business interests fight off attempts to list the bird as an endangered species, arguing that activists had overstated its needs. Loggers claimed that hundreds of thousands of jobs hung in the balance. Environmentalists claimed that 1.5 million acres (607,000 hectares) of "virgin forest" were at stake. Ultimately a compromise forbade the sale of timber from spotted-owl habitat, although it reduced the size of the lands officially designated as habitat. By all accounts the decision balanced more on political interests than environmental concern. But the owl and its habitat do appear to have been saved.

Zoos: Ups and Downs

The golden lion tamarin stands a good chance of staging a comeback thanks to years of propagation in zoos.

In an age of rampant habitat loss, zoos and zoo researchers offer a last, often best, hope. This year saw some victories. The golden lion tamarins, now being reintroduced in Brazilian nature reserves, might have gone extinct by now if they hadn't been shipped to zoos over the years. Black-footed ferrets, once believed extinct, have been saved by captive breeding; the first of these animals will soon be set free on their former Wyoming range. The world's only captive flock of whooping cranes, at the Patuxent Wildlife Reserve Center in Maryland, has more than doubled in the past 10 years.

Research departments funded by zoos and reliant on zoo animals continue to churn out important treatises on genetics, disease, and captive propagation. In 1989, unfortunately, the good news was tempered by two highly publicized deaths. In Washington a tiny panda cub born at the national zoo died after succumbing to an infection. Perhaps the more startling event occurred in San Diego, where a killer whale died after colliding with another orca in the midst of a show at Sea World. The tragedy renewed criticism from organizations opposed to the use of animals for amusement purposes.

Controversy at Yellowstone

Lovers and haters of hunting took turns on the soapbox last year. Using radios, four-wheeled trucks, and high-powered rifles, supervised hunters picked off hundreds of bison that strayed out of Yellowstone National Park into public and private lands in Montana. Drought and fires in the park had caused many of these animals to wander beyond their legal territory onto lands owned by ranchers. The kill rate in this legal "hunt" was 100 percent. Animal-rights groups argued that park officials were obligated to try harder to keep the bison from roaming.

Even for those who denounced the hunts as cruel, the alternative seemed just as heartless. Inside Yellowstone, native elk and bison from the two largest herds in North America died off at rates unmatched since record-keeping began. In one town bordering the park, elk were dropping dead of hunger in front yards; sickly elk were hit by cars on a daily basis. Critics said the park was remiss for allowing the herds to grow so large. One issue concerned the tricky question of whether these die-offs were "bad," or simply nature's way of regulating itself.

A related issue concerned the relationship between wildlife and public lands. In September 1989, a report prepared by the General Accounting Office (GAO) concluded that wildlife on federal preserves was being harmed by a whole range of human activities from waterskiing to military dogfight practices.

Much controversy arose when hunters picked off elk, bison, and other large animals that wandered from Yellowstone National Park in search of more plentiful food.

Photos: © Phil Huber/Sports Illustrated

Everglades Agony

Florida's Everglades has suffered at the hands of man for decades: slowly but surely, the delicate marshland has been losing its wildlife while gaining pollutants and unwanted species. The National Audubon Society estimates that 90 percent of the wading birds in these waters have disappeared since the 1920s. In recent years, agricultural runoff triggered choking algae blooms and explosions of exotic weeds. The degradation may have peaked in 1989. Fires fanned by drought swept thousands of acres of the habitat. Rising mercury levels in the water even led to bans on fishing and alligator hunting. The Wilderness Society described the Everglades National Park as verging on "extinction."

Restoring the Everglades presents a nearly impossible complex task. Southern Florida is now home to 4.5 million people, all of whom need water. Its economy is partly anchored by an agricultural industry built largely on reclaimed lands. But the blooms and the fires and the trends toward extinction could soon force a change. In January 1990, the governor of Florida announced a plan to dismantle the 1,400-mile (2,250-kilometer) network of levees and canals leading south from Lake Okeechobee. His plan carried the distinction of being the first time in American history that a project built by the U.S. Army Corps of Engineers would make way for a natural environment.

Pollution and fire have wreaked havoc with the delicate Everglades ecosystem.

Keith Meyers/NYT Pictures

Saving the Elephants
by Jane Perlez

The flat, khaki grasslands around Nairobi recede into the blue distance as the light plane piloted by the new director of Kenya's wildlife agency pushes north, past the snowy slab of Mount Kenya to the once-magnificent Meru National Park. Today the park lies in tatters, most of its rhinoceroses and many of its elephants killed off.

Dr. Richard E. Leakey has an important passenger on board, a veteran game warden who is being brought out of retirement to rescue Meru's dwindling animal population and its majestic riverine forest and acacia woodlands from fearless poachers, enfeebled rangers, and years of official neglect. As Leakey and his warden, Peter Jenkins, alight on the dirt airstrip, they are ushered into a shiny new Land-Rover, a recent gift to Meru from Leakey's office, and the only sign of anything that works.

For the next hour, the two men grimly inspect what used to be a resplendent park headquarters: a fleet of rusted trucks, including a 1948 Ford, encrusted in vines; a paralyzed bulldozer; three road graders, only one of which functions; an empty spare-parts room; and an armory equipped with World War I rifles.

The local warden is so overweight and out of shape he has difficulty keeping up with the

© Zig Leszczynski/Animals Animals

Africa's elephant population has declined by an alarming 54 percent in just 10 years. In Kenya, only 18,000 elephants hold on, compared to 65,000 a decade ago. Fortunately, one glimmer of hope remains: anthropologist Richard E. Leakey. As the recently appointed director of the Kenya Wildlife Service, Leakey has moved boldly to clean up the parks, inspire the demoralized game wardens, eliminate the ivory trade, and encourage tourism.

energetic Leakey as he makes his rounds. No wonder, Leakey mutters, that last November six priceless white rhinoceroses, supposedly protected inside a sanctuary within Meru, were brazenly gunned down by poachers, who chopped off the horns with axes, only yards from the warden's house.

Worldwide attention focused on Kenya's parks in 1989 when poachers turned their automatic weapons on tourists. In July of that year, as she traveled in a convoy of minibuses trundling between Tsavo and Amboseli national parks, a Connecticut tourist with the Audubon Society was killed by gunmen. That same month two French honeymooners on a self-drive safari were killed on one of the park's dirt tracks after they stumbled on poachers carving up a zebra. And in August, at Kora National Reserve, adjacent to Meru, George Adamson, the renowned friend of the lions, was shot to death at midday, a mile from his camp.

Years of Neglect

The slaughter of Kenya's elephants has been going on for years. Inside Tsavo, the country's largest park, the detritus of the poachers' handiwork—bleached white bones and bloodied carcasses—has become almost as common as the sight of live elephants. The sight of Warden Jenkins at Meru, however, was a distinct rarity, and evidence of what is being billed by Kenya as a new era in wildlife management.

The government pledges to reverse the course it has followed since the mid-1970s, when men of skill and experience—many of them, though by no means all, whites like Jenkins who had trained in the colonial days—were shunted aside in a political housecleaning. At the same time, the independent government agency that oversaw Kenya's national parks and managed their revenues was abolished, replaced by a department beholden to the general treasury. Under the new arrangement, the government could siphon off the steady income the parks generated in tourism. Little was returned to the parks, and before long the wildlife began to suffer. Poaching—in many cases by disgruntled, underpaid park employees—became commonplace, as did the theft of park entrance fees and equipment.

Corruption was widespread, Dr. Leakey and others now contend, because it was given a helping hand by senior officials in the government. One measure of Kenya's acknowledgment of its wildlife crisis may be that today the same government that presided over years of gross mismanagement grudgingly accepts such criticism. President Daniel arap Moi, in power since 1978, realizes that Kenya's vital tourist trade, not to mention much of its heritage, will be irrevocably eroded if its wildlife all but disappears. In May 1989, he turned to Leakey, internationally known for his fossil discoveries of mankind's origins, for help. He promised the high-profile paleontologist full political support as director of an agency once again in control of park revenues. After much delay the Kenya Wildlife Service was finally established by Parliament in November 1989.

Poaching, Population, and Proposals

Can Leakey save the wildlife? And—a critical corollary for the economic well-being of East Africa's most prosperous nation—can he persuade American and European tourists that it is both safe and scenic to come on safari to Kenya?

The most pressing and emotionally charged problem is rampant poaching, which has been spurred in the past decade by the soaring value of ivory. Only recently, with the adoption of a worldwide ivory ban at an international conference in Switzerland, has the price of ivory tusks dropped. Elephants are notoriously difficult to find and count, especially those that inhabit the densely canopied forests of central Africa, but it is generally agreed that the African elephant population has declined from 1.3 million 10 years ago to 600,000 today. In Kenya the elephant's decline has been even more extreme—from 65,000 a decade ago to 18,000 today.

But there are other formidable obstacles to saving the animals, ranging from the corruption in the wildlife department to basic issues of environmental planning. The outside world tends to view Kenya as the world's premier wildlife haunt, but the country has a rapidly expanding agricultural base and the highest population growth rate in the world. Kenya's biggest predicament may be how to manage the competition for land that has arisen between the burgeoning human population and its major foreign-exchange drawing card, the wildlife.

Leakey already has put forth some audacious proposals. He envisions running the individual parks as divisions of a large corporation, each park accountable for its revenues and expenditures. Once the profits flow, he plans to follow the example of Zimbabwe, which fosters support for wildlife preservation by giving farmers whose land adjoins a park a share of its proceeds.

Long gone, he emphasizes, are the days when rich visitors on safari could traipse around the landscape wherever they fancied—an observation hardly lost on a visitor to Kenya's parks, who now has to compete with other visitors for a view of the animals. On a recent day in the popular Masai Mara National Reserve, 24 minivans jostled each other for a glimpse of a few lions lounging in the shade of a tree. The quiet of the wilderness was ruptured by the exclamations of more than 100 people and the repetitive clicking of their cameras. At night, tourists often watch game attracted by salt placed at artificial ponds, a setting not unlike a well-landscaped zoo.

Fenced-in or Free-Ranging?

To Leakey, this concentration of wildlife is inevitable. The human population has doubled in the past 20 years, and agriculture has consumed much of the land where animals once roamed. With the competition for land between man and wildlife becoming steadily more intense, he envisions an ever-more-zoolike parks system, with hundreds, perhaps thousands, of miles of fencing.

Animal enclosures will greatly benefit farmers, especially in mountainous regions like Mount Kenya National Park and Aberdare National Park, where elephants regularly wander onto nearby tea plantations. To those purists who recoil at the thought of interfering with animal migrations, which these enclosures would likely do, Leakey's response is bluntly pragmatic: "Anyone who says you can't 'manage' the park is talking another century."

Ever one for flying trial balloons, Leakey seemed at first positively infatuated with fencing. He talked of putting one big fence around Tsavo, another around the Masai Mara and Serengeti ecosystem. He even insisted that the vastness of these enclosures would mean the spectacular migration of the wildebeest could continue unimpeded. In this famous annual peregrination, 1.5 million wildebeests travel in densely packed herds over the plains from the Serengeti in Tanzania, across the Kenyan border into the Masai Mara, and back again.

Leakey has since modified his plans for fencing—apparently after some acerbic advice from biologists on the consequences of such human interference. For the moment at least, he has dropped the notion of making the Masai Mara and Serengeti the world's biggest enclosed wildlife park. But "managing" with fences is still central to his plans: "That we have to constantly interfere," he says, "is a given." The days of free-ranging wildlife are over for Kenya. "It's not realistic in America," he says. "And if you don't accept it in your country, you can't expect us to accept it here."

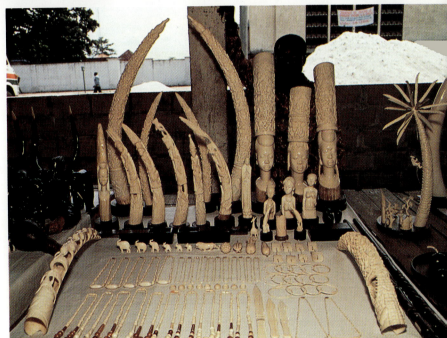

Ivory poachers simply cleave off the elephant's tusks and leave the animal to die (above). The tusks, some weighing up to 200 pounds, sell quickly on the black market. Ivory carvers (upper right) sculpt intricately detailed figurines, jewelry, and other items from the tusks. Carved ivory objects eventually find their way to shops in Africa and the Far East that cater to the tourist trade.

Top left: © Ted Wood/Picture Group; top right: © William Campbell/Time Magazine; above: © P. Robert/Sygma

Raising Morale

A tall, energetic man, weathered in the ways of Kenya's byzantine politics as well as the more gilded avenues of international fund-raising, Leakey, 45, is confident he can carry off the challenge. He is also sure, almost cocksure, that he has the backing of the top powers in the government, specifically President Moi. But if any of his proposals are to work, he'll have to raise a substantial amount of money—about $150 million, he estimates—much of it from the international economic community. Given that Kenya's foreign-exchange reserves amount to only about $200 million, some Western observers think he is overly ambitious. But with the World Bank showing interest in wildlife as an economic resource for Africa, Leakey remains hopeful.

A good deal of money should become available now that the Wildlife Service controls its own budget. The parks and Kenya's Indian

WILDLIFE 351

Both photos: © Louise Gubb/JB Pictures

Ocean resorts generate yearly foreign-exchange earnings of more than $300 million. No longer will Kenya's parks be expected to run on a feeble $7 million annual budget. The Wildlife Service will also be raising the park entrance fees. It now costs 80 Kenya shillings, or about $4—"the cost of a tot of whiskey,"—to enter a park for an indefinite stay.

Touring his new fief, Leakey sometimes assumes the manner of head prefect at an English boarding school. "Why do you have those South American exotic plants inside there?" he recently chided a nervous-looking gardener about the shiny, decorative shrubbery inside one tourist lodge. "It's inappropriate; it's like importing tigers. Take them out and put in local varieties," he ordered, adding with a smile: "Mine is a cause of ethics; yours is of beauty. One is higher than the other."

Leakey hopes potential foreign donors will be swayed by the improvements already visible at Tsavo National Park, where the slaughter of two or three elephants *per day* has been virtually halted since he took charge. "The rangers were totally demoralized," he explained recently, aloft in his Cessna, his usual means of transportation. "They had very old equipment, little ammunition, ragged clothes, and gym shoes. They were getting no allowance, which they are supposed to get when they go out on patrol.

"We raised some funds, reequipped them with automatic rifles, ammunition, boots, uniforms, and daily allowances. We now have small, effective groups of rangers, 10 or so in

The campaign led by Dr. Leakey (below) to crush the illegal ivory trade has greater support now that the U.S. and 102 other nations have declared the elephant an endangered species and banned all trade of ivory products. The Kenyan government showed further support by burning 12 tons of elephant tusks taken from poachers (above).

each group, who set off on foot on planned operations. They carry radio equipment—if a poacher's footprint is seen, they can radio base, and directly an aircraft can come overhead."

In addition to two new helicopter gunships, Leakey's park rangers are now backed by a network of informers—former poachers, many of them Somali—who are rewarded with up to $500 for information that turns up poachers and/or their weapons and contraband ivory.

The rangers' most powerful weapon may be a controversial "shoot-to-kill" policy President Moi introduced in late 1988 as a means of combating poachers. The policy has been strongly criticized by several human-rights organizations. According to the Wildlife Service, more than 60 poachers have been killed or captured since the reinforced patrols started last May, and 500 tusks have been recovered.

The poachers are tenacious—in early October 1989, in a major setback, 11 elephants were found slaughtered just outside the perimeter of Tsavo. But that month also marked an event Leakey hails as the biggest triumph so far in the battle with poachers: the shooting, by a patrol, of Mohamed Hussein Omar, reputed to be the head of a poaching gang that had been plaguing Meru National Park. Ballistics tests showed that the bullets from Omar's G-3 automatic rifle (a government-issue model, and thus poacher's booty) matched the bullets that killed the two French tourists earlier in the year.

Ambivalence About Tourists

But Leakey must now contend with a by-product of his success in curbing the attacks on elephants—an increase in the danger for people. Robbing both tourists and Kenyans appears to be emerging as an alternative to shooting elephants. "In three weeks, as a poacher, you could get 10 good tusks and get $1,000 to $1,500," he observes. "Suddenly you can't do that because you're chased with planes overhead. So you do something else."

The tourists, on whose behalf much of the conservation work is being undertaken, are also contributing to the wildlife crisis. Amboseli National Park, some 150 square miles (390 square kilometers) of plains, swamps, and marshes spread below the glistening snowcap of Tanzania's Mount Kilimanjaro, is the most heavily visited of the 16 major parks in Kenya. Its compactness, accessibility, and density of elephants have made it a regular stop for the safari companies that guarantee their clients the "big five"—elephant, buffalo, leopard, lion, rhino—during five days of hard driving from park to park in zebra-striped minivans. To the old Kenya hands, Amboseli was a favorite haunt, its setting unsurpassed. No longer. "Between the cars and the animals, they've devastated it," says David Allen, a private safari guide and former hunter and game warden. "I don't take clients there if I can avoid it."

Leakey recently flew down to Amboseli, an hour's journey south to the Tanzanian border from Nairobi, over the Athi Plains and the floor of the Great Rift Valley, the 4,000-mile (6,400-kilometer) scar that slices through East Africa. There he met Dr. David Western, a fellow Kenyan who is the director of Wildlife Conservation International, the research and conservation division of the New York Zoological Society. With Western was Amboseli's new warden, Francis Mukungi, an energetic, perennially smiling member of the wildlife department.

The three drove to a nearby tourist lodge through a worn, dry, and barren landscape, with trees stripped of their bark, trunks broken and splattered across the yellow earth. All this was elephant damage. The harm done by tourists could be seen in the rutted and pockmarked roads. Safari vans tore around in swirls of each other's dust. Deep tire marks veering off into the grasslands attested to the fact that drivers often break the rules—all vehicles must stay on roadways—to offer tourists better views.

Amboseli reflects the harsh imperatives of wildlife management in an era of limited resources. In the mid-1970s, poachers decimated herds of elephants on the park's periphery, causing the survivors to crowd into Amboseli for sanctuary. Today 720 elephants live in an area too small to sustain them indefinitely. Thanks to their clamorous stomachs, the tree cover has gradually given way to savanna, pushing out the giraffes, monkeys, and bushbuck that thrived on the old vegetation. There is no place left for Amboseli's elephants to go. If they are turned out of the park, poachers will attack them. Instead, says Leakey, some of these elephants may have to be destroyed, or culled. (This form of wildlife management, historically repugnant to Kenyans, recognizes the elephants' sophisticated social organization: entire families are culled so as to prevent orphaning the young and traumatizing older group members.)

Another remedy, abhorrent to the tour operators, but presumably appealing to the tourists, will be to limit the number of vans entering what is fast coming to resemble a desert moonscape. But overshadowing all these plans is the legacy of entrenched government indifference. On his periodic inspections of the parks, Leakey finds himself the official representative of a ruling power whose credibility is badly tarnished and must be quickly restored.

An Anomaly for Kenya

Kenya's wildlife directorship came to Richard Leakey, as have many things in his career, somewhat unexpectedly. In May 1989, perhaps spurred by the international outrage over news photographs of bloody elephant carcasses, President Moi ousted Dr. Perez Olindo, a black Kenyan with a respected scientific background but little management experience. As Olindo's replacement, Leakey was at once a flamboyant and popular choice. He knew far less about living creatures than about fossils, but had strong management skills, acquired during 20 years of building the National Museums of Kenya into a research institution that, among other things, supported research into man's African origins.

Leakey's experiences with wildlife—aside from holding the title of chairman of the East African Wildlife Society, which gave him a springboard from which to deplore the state of Kenya's wildlife from time to time—came in his youth, as a safari guide running his own business. In his autobiography, *One Life,* he recounts his effort one evening to dislodge a herd of elephants feeding around a tree house in which he and his clients planned to sleep. He shot a heavy-gauge gun into the air to scare the elephants off, but was practically knocked off his feet by the blast—and, worse, enraged one elephant, who came charging down the path. "I fled back to where my party was," he writes, "arriving breathless and very red in the face."

Yet when Leakey assumed the wildlife post, there were sighs of relief from even the most skeptical old Kenya hands. The tour operators welcomed his reputation as a firm manager, and liked what they saw as his independence from the tribal politics that seemed to discourage decision-making under Olindo. His fund-raising experience, his British Broadcasting Corporation (BBC) television series "The Making of Mankind," which has appeared many times over in the United States, as well as his fossil work and its attendant international lecturing, endowed him with the kind of clout that many inside and outside Kenya felt was essential for the job. And there was evident approval from ordinary Kenyans—teachers and students in particular—who started writing to the wildlife department for the first time, and visiting the tourist-laden parks themselves.

Leakey is a bit of an anomaly for Kenya, a fiercely patriotic white man who broke from the colonial ranks by opting for Kenyan, rather than British, citizenship at independence in 1963. He is capable of trading on that distinction, taking

Conservation also has an economic incentive: the important source of revenue that tourism represents. Governments in tropical Africa realize that once their wildlife disappears, tourists will go elsewhere to spend their money.

© Mark W. Boulton/Photo Researchers, Inc.

advantage of the fact that he can get away with behavior others in the Kenyan political milieu cannot. Before his appointment, he was not so subtly campaigning for the post by starting a bruising public fight with the then minister of Wildlife and Tourism. "Nothing was being done [by the government] about the poaching," he says. "People were coming to me saying I was the only one who could create a crisis." (The minister has since been dismissed.)

Leakey is not shy about his ambition, and his talk can edge toward the grandiose. He likes to mention his ability to get President Moi on the telephone whenever he needs to, and he delights in telling anecdotes about the cabinet meetings he now attends. Recently, he says, he told the ministers present during one informal meeting with the president that there was no need to bob up on their feet every time their leader had to leave the room to make a telephone call. "They saw my point," he notes dryly.

The best-known of the three sons of the paleontologists Louis and Mary Leakey, and a distinguished paleontologist himself, Richard Leakey grew up in Kenya in a household that held few of the views of the colonial white settlers. His father was a fan of Jomo Kenyatta, the country's first president. In his own words, Richard became "very nationalistic" early on, as did his brother, Philip, who today sits as the only white member of Kenya's parliament.

By the time independence came in 1963, the senior Leakeys were well on their way to becoming international celebrities with finds of early man at Olduvai Gorge in Tanzania. It did not take long for young Richard to follow suit. In the late 1960s, he joined a fossil expedition led by his father to southern Ethiopia, made at the invitation of Emperor Haile Selassie. On a flight there, Richard spotted ancient lake sediments, likely terrain for fossils, on the eastern shore of Lake Turkana in northern Kenya.

His father was skeptical, but with the bravura that he had learned from living in a prominent, well-traveled family, 24-year-old Richard persuaded the National Geographic Society's research committee in Washington, D.C., to finance an expedition and a base camp at the lake. His hunch was right. Leakey and his second wife, Meave, a zoologist who completed her doctorate with Louis Leakey, have spent a large part of the past 20 years at the camp, known as Koobi Fora, uncovering numerous skulls of early man, including *Homo erectus,* representing the earliest record of that species in Africa.

© Verna R. Johnson/Photo Researchers, Inc.

Superstitions involving its horn have also made the endangered white rhinoceros a favorite victim of poachers.

Just before his father's death in 1972, Richard made his most significant find at Koobi Fora: the "1470" skull, so-called for its museum number, of the nearly 2-million-year-old man.

A Manager First

Leakey may lack direct experience with wildlife, but the important credentials for today's crisis, he maintains, are management skills. "I manage people; there's a lot of expertise around," he says. "We can buy it." He is not enamored of environmental studies; he thinks there are probably enough of those on everyone's shelves to last a lifetime. Instead, he has taken a characteristically unfamiliar route—enlisting the services of a team of accountants from Price Waterhouse, with financing from the U.S. Agency for International Development (AID), to review the Wildlife Service's accounts.

This businesslike approach extends to a startling honesty and pragmatism about the current state of the parks. As he headed to the U.S. to pursue discussions on aid with the World Bank and AID, he was asked by a major tour operator in Kenya, Abercrombie & Kent, to issue a statement abroad stating that tourism in Kenya is safe. After some hesitation, Leakey agreed to make a statement. But he remains candid about the possibility of danger.

"I think it is safe," he says. "But it is always a risk in a country like this, with remote areas and large unemployment. The safari companies have been advised not to use certain areas and park routes. We want to avoid giving guarantees. I don't want to end up being sued."

WILDLIFE 355

© Michael A. Smith

The Day of the DOLPHINS

by Justine Kaplan

In our own oceans lives a sentient being, a creature perhaps as curious and communicative as we are. Torpedo-shaped and alien as Martians, 66 species of dolphins, porpoises, and other toothed whales inhabit the Earth's waters, detecting sound through their jaws and breathing from openings on the tops of their heads. It is perhaps irresponsible to draw comparisons between man and a 300-pound (136-kilogram) hairless marine mammal with skin like a wet inner tube. But stare into the eye of a dolphin, and some bit of information is exchanged; some mutual yearning happens. There is a reflective glint, a hint of wisdom.

From the beginning of time, the dolphin's remarkable abilities have been extolled by poets, philosophers, and historians as well as by the ancients, whose legends tell that dolphins were really men transformed by the gods. Only in the past decade, however, have researchers in the field of cetacean behavior begun to determine how intelligent these aquatic sages truly are. Scientists analyzing the dolphins' communication systems are working to decode the animals' verbal and nonverbal signals. Researchers are attempting to determine how echolocation—the complex sonar communication system dolphins use to navigate their undersea world—can be used to protect them from their greatest threat, the tuna net. Cetacean scholars from Florida to Australia are studying the dolphin's complex familial and social relationships. And in the laboratory, neurologists are attempting to unravel the mystery of the dolphin brain, an organ quite different in function from man's, but with a creativity center larger than our own. While most of the research focuses on understanding the communication systems and minds of these animals, more recent—and perhaps riskier—studies include using dolphins as an adjunct to treating a variety of human ills.

A Kiss from a Dolphin

Every Tuesday at the Dolphin Research Center in Grassy Key, Florida, psychologist David Nathanson relies on six Atlantic bottle-nosed dolphins to help bolster the speech and memories of mentally handicapped children. Nathanson is finding that the children pay closer attention and learn up to 10 times more quickly in these surroundings than they do in classrooms, where their usual reward is verbal praise and a hug or a kiss. "They prefer a kiss from the dolphins to a kiss from me," says Nathanson. During some lessons, pictures displaying simple words such as *bus* or *dog* are tossed to the dolphins, which bring them to the children. If the child pronounces the word correctly, the reward is a swim with the dolphins.

"Dolphins seem to sense these kids are handicapped," Nathanson adds. "They perceive these children as more helpless than people they usually swim with, so they're gentler."

A second facility devoted to human-dolphin interaction is Dolphins Plus, a private "dolphinarium" in Key Largo, Florida. In the summer of 1987, the owners turned over the facility for eight days to Betsy Smith, Ph.D., an associate professor of social work at Florida International University in Miami. Smith, who has been working in animal-assisted therapy for 15 years and was the first to try dolphin therapy in 1971, wanted to test her theory that autistic children who had not benefited from other types of treatment might become more sociable and communicative by being with the dolphins. Ten years ago Smith began studying the responses of neurologically impaired children to dolphins and found that those children with autism responded most dramatically. "Dolphins, who have no preconceived expectations, approach an autistic child as they would any other—for spontaneous interaction," says Smith.

The dolphin's playfulness, intelligence, and affinity with humans have made it the most beloved of sea mammals and a favorite subject for wildlife research.

While the children's social skills appeared to improve after spending time with dolphins, Smith admits she has not proved that the dolphins, rather than the therapeutic effects of water, are responsible for her success. But she says the children who spent time with the dolphins appeared to be more energetic and motivated than a control group that was taken to swim at another beach. Questionnaires and interviews with parents have shown that the "dolphin kids" demonstrated long-term behavioral and social changes, and she plans to test these results further in a follow-up study.

Dolphin Cognition

As much as it needs more raw data, the scientific community also needs concrete evidence to support the various claims concerning dolphin cognition. Most cetacean scientists agree that the person who has given scientific credibility to the field is psychologist Louis Herman, director of the University of Hawaii's Kewalo Basin Marine Mammal Laboratory. "Herman's a real titan in the area. His work is the very backbone of the field of dolphin cognition," says Kenneth Norris, a field biologist at the University of California at Santa Cruz. "He has almost singlehandedly defined the capabilities of the dolphin mind." For the past ten years, Herman and his colleagues have worked with two Atlantic bottle-nosed dolphins, Phoenix and Akeakamai, methodically testing their ability to understand and execute commands in two kinds of artificial language. Phoenix's language consists of electronically generated computer whistles. Ake's is based on hand and arm gestures. Each language has a "vocabulary" and a set of rules governing how the sounds or gestures are arranged in sequences that form thousands of sentences. Using these languages, Herman has shown that dolphins understand the meanings of the words in their languages and, even more important, how word order affects meaning. This ability is considered to be at the core of most human languages, a trait many linguists and philosophers would argue is a sign of intelligence.

Herman has discovered, for example, that dolphins can differentiate between phrases such as "Pipe fetch surfboard"—which translates to "Get the pipe and take it to the surfboard"—and "Surfboard fetch pipe," which means "Get the surfboard and take it to the pipe."

Some scientists have argued that it is misleading for Herman to call what the dolphins have learned "language." David Premack, a former ape researcher at the University of Pennsylvania, who has retired from his animal language work, claims Herman's "free use of a sentence" is a problem. Human language, he argues, consists of abstract concepts, not just objects and actions. And a decade ago Columbia psychologist Herbert Terrace published statements that ape-language researchers were incorrect, that animals learned not language but behaviors—behaviors that earned rewards.

Herman retorts that "the fact that an animal gets a reward does not invalidate the involvement of language." Furthermore, Herman contends that dolphins "develop an understanding of the words of their language at the level of a concept." For example, *under* means passing beneath, and dolphins will raise an object from the tank bottom to swim below in response to *under*. He has also demonstrated that dolphins understand references to absent objects. When asked "ball question," which means "Is there a ball in the tank?", Ake searches the pool and responds on a "yes" or "no" paddle. When the dolphin presses the "no" paddle, it implies she has understood the sign, formed a mental image of the object referred to, and deduced the ball is not there. This ability—called referential reporting—has previously been documented only in apes and man.

Signature Whistles
Scientists have known since 1965 that dolphins have distinctive signature whistles, which they use to identify themselves. Work conducted by Peter Tyack, an assistant scientist at Woods Hole Oceanographic Institute, showed that the animals were actually imitating one another's whistles. "Learned mimicry is rare in the animal kingdom, but dolphins can be trained to imitate particular sounds, and wild dolphins appear to use this skill to imitate each other's whistles, perhaps to initiate social interaction," says Tyack. With support from the Office of Naval Research in Arlington, Virginia, Tyack is studying the social function of the whistles by recording the sounds of captive dolphins at the New England Aquarium and Chicago's Brookfield Zoo. Tyack is also analyzing the signature whistles of a wild-dolphin population followed in Sarasota Bay, Florida, for the past 19 years by researcher Randall Wells.

Five months out of the year, Wells, along with guest researchers and volunteers from Earthwatch, a nonprofit research organization, tracks the Sarasota dolphins to understand how they divide themselves into subgroups. The dolphins are captured for short periods of time for identification and sampling; during that time, Tyack records the whistles of each dolphin handled, including those of pairs that have bonded together, such as mothers and calves.

Wells has recently expanded his population studies to the Tampa area. Since he began his studies in 1970, Wells has identified nearly 600 individual dolphins from Florida's central west coast, and has determined that the populations segregate themselves into groups. It appears that adult females with calves form bands that may include three generations of females as well as some unrelated females. It has also been established that the adult females seen briefly accompanying calves may be, not their own mothers, but "baby-sitters." As dolphins mature, Wells and his colleagues have found, they join sex-segregated groups. Within these groups, males form pair-bonds with other males for years at a time. A similar study by a group of University of Michigan graduate students at Monkey Mia, a remote bay in Australia, has shown that the male pairs often herd or "kidnap" females for mating. This bonding, which Wells says may be as tight as female-calf bonds, may give male dolphins the group dynamics they seem to need, as well as protection from sharks.

Dolphin Sonar
The greatest threat to dolphins, however, is the tuna net. While the Marine Mammal Protection Act has been amended to limit the number of dolphins "accidentally" killed by fishermen, enforcement is lacking. Scientists studying dolphin sonar (*so*und *na*vigation *r*anging) as well as their schooling systems are attempting to develop ways to save them from the nets. The sonar, or echolocating, blasts are actually short bursts of sound pulsing through the fatty tissue of the dolphin's forehead, or mellon, where they are focused into a beam that travels through water 4.5 times faster than through air. When the sound is sent, it bounces off a target and returns an echo containing information on the surrounding environment. The dolphin hears through fat deposits in its jaw, from which sound is transmitted to the inner ear and then to the brain. A broad band of low- and high-frequency sound emissions, which bounce off or go through fish and other animals like an X ray, lets dolphins navigate in turbid waters and locate and identify objects well out of visual range, up to 2,600 feet (800 meters) away.

There has been some speculation that dolphins may use their sonar to project images into other dolphins' brains. Perhaps sonar also gives dolphins an understanding of the humans that enter their environment. Pregnant women have reported that when they are in the water with dolphins, they can feel blasts of energy directed at their wombs. The echolocating beams are gentle, soothing vibrations that feel like energy pulsing through a machine.

Ken Norris, who identified the dolphin sonar 40 years ago, believes that some of the sounds are powerful enough to kill fish. "When the dolphin gets near a fish, there's a loud bang 200 to 500 times as long as a regular echolocation click." In his lab, Norris has killed anchovies with simulated feeding sounds.

The Net Threat

Even with such a fine-tuned sonar system—a dolphin can detect a vitamin capsule from the far end of a large pool—dolphins are still unable to sense fishing nets. Given that deficiency, Norris has proposed a way to release dolphins from tuna nets based on his studies of the mechanics of the dolphin school. In vast regions of the tropical Pacific, prized yellowfin tuna are often found in the company of spotted, spinner, and

The mysterious means by which dolphins communicate has fascinated humans since ancient times. A dolphin emits bursts of sound waves through the fatty tissue of its forehead, or mellon. The sonar beam bounces off objects and returns as an echo to the dolphin, giving the creature information about its environment. Researchers can pick up the dolphin's sounds by attaching a microphone to the mellon by means of a suction-cup device (above). Researcher Diana Reiss (right) has designed a keyboard through which dolphins can "ask" for a ball, a rub, and so on, by pushing the appropriate key.

common dolphins, an association that has helped to build a billion-dollar fishing industry. Fishermen locate schools of tuna by spotting dolphins, then herd the tuna and dolphins into a bunch and drop a large net around them, which they shut at the bottom and winch in. Although they attempt to release the dolphins while keeping the fish, the dolphins become disoriented and entangled in the nets, where they suffocate and drown. Approximately 125,000 dolphins are killed by fishermen each year.

By understanding the dolphin schooling system, which he calls a "magic envelope," Norris believes a method can be developed to release dolphins from the nets. One function of the school, says Norris, is the "chorus-line effect"—an intricate system that allows dolphins to evade sharks. As each dancer on the chorus line anticipates the kicks of his or her neighbor, Norris explains, so does each dolphin foresee the moves of the others in its school and then acts in concert with them.

Caught in a tuna net, dolphins are "stripped of this magic envelope," says Norris. "Their school system does not work when they are pressed together. To release them, we have to design a way to let that magic envelope out intact and maintain the spacing of individuals."

In Hawaii in the late 1970s, Norris tested his theory by encircling schools of dolphins in nets with 12-foot (3.5-meter) openings—and found the dolphins wouldn't go out. At 20 feet (6 meters) some of them did. Norris and his colleagues are testing the configurations of net openings to determine the size of the opening required to let the entire school escape.

Big Brain

It would seem that a creature with a brain approximately the same size as man's could figure out how to avoid such trouble. Its keen sense of hearing did, after all, evolve to compete with the shark's acute sense of smell.

It's a strange combination of abilities and inadequacies that has led to the fascination with the dolphin's big brain. In humans the part

Above: © Joseph A. Thompson/Seavision Productions; below: © Dawson/Greenpeace

Tens of thousands of dolphins die each year after becoming entangled in tuna nets (above). The carnage has not gone unnoticed: Greenpeace and other environmental groups continue to stage daring rescues of entrapped dolphins (left). The United States and other countries are pressuring the tuna industry to develop nets from which the dolphins can readily escape.

of the brain known as the neocortex is considered the seat of creativity and thought. In man the total cortical surface is 96 percent neocortex; in the dolphin, 98 percent. But despite the similarity in size between the neocortices of man and dolphin, they are quite dissimilar.

"What the dolphin neocortex has done evolutionarily," says Peter Morgane, senior scientist at the Worcester Foundation for Experimental Biology, "is get bigger, but it doesn't seem to have gotten more complex. It has a very ancient type of organization, and we don't see this type of arrangement in any modern land animal. The dolphin has kept the brain it had when it first went into the water some 50 million years ago." Qualitatively, the dolphin neocortex bears more resemblance to the brain structure of the hedgehog—whose cortex evolved more than 100 million years ago—and of the bat—another ancient brain—than it does to those of modern land animals.

On land, mammalian neocortices evolved, becoming more specialized. These cortices are thought to be crucial to learning and emotion. For instance, the smell of pine can evoke thoughts of Christmas past. In the dolphin brain, the somatosensory areas—vision, hearing, touch—remain more generalized than those of land mammals. The last stage of cortical evolution—the formation of primary sensory and motor regions—did not take place.

In 1986 Harry Jerison, a microbiologist at the University of California Medical School, proposed that "motivational" functions such as intimacy and feelings may be more active in the dolphins' neocortex than in humans', and their thoughts more emotionally charged.

He also theorized from research on bats that huge areas of dolphin brains are devoted to a sophisticated echolocation system. Given the enlarged dolphin neocortex, he proposed, the dolphin echolocation system may have evolved to the point where it creates the dolphin's "reality." Dolphin echolocation may operate as visual processes do in humans, "creating both perception of the outer world and of the self."

Shared Perceptions?

It has already been demonstrated that bats can intercept one another's signals. This kind of interception could mean they could share perceptions about the external world. "Within its perceptual world, the dolphin may not have constructed an individual but a communal sense of self," says Jerison.

Norris is not willing to accept this group-consciousness idea. "There is no question dolphins pick up a lot of information by listening to each other's echoes," he says. "And I think the idea is engaging, but there is no proof. We do have evidence that when they echolocate, they avoid spraying each other with sound. They have a thing we call dolphin echolocation manners: they simply shut down their gear. They have developed a window to spray the environment, but not each other, with sound."

He is also unconvinced that dolphins have language. Instead, he hypothesizes, there may be a more ponderous, multilayered, emotionally based communication system. "It's more like music," says Norris. "There is a lot of rhythmicity, multiple messages carried in beats," such as a single whistle that identifies one dolphin to another and at the same time tells that dolphin that there is trouble.

Some researchers, however, describe Jerison's notions of a communal consciousness as "philosophical" and "ethereal." Jerison seems to be proposing that the dolphin sees the world differently than man does—"which it may," Morgane says, "if it has a different brain."

Researchers from SETI, the Search for Extra-Terrestrial Intelligence, would probably agree, which is why a number of space agencies have taken a keen interest in the research of Diana Reiss, an animal- and human-communications professor at San Francisco State University and director and founder of Project Circe at Marineworld Africa, USA in Vallejo, California. Reiss's work decoding dolphins' communication systems seems similar to the quests for extraterrestrial intelligence. In a 1987 paper, Reiss spelled out the problems the two quests share. "How do we detect signals and recognize patterns when we are unsure of the nature of the signal?" she asked. "How does an observer of one species penetrate the communication system of another when we are really deaf and blind to that system?"

During her eight-year study of four bottlenosed dolphins, Reiss has concentrated on analyzing both vocal and nonvocal signals. She constructs "ethograms"—a behavioral code—by watching and videotaping the dolphins, tape-recording her observations, then sorting their behaviors into categories, searching for a relationship between the dolphins' sounds and their other behaviors.

Reiss has also designed an underwater keyboard of nine keys that the dolphins can manip-

The U.S. Navy sponsors much important dolphin research. Trainers (above) have taught dolphins to guard submarines, retrieve objects lost overboard, and detect military equipment in murky waters (left).

Top and above: Official U.S. Navy Photographs

ulate. Each time a dolphin pushes a key with its beak, a computer records it and produces a sound specific to that symbol. Then the dolphin gets a specific object or activity: a ball, a rub, and so on. Soon after Reiss installed the keyboard, the dolphins taught themselves to use it, imitating the computer whistles they would hear after pushing the keys, and incorporating those whistles into their own vocal repertoire. This suggests that they were learning to associate key whistles with visual signals.

Navy Research

Not all research has been so benign. Currently most financial support for dolphin studies comes from the U.S. Navy, an organization recently accused of abusing the marine mammals in its program. In the past four years, $30 million has been spent on work the Navy refers to as research into "marine biological systems" with dolphins, sea lions, and beluga whales. During the past 30 years, the Navy has trained 240 animals at secret facilities in Hawaii, Key West, and San Diego, and has published more than 200 papers on its research into the dolphin's sonar, diving, and retrieval abilities and anatomy, as well as the electrical activity of the dolphin brain. But there is much public interest in what the Navy will not disclose, which appears to be plenty.

Navy officials admit that the dolphins were sent to Vietnam for research into the animals' ability to detect military equipment in the murky waters. They have also revealed that dolphins were in fact involved in recovering torpedoes, detecting mines, and locating hostile frogmen in the Persian Gulf. And the Navy has publicly confirmed its plans to use 16 dolphins to guard the Trident nuclear submarine base in Bangor, Washington—an action 15 environmental and animal-rights groups are trying to block.

The fact remains, however, that Navy funding is essential for much of the research on dolphins to continue. And with recent cutbacks in the Department of Defense budget, even those funds are waning. Grants from the National Science Foundation (NSF) and other research organizations are coveted, and a university affiliation for a marine-mammal scientist is rare. "The peer-review process at NSF has become very conservative in their support for new or risky projects. The Office of Naval Research specializes in funding this kind of innovative research," says Woods Hole's Peter Tyack.

Swimming with Dolphins

Excluding the Navy, 641 permits have been granted to date by the National Marine Fisheries Service (NMFS) for purposes of research and public display in the United States, and last year the Marine Mammal Protection Act was amended to require that "dolphin swim" programs contain some kind of conservation or educational element.

The Marine Mammal Commission is currently reviewing the four permits granted to these programs, where guests pay up to $55 to swim with dolphins. The original permits were granted "with reservations," says John Twiss, executive director of the Marine Mammal Commission. "We approved the NMFS permits, but recommended that these swim programs be conducted on an experimental basis until the end of 1989, when preliminary data were available to evaluate them." The last "swim" permit to be granted was to the Hyatt Waikaloa on the Kona Coast of the island of Hawaii, a megaresort with the largest captive habitat for dolphins in the world. At the Hyatt's Dolphin Quest, scientists and trainers are merging research, education, and entertainment; and there is no question that it will be a profitable venture.

The habitat is a 5-acre (2-hectare) natural seawater lagoon, open to the ocean and teeming with tropical-reef fish and mullet—fish the dolphins eat naturally in the wild. The space is so large, in fact, that the dolphins can choose not to participate by swimming away.

The popularity of this program is obvious on examining the daily waiting lists for the $55 dolphin encounters. They number in the hundreds for the 40 spaces available under a lottery system. And of the 30 minutes allotted to the experience, only 10 are actually spent with the dolphins in the water; the other 20 consist of a lecture on dolphin behavior and anatomy. But the swimmers still come.

Without question, swimming with dolphins is an uplifting experience. Despite the intrusion into their captive environment, they appear to unconditionally embrace humanity. Our communion with them is real, not imagined, reiterated time and again by the myriad scientists devoted to comprehending their "alien" intellect. "We're not on a pedestal someplace, alone and magnificent. Dolphins are one branch of a tree we are all a part of," says Norris. "We are of them, they are of us, and the more we know about them and the other animals, the fewer the barriers there will be between us."

"Dolphin swim" programs have skyrocketed in popularity. The dolphins seem to enjoy their excursions with humans.

© M. Timothy O'Keefe/Bruce Coleman Inc.

America's Other Eagle

by Gary Turbak

Though wise in so many ways, America's Founding Fathers were perhaps poor judges when it came to birds. By an act of Congress in 1782, they chose to adorn the Great Seal of the United States with the bald eagle, an undeniably beautiful bird, but also a mere scavenger and catcher of fish.

Presumably, the members of the Continental Congress sought a symbol of courage and independence to match their hopes for the infant republic; undoubtedly, they wanted a bird embodying the splendor of the new nation and its magnificent virgin lands. So how come they overlooked a choice as good as gold—the golden eagle, that is?

America's "other eagle" is more aggressive, more numerous, more skilled as a hunter, and even a bit larger than its white-headed cousin. Yet it has traditionally been overshadowed by the national symbol. Many Americans, particularly people in parts of the East, where the golden eagle infrequently flies, may not even know the bird exists. Meanwhile, in some areas of the West, the creature's taste for mutton long ago earned it the enmity of sheep ranchers. Today a symbol of the country's wide-open spaces, the magnificent raptor is also a lightning rod for controversy.

Weighing up to 13 pounds (6 kilograms) and with a wingspan of 7 feet (2 meters) or more, the golden eagle is one of the world's largest birds of prey. In North America the only larger raptor is the nearly extinct California condor, now confined to zoos. With that great size comes power, "truly awesome power," in the words of Mike Kochert, research leader at the Snake River Birds of Prey Area in Idaho. "Some smaller raptors may be more aggressive, but for pure energy on the wing," he says, "the golden eagle is tops."

Stable and Healthy

Soaring across stretches of North America, Central America, Europe, Africa, and Asia, the species is most abundant in places—like the American West—that provide unrestricted, panoramic views (for spotting prey) and consistent breezes (for easy soaring). Goldens breed in the seventeen westernmost states and in nearly all of the Canadian provinces. Biologists put the United States population at about 63,000 birds, including more than 17,000 breeding pairs. "The golden-eagle population is stable and healthy," says Kochert.

One reason for that success is that the species escaped the toxin scourge (of DDT and other chlorinated hydrocarbons) that severely reduced populations of bald eagles and other raptors. Predators suffer mostly from environmental poisoning when their prey sits at the top of a long food chain of other animals. Golden eagles, however, mainly consume grass-eating rodents (comparatively low on the food chain), and seldom receive high concentrations of poisons. They feed infrequently on waterfowl and fish, two types of animals known to accumulate toxins.

The big birds mate for the first time in their fourth year, and the bond often lasts for the eagles' entire 15- to 20-year lifespan. In late winter the golden pair claims a territory of several square miles, and prepares a grass-lined stick nest situated on a rocky ledge or high in a tree. Sometimes the nest becomes huge with reuse. One nest, on a high bluff near the Sun River in northwestern Montana, stands 21 feet (6.5 meters) tall—and is still growing.

The eagles' creamy buff eggs (usually two) are laid in early spring; after 45 days of incubation by the female, the chicks emerge wobbly and pale gray. In a few days, their down turns snowy white, and three weeks later, feathers begin to appear. By the time the eagle chicks are five weeks old, they are able to move about the nest. They may even attempt to rip meat from

The magnificent golden eagle—North America's largest bird of prey still in the wild—continues to thrive, particularly in those areas of the western states that offer wide, sweeping views (left). More than 17,000 pairs of golden eagles nest in the U.S. alone (right).

Left: From THE AMERICAN WILDERNESS: The Snake River Country Photograph by George Silk © 1974 Time-Life Books, Inc.; right: © Robert Balbo/Animals Animals

the one or two prey animals the male brings to the aerie each day.

The young take to the air for the first time when they are about two months old. Sturdy feathers have replaced their down, and rigorous flapping practice inside the nest has prepared their muscles for the inaugural flights. Clumsy at first, the fledglings soon acquire the skills that make the golden eagle an able flier.

Feathered Cannonball

In mating and territorial displays—and maybe even just for fun—soaring adult golden eagles sometimes close their wings and plummet like spent rockets. Before nearing the ground, the speeding birds pull out of their dive and climb—to do it all again. Observers have seen eagles add a prop to this diving game, dropping a dead grouse from a great height, then catching it before it hits the ground.

Cruising the Idaho back roads near the Snake River Birds of Prey Area, biologist Kochert once happened upon a rough-legged hawk (about one-fourth the size of a golden) sitting on a jackrabbit it had killed. Kochert paused to watch the hawk feed, then spied a golden eagle soaring high above. "When the eagle saw the hawk and the dead rabbit, it tucked its wings and began to plummet straight down like a feathered cannonball," recalls the scientist. "When the

Golden eagles use twigs and sticks to build a nest, or aerie, on a rocky ledge (above left) or high in a tree. Two eggs are laid in early spring and incubated for 45 days. During their first two months, the eagle chicks remain close to the nest, enjoying the devoted care of their parents (left). As early as four weeks, some golden eagle chicks begin to display aggressiveness (below).

366 WILDLIFE   Top left and bottom right: © W. Perry Conway; lower left: © Jose L. G. Grande/Photo Researchers, Inc.

With its extraordinary vision, a golden eagle can spot its next meal at great distances and, in a matter of seconds, silently swoop down (above) on its victim and make the kill. Golden eagles prefer grass-eating rodents to the fish and waterfowl sought by other birds of prey. A battle unfolds when one golden eagle tries to steal another's freshly killed hare (right); the victor wastes no time establishing ownership over the fallen prey (bottom).

hawk saw the eagle dive, it began furiously pumping its wings to get out of there." The hawk escaped while the eagle dined on rabbit.

The eagle's visual acuity is truly astonishing. Biologist Michael Collopy, who studied golden eagles in Idaho, recalls observing a Snake River golden eagle in 1979, as it climbed higher and higher into the western sky. "Eventually it became just a speck," says Collopy. "It must have been several thousand feet up."

Suddenly the bird ceased its lazy circling, tucked its wings, and dove almost vertically toward Earth. One hundred feet (30 meters) or so above the desert floor, the eagle eased out of its kamikaze dive and swept with full force into an ambling jackrabbit, killing it instantly. "There's no question that the eagle had its eye on that rabbit from the very beginning," says Collopy.

Controversial Predators

Golden eagles occasionally attack larger prey. University of Calgary biologist Elden Bruns once witnessed an eagle attacking a pronghorn on the Alberta prairie. After four failed stoops on a fleeing herd of 60, the golden landed atop a running pronghorn and sank its talons into the animal's back. While the doomed pronghorn bucked and ran, the eagle spread its wings for balance and opened and closed its talons in an attempt to pierce a lung or artery. After nearly 20 minutes with the eagle on its back, the animal collapsed. Bruns estimated its weight at 70 pounds (30 kilograms), several times that of the eagle that began feeding on the carcass.

Because they can kill game animals and small livestock, goldens long were targets of bullets, traps, and poison. As recently as 1948, some states paid a bounty for dead golden eagles, and, until outlawed in 1962, aerial gunning was common in parts of the West. By one estimate, airborne shooters in western Texas and eastern New Mexico killed 20,000 golden eagles in the 1940s and 1950s.

"Many ranchers in the West still consider the golden eagle very controversial and destructive, but the problem is often greatly exaggerated," says Lynn Greenwalt, former director of the U.S. Fish and Wildlife Service. In fact, golden eagles can benefit ranchers. Ten jackrabbits consume about the same amount of valuable forage as one sheep, and a nesting pair of the eagles can kill several hundred hares a year. But under certain circumstances, as biologists learned in 1974, goldens can be voracious consumers of mutton.

That June, ranchers near Dillon, Montana, summoned E. V. Cofer, a Fish and Wildlife Service law-enforcement agent and biologist, to investigate golden-eagle predation on lambs. Like most other scientists, Cofer doubted they killed many lambs. He was astonished to find dozens of carcasses marked with telltale talon punctures, and to count—in one day on one ranch—65 eagles. "It was unlike anything I had ever seen," he recalls.

Two Montana ranches were virtually under siege, sometimes losing nearly 100 lambs a day to hundreds of golden eagles gathered in the area. "I could drive around those ranches, stop almost anywhere, and find five or ten lamb carcasses," says University of Montana biologist Bart O'Gara, leader of the Montana Cooperative Wildlife Research Unit.

Eagle Relocation

To rectify the situation, biologists conducted the largest eagle-relocation project in history. Using padded, weakened leg-hold traps, they captured hundreds of eagles and moved them to Colorado, Wyoming, and other parts of Montana.

But the relocation project seemed to have little effect on predation; lamb production fell to an all-time low. "There were so many eagles that nothing we did made much difference," says O'Gara.

Then nature accomplished what human intervention could not. Eagle numbers fell, predation waned, and lamb production improved. Later, researchers pieced together what had happened. First, a plentiful supply of jackrabbits had helped eagle numbers rise. Then, in 1972 and 1973, jackrabbit numbers plummeted in the West. In addition, unusually cold, wet springs kept young ground squirrels in their burrows, and hordes of immature—and hungry—golden eagles began preying on lambs. Later, when jackrabbit numbers rebounded and ground squirrels became more venturesome, most of the Montana raptors lost their appetite for mutton. "The whole situation was a fluke of nature," says O'Gara.

Golden eagles continue sporadic livestock predation in several western states. Some west Texas ranchers claim to lose 5 percent of their annual lamb crop to the birds. "It can be a serious problem," says Bill Sims, executive secretary of the Texas Sheep and Goat Raisers Association, "and ranchers would like the right to shoot the offending eagles."

Some ranchers already shoot eagles—quite illegally. But few of the eagle killers are caught. Several years ago, during a conversation with Texas ranchers, Bart O'Gara lamented that most eagle leg bands are never returned. The bands, he explained, could add to understanding of eagle migrations and predatory habits. A rancher later sauntered up to O'Gara and asked if he wanted some bands from golden eagles. "I got a 2-quart jar full of 'em at home," drawled the rancher.

Most eagle watchers downplay the wrath of ranchers. "In recent years, eagle persecution has become more verbal and symbolic than real," says Lynn Greenwalt. "The days of wholesale eagle slaughter are long gone. It's likely that the sheep rancher may disappear from the West before the golden eagle does."

Even if ranchers and eagles enjoy no more than an uneasy truce, that may be enough. "Because it's largely a Western bird, the golden has plenty of room to roam, and its future looks bright," says Greenwalt. Passed over long ago for that choice spot on the Great Seal, America's other eagle should nevertheless grace our nation's skies for a long time to come.

© Bianca Lavies

The Disarming Armadillo

by Jim Watson

On the surface, the armadillo hardly seems like celebrity material. Its body looks like a conglomeration of some of nature's most bizarre features: the figure of an overgrown sow bug, possumlike ears, turtle feet, lizard skin, and a dinosaur's tail. Draped over its back is a thick, medieval-looking cape called a carapace. Patches of coarse hair cover its armored body, and scent glands at the base of its tail give off a pungent, musky odor. "You could say it's not an animal we humans can easily relate to," says armadillo expert Eleanor Storrs, a professor of biology at the Florida Institute of Technology in Melbourne. Even so, a remarkable discovery she made 18 years ago thrust the creature from the shadows of anonymity into the glare of scientific fame: the armadillo is one of the few creatures besides people that can contract leprosy.

The discovery toppled a barrier that had stalled research on the disease for more than a century, and in the process pushed the armadillo into a position of importance. Now, however, many researchers believe the leprosy connection may just scratch the surface of the animal's trove of secrets.

"It's really a remarkable animal, and there's still so little known about it," says Storrs, who first became fascinated by the crea-

The nine-banded armadillo has ears like an opossum, skin like a reptile, a tail like a dinosaur, and a thick medieval-looking suit of armor called a carapace. When startled, the remarkable little creature leaps into the air.

WILDLIFE   369

Unlike any other mammal, the armadillo always gives birth to four identical young (upper left). An adult armadillo consumes more than 200 pounds of insects per year (above). Armadillos ford rivers by inflating their stomach and intestines and simply floating across (left).

Top left and above: © Bianca Lavies; top right: © Jeff Foott

tures nearly three decades ago. Since then the 62-year-old biochemist, who is director of the school's Comparative Mammology Laboratory, has found that under the armadillo's primitive upholstery lies a shy but intriguing animal with a vast repertoire of unusual traits. The more she peers under the armor, she says, the more remarkable it becomes.

An Ancient Order

Of the 20 or so types of armadillo, Storrs has focused on *Dasypus novemcinctus*, the nine-banded species, which ranges from Argentina to the southern United States, and is the only variety found in this country. Although it can't roll up in a ball like its three-banded cousin, the nine-banded armadillo can leap straight up in the air (a habit that often proves fatal in encounters with cars). Like a minisubmarine, it can fill itself with air and float, or expel the air and sink.

Unlike any other mammal, the armadillo produces four genetically identical offspring almost every time it gives birth. In addition, the nine-banded species practices family planning: a unique time-release mechanism allows it to postpone delivery until conditions seem safe.

Armadillos, anteaters, and tree sloths are all that is left of Xenarthra, an ancient order of animals that flourished 55 million years ago in South America. The animals made their way into what is now the United States when the land bridge between the two continents formed, only to vanish mysteriously some 10,000 years ago.

Among the survivors in Central and South America was the nine-banded armadillo, which weighs from 6 to 10 pounds (2.5 to 4.5 kilograms). In the 1850s a group of the creatures appeared in Texas and swiftly spread through the Gulf states toward the Atlantic.

To cross the Mississippi and other large rivers, the animals may have hitched rides on logs, says Storrs. When they came to a creek or narrow stream, they probably just floated across, buoying themselves by gulping air into their stomachs or intestines. Or they might have used a technique she and her husband, biochemist Harry Burchfield, observed while stalking an armadillo in Texas. The creature scurried into a stream and sank to the bottom. After about five minutes, it strolled out on the other side.

Texas Horde

In 1922, while the migrants were making their eastward journey, a pair of captive nine-banded armadillos escaped from a small zoo near Cocoa, Florida. Soon the state was crawling with tens of thousands of the creatures, all thought to be descendants of this Adam and Eve.

The Texas horde reached the Florida panhandle in the late 1970s, and probably met up with their East Coast kin within the past five or six years. Today an estimated 30 million armadillos roam the southern United States. Reactions to their arrival have been mixed. One armadillo can eat 200 pounds (90 kilograms) of insect pests a year, so some people consider it an ally. But homeowners whose yards have been turned into Swiss cheese by the restless burrowers are less hospitable.

Using their heads as shovels, armadillos carve out passageways to underground dens. They come out to feed in late afternoon. Their eyesight is poor, but a superb sense of smell allows them to locate insects as deep as 6 inches (15 centimeters) underground.

Time-Release Family Planning

The armadillo might not win many animal intelligence contests, but, as Burchfield points out, "it's been very successful." The creature owes its staying power in part to the routine consistency of its quadruplet births. Only the nine-banded armadillo and a few close cousins in South America regularly produce litters of four pups from the same fertilized egg.

"Basically," says Burchfield, who taught biology at the Florida Institute of Technology, "the armadillo produces only one ovum per year, which normally would limit its chances of reproduction." However, the development of polyembryony (multiple embryos from one egg)

Both photos: © Jeff Foott

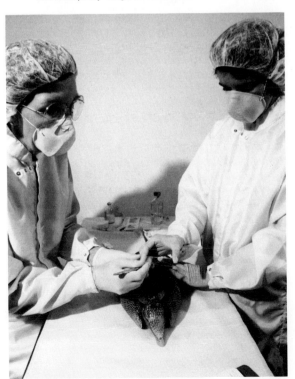

The armadillo has the further distinction of being one of the few creatures besides humans that can contract leprosy. Researchers now maintain laboratory colonies of armadillos in order to grow the leprosy-causing bacillus and to test antileprosy drugs and vaccines.

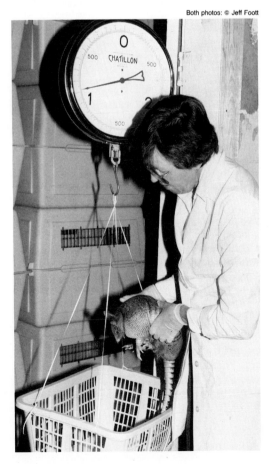

"increases the litter size by a factor of four, a much more favorable position."

Another startling reproductive talent plays perhaps an even larger role in the species' success. Armadillo females can suspend their pregnancy for up to two years, an ability unheard of elsewhere in the animal world. Scientists have known for years that a small number of mammals can delay for several months the implantation of the embryo in the uterine wall. But only recently has Storrs begun to appreciate the armadillo's "superdelay."

At first, when captive armadillos at her laboratory began giving birth long after their last contact with males, "we figured we must have made a mistake or overlooked something," says Storrs.

She began to suspect a delay in March 1986, when five of the seven females she had sent to a lab in England gave birth. The animals had not been near males since their capture in November 1984. Says Burchfield: "The nicest part about it was that it didn't happen in our laboratory—no one would have believed it."

After the egg is fertilized, the early embryo—a free-floating ball of cells called a blastocyst—embeds itself in the uterus and begins to grow. But sometimes, says Storrs, "instead of remaining free-floating for two or three months, as is normal, this blastocyst goes on free-floating for an additional 12 months, or even 24 months, before implanting."

She believes the delay could be a reaction to stress, which might cause other species to abort. By postponing the implantation, she suggests, the armadillo can wait out the stressful conditions and give its litter a better chance of surviving.

Potential Lepers

At Storrs' lab in Melbourne, the walls are stacked with hundreds of "convertible pet apartments," each occupied by one armadillo. Eventually these animals become part of a crucial leprosy study. Their bacteria-laden tissues are harvested, frozen, and sent to research labs worldwide.

Thanks to the armadillos, scientists have made great strides toward conquering the ancient disease, which still affects some 15 million people worldwide, including 4,000 in this country. First identified 3,000 years ago, leprosy in humans is seldom fatal, but produces skin and nerve lesions that, left untreated, can cripple or severely disfigure.

Fighting leprosy was the farthest thing from Storrs' mind in 1963, when, as a newly-wed biology student at the University of Texas, she became intrigued by the regularity of armadillo quadruplets. Working on her doctoral thesis, she began studying captive animals. Then, in 1968, after the couple moved to Louisiana, Storrs got a grant from the National Institutes of Health (NIH) to study armadillo reproduction. But a casual remark changed the course of her research—and her life.

That year, members of the U.S. Leprosy Panel visited Burchfield, then a biochemist with Gulf South Research Institute. They wanted to know about a technique he had suggested for testing the antileprosy drug dapsone.

During their visit, they mentioned some of the problems with studying the disease. Though the leprosy-causing bacillus was discovered in 1873, no one could make it grow in the laboratory. And because there was no animal "model," scientists had to rely on tissue removed from people.

The visitors pointed out that leprosy attacks cooler areas of the body such as the nose and ears. From her work, Storrs knew that the armadillo's body temperature, at 92° to 95° F (33° to 35° C), was lower than a human's. Playing on a hunch, she blurted out, "Has anyone ever tried the armadillo?" As it turned out, scientists had tested everything from fish to buffalo to monkeys, but never armadillos.

In 1969, with a grant from the Centers for Disease Control (CDC), Storrs set out to test her theory. She chuckles about it now: "We didn't have any background showing that armadillos would develop leprosy, and the odds that they would were pretty remote. It was a shot in the dark."

The shot hit its mark two years later, when an armadillo she was working with developed the disease. Laszlo Kato, director of leprosy research at Catherine Booth Hospital in Montreal, recalls his reaction: "Eureka! For decades we prayed for an animal model. Then this shy, kind person came along and said, 'Here it is.'"

Storrs took her armadillo colony to Melbourne in 1977, and she now maintains it with the help of grants from the World Health Organization (WHO) and other groups. Before her discovery, scientists were no closer to a leprosy vaccine than were the physicians of the Middle Ages. Now the best hope of wiping out one of mankind's oldest scourges lies with this scaly throwback to the days of dinosaurs.

Return of a Reptile

by James Gorman

I find Frank Godwin out by the alligator pens. Godwin, 50, is the owner of the Gatorland Zoo in Orlando, Florida. He displays his 4,500 alligators for the entertainment of tourists, and he also markets their hides and meat. One of Gatorland's biggest attractions is the Gator Jumparoo, which features large alligators leaping out of a pond to snatch chicken carcasses suspended from a pulley-and-chain device that stretches over the water. When the chain is shaken to attract the alligators' attention, the footless and featherless chickens, suspended by their wings, seem to dance on air. This may not be the most grotesque thing that goes on in Florida, but it's certainly close.

It is after witnessing the Jumparoo with about 100 other people (some of whom were munching Gator Bites, little bits of breaded, deep-fried alligator flesh) that I seek out Godwin. He is supervising the installation of netting over some of the outdoor pens. "Alligator-husbandry techniques are being refined all the time," he says, after explaining that the netting is a solution to a problem that had been plaguing

The numbers of once-rare American alligators have risen dramatically in Florida and throughout the South.

his alligators. For several weeks they had been showing signs of stress, which, in the alligator world, means that their heads were cut and scarred. Godwin knew that they were chewing on each other, but he didn't know why.

The reason for his ignorance was that the crocodilian carnage was taking place while Godwin and the tourists were watching the Jumparoo. It seems that the bell that summoned spectators to the big event was also a signal to the 200 wood storks in the area that the coast was clear: without human interference, they could try to grab the chopped fish Godwin had left in the alligator pens. While the prize performers were earning their keep in the Jumparoo, the rest of the gator herd back on the ranch was being mightily provoked by the storks. "The alligators were getting so excited that they'd jump up for the birds, and what they did, they grabbed each other by the head," says Godwin, who finally saw the spectacle—the flapping wings, the diving birds, the snapping alligators catching mouthfuls, not of wood stork, but of each other. "That was absolute chaos. I've never seen such a thing."

Renewable Resource?

Once hunted into scarcity, then protected as an endangered species, the alligator is now back in force—in the wild, on farms, and in fine stores everywhere. The state of Florida recently held its first public (although limited) alligator-hunting season since 1962. Six thousand hunters applied for permits, and 238 were chosen by lottery. All but eight coughed up the $250 for a license. The hunters took a total of 2,988 alligators from 28 designated areas throughout the state. Though hunting was allowed only at night, no major injuries were reported—perhaps because no firearms were permitted. One man did shoot off the tip of a finger with a bang stick, a long tube that holds, at the business end, a .44-caliber or .357-magnum cartridge. The bang stick, developed for killing sharks, is not considered a firearm (pronounced "faaarm" by true Floridians), because the bullet is fired only when the stick is pressed against the target.

In addition to the population in the wild, there are now more than 75,000 captive alligators on 48 farms. Last year the farms produced 7,529 hides and 71,099 pounds (32,250 kilograms) of meat. As Godwin's son Mike, who works the Jumparoo, says just before he induces an animal named George (I'm taking his word for it) to pluck a chicken carcass from his hand, "The American alligator is now considered a renewable resource."

The high point for crocodilians, of which alligators are one variety, was the Mesozoic era, the Age of Reptiles, which occurred between 225 million and 65 million years ago. One crocodilian ancestor, *Phobosuchus,* grew to 50 feet (15.25 meters) in length and may well have eaten dinosaurs. Of course, 65 million years ago, the dinosaurs became extinct. It was all downhill from there for Florida's crocodilians. First the mammals took over, and then, before you knew it, people appeared—the Indians, and then Ponce de León, Marjorie Kinnan Rawlings, Don Johnson. The airboats arrived, and the condos. Swamps were drained, and the words *Miami* and *vice* became inseparable. It got so that when people said Florida was full of reptiles, they weren't talking about alligators.

The resurgent reptiles have proved to be a powerful tourist attraction. At the Gatorland Zoo in Florida, visitors flock to see alligators leap up to snatch a chicken carcass during an event called a "Gator Jumparoo."

© Kevin Kolczynski/Sports Illustrated

Territory-hungry alligators have invaded swimming pools, golf courses (above), and other places where humans consider carnivorous reptiles unwelcome. Florida now permits a limited alligator-hunting season (below).

The growth of the human population tended, until recently, to be accompanied by a decrease in the alligator population. There was extensive hunting of alligators in Florida as early as 1800, when a 6-foot (1.8-meter) hide went for $7. By 1929, according to one estimate, 190,000 hides were being taken out of Florida annually. The hunting of alligators was first restricted in Florida in the mid-1940s, and then, in 1962, it was completely forbidden, though poaching continued. In 1967 the alligator was put on the Endangered Species List by the U.S. Government, and in 1970 amended federal regulations finally shut down the interstate movement of alligator hides, thereby effectively stopping the hunting.

The alligator population bounced back, and fast. In 1977 the U.S. Fish and Wildlife Service changed the reptile's status in Florida to the less serious "threatened" category. In 1985 the alligator was taken off that list (for Florida), and given the formal status of "threatened by similarity of appearance," a quirky designation that allows the government to inspect skins whenever they deem it appropriate, to make sure that unscrupulous trappers are not passing off the skins of the endangered crocodile or the much less valuable caiman as hides of the alligator. According to a Fish and Wildlife spokesman, the alligator is now "completely recovered throughout its range."

Just Hiding?

Today most alligator people, from trappers to game-and-fish officers to zoologists, will tell you that although the alligator may have been in some trouble, it was never in danger of disappearing. It was simply hiding from hunters. Once the hunting stopped, the alligators reappeared quickly, then proceeded to reproduce. In Florida in the 1970s, the alligator and human populations were exploding at the same time, with the result that confrontations between the two species began increasing.

Usually alligators consider human beings too big to eat, but they (the alligators, that is) are unpredictable. Sometimes they change their minds. In 1973 there was a fatal attack on a 16-year-old girl swimming near Osprey. And in each of the past two years, there has been a fatal attack by an alligator on a human being. The incident last year, in which an alligator killed a four-year-old girl in Englewood, was particularly gruesome. But despite the publicity such attacks generate, alligators do not wreak the havoc one might imagine. Since 1948, when the Game and Fresh Water Fish Commission, Florida's alligator authority, began keeping statistics, there have been only five confirmed fatal attacks on human beings by alligators.

Pets are another matter. Most dogs and cats are just the right size for alligators, and there have been numerous reports of attacks on pets. For the past decade, Game and Fresh Water Fish Commission biologist Michael Delany has been looking in alligator stomachs to find out what gators eat. The answer is: mostly insects when the alligator is small, and fish and turtles when it gets bigger, with the odd bird or mammal thrown in. But an alligator will eat just about anything. Delany has a plaque in his office displaying assorted "junk food" that he took out of alligator stomachs: besides beer cans, bottles, fish lures, outboard-motor pull cords, bird bands, and bands from the legs of baby alligators, there were also two tags from dogs, which pretty much sum up the fate of unleashed pets in alligator territory. One reads *My Name is Blackie*. The other reads *Leave Me Alone and I'll Go Home*.

But the alligator confrontations with the human world were not usually deadly. In the 1970s gators began making a nuisance of themselves in yards and swimming pools and shopping-mall parking lots, leading the Game and Fresh Water Fish Commission to set up a program in 1978 to use trappers to kill the so-called nuisance alligators. Last year the program's 50 or so trappers killed 4,464 alligators in Florida.

One of the regional coordinators of the nuisance-alligator program is Lieutenant Mitch Brown of Lakeland. Brown is a very serious law-enforcement official, which means he wears a gun belt and doesn't tend to focus on the humor in his job. His demeanor doesn't suggest

Below and facing page: © Kevin Kolczynski/Sports Illustrated

Alligators tend to eat fish, turtles, birds, and small mammals—including an occasional household pet. But, as the contents of an alligator's stomach show (left), they'll eat just about anything that comes their way.

Baby alligators grow twice as quickly in the controlled environment of a hothouse as they would in the wild.

unfriendliness, mind you; it just inspires the desire to be law-abiding and to avoid committing any "crocodilian offense." (In Florida this means doing something nasty to a reptile, not behaving like one.) I ask Brown about the thousands of calls his office receives every year about nuisance alligators. He says, "We get complaints about alligators on golf courses, in carports and swimming pools, backyards and front yards, under people's cars, and on their porches."

One 10-foot (3-meter) alligator tried to eat the concession stand at a Little League baseball game. Another refused to move from an airport runway, making it a tad difficult for planes to land. Another, only 3 feet (1 meter) long, was found in a fenced-in area around the electric meter of a house. The gator had been placed there by the homeowner, who was behind on his electric bills. The plan didn't work: the man's power was turned off anyway!

And then there was the golf-ball thief. "We had a person not too awful long ago that was on a golf course," Brown tells me. "He was trespassing. He was actually in the pond. He was on his hands and knees in shallow water, late in the evening, almost dark, and he was attacked by a six-foot (1.8-meter) gator. He wasn't injured to any great extent at all, but he was bitten. He was in a place where he should not have been. Wrong time of day. Wrong place. Doing the wrong thing." Was that alligator considered a nuisance? Brown pauses and then says, "I think we left that alligator."

The contract trappers, some of whom go back to the days of open alligator hunting 30 to 40 years ago, pay the state $250 for a license and a $30 tag fee for each hide, the same fees charged for the public hunt. The money goes to the Game and Fresh Water Fish Commission for its alligator-management and -research programs.

Alligator Husbandry

If nuisance-alligator hunting is a bit nostalgic, a nod in the direction of Florida's past, alligator farming is just the opposite. Like the glittering city of Miami, it's a glimpse of Florida's future. The raising of other animals—goats, chickens, cattle—reaches deep into the human past. But reptile husbandry, certainly on any large scale, has no such tradition. It's the Monty Python version of farming—something completely different.

It is also something that is growing almost as fast as the wild-alligator population. Twenty years ago there were five farms in Florida on which alligators were raised in the hope that one day their skins could be marketed. Now there are close to 50. A similar trend has developed in Louisiana, the state with more alligator farms (55) than any other. Alligator farmers nationwide have organized into associations, hired lobbyists and sales representatives, and made contact with people who are farming crocodilians in such other countries as Zimbabwe, Zambia, Botswana, and New Guinea. The impetus behind this surge is the increasing scarcity of crocodile and alligator skins, caused by the reduction of poaching and, to a lesser extent, by the erosion of crocodilian habitat worldwide. In the 1960s, 500,000 alligator and crocodile skins entered the world market per year; now it's 150,000, making them more valuable.

Often, when something hasn't been done before, there are good reasons, and nowhere is this truer than in the raising of large, carnivorous reptiles. On an alligator farm, there are no cute red barns, no majestic silos, no homey feed bins full of corn. Instead, there are coolers, frozen fish, and sometimes big, thawing mounds of nutria carcasses. Nutria are rodents, something like muskrats, native to South America. They were introduced in Louisiana around the turn of the century, where they flourished, as did the alligators that ate them.

Edwin Froehlich was one of Florida's first alligator farmers. Today, in the town of Christmas, he has 40 concrete outdoor pens housing about 4,000 alligators. He also has five breeding ponds under construction, and a building that contains several coolers, his office, and a spotless butchering room with stainless-steel tables and a red tile floor.

He used to feed the gators a lot of nutria, ground up for the little ones and quartered for the big fellows. "They like to sink their teeth into something once in a while," he says. "It helps their appetites." But recently he has been switching over to beef and alligator-feed pellets because of the scarcity of nutria meat. His gators gulp more than 150,000 pounds (68,000 kilograms) of the mixture in a year.

The table manners of captive gators are no worse than those of, say, swine. Godwin's gators eat fish, and some of the small ones, lined up and jostling at a long mound of food, *do* have the air of hogs at the trough. Froehlich's small gators receive their meals in conical heaps.

The carnivorous nature of the alligator not only reduces the pastoral appeal of this sort of farming, but also makes for logistical difficulties. As Froehlich points out, "You can't drive them." It would be unwise, for obvious reasons, to introduce Border collies to move an alligator herd. When Froehlich needs to shift his animals, he picks them up one by one, by hand, tapes their mouths shut, and loads them on a wagon, which he pulls to the next pen. Then he picks them up again. Froehlich, who still has all his fingers, says, "You have to handle them until they're too big to handle. Then you leave them until they're ready to be killed."

An even bigger problem is reproduction. Not that the alligators have much trouble mating. They mate just fine. As a display at the Gatorland Zoo says: "During courtship 'gators show affection by swimming together, climbing across each other, and gently rubbing snouts." However, after mating in captivity, only a quarter or a third of the females lay eggs, only about a third of the eggs laid are fertile, and only a third to two-thirds of those will actually hatch. So alligator farmers also get eggs from the wild, which they can incubate and hatch with much greater success.

Below and facing page: © Kevin Kolczynski/Sports Illustrated

Some alligator farmers feed their livestock chopped nutria meat (left). Alligators in captivity devour tons of food per year.

Alligator farming pays off at skinning time. Alligator skin is still used for wallets, purses, and other items.

A final difficulty with reptiles is that—unlike cows, goats, and chickens—they are cold-blooded, with the result that when the weather turns cooler, the gators eat less and grow less. It's nerve-racking to tie up money in an animal that grows only eight months of the year. And, says Froehlich, it can easily cost $1 million to establish a farm with 200 breeding adults, the number needed to make a go of it.

To speed growth, farmers have turned to hothouses. In the wild, alligators take six or seven years to reach the 6-foot (1.8-meter) skinning size. Godwin, who grows them to the 3- to 4-foot (1- to 1.2-meter) stage in hothouses, says he gets them to 6 feet in about 2.5 years. His hothouses have windows, like greenhouses.

Froehlich has gone Godwin one step better, or perhaps weirder. He has built two hothouses with 12 tanks each. The hothouses are without windows. Their walls and roofs are of galvanized tin, and they are thoroughly insulated. The concrete floors have built-in heating pipes. In the airtight buildings, the alligators are kept in darkness and silence (except for feeding times) at a temperature of 88° to 89° F (31° to 32° C). When alligators are raised in the hot, quiet dark, says Froehlich, "they do real well. They're not as nervous. Nothing disturbs them."

There have, however, been some reports that alligators raised in hothouses have been upset by the variety of new stimuli when they are taken outside. In particular, says Froehlich, "the sound of the human voice petrifies them."

He is waiting to see how his current crop of hothouse alligators does, but in case there is a problem, he's ready for the next batch. His hothouses include a system that will allow him to pipe in a local country-music station with a lot of chatter. This seems only appropriate: the hides of some of his gators will no doubt become cowboy boots. If the alligators make a comfortable transition from the silence of the hothouse to the world of the outdoor pen, Froehlich's next crop of gators will continue in quiet. If not, says Froehlich, they "will be hearing country music 24 hours a day."

Tourist Reptile

People who spend time with alligators seem to develop a certain fondness for them. But it's not love that moves the public to keep them around. Nor do I think it's the economic incentive, although they do make beautiful shoes. I think the alligator's biggest selling point is that it's a great tourist reptile. It's large and fearsome, and it can be deadly, but usually it isn't. The alligator experience for most tourists is a titillating shiver of fear, not bodily harm. Godwin says: "People in the state of Florida are the most fortunate people in the world that the alligator is not an aggressive animal." I think he has it wrong. Alligators may have big teeth, but let's not kid ourselves. We're the ones with the airboats and the bulldozers. And, unlike the alligator, we are consistently aggressive. It's the alligator that needs the protection, not us.

IN MEMORIAM

ADAMSON, GEORGE (83), Indian-born wildlife conservationist who, with his wife, Joy Adamson, gained fame for his work with Elsa, the lion star of the film *Born Free*. He spent much of his life studying lions and working to preserve their habitat. He was an outspoken critic of Kenya's lax policies on elephant poaching, a stance that ultimately led to his murder in an ambush a few miles from his camp; d. Kora Rock, Kenya, Aug. 20.

ANDREASEN, GEORGE (55), U.S. orthodontist who introduced nitinol, a nickel-titanium alloy used in arch wires to straighten teeth. Commonly called "memory wire" for its ability to resume its original shape after bending, nitinol wire straightens the teeth faster than stainless-steel wires, and without the periodic adjustments; d. Iowa City, Iowa, Aug. 11.

APPLEMAN, WILBER (83), U.S. engineer who invented the electric hospital bed, a motor for the Jacuzzi whirlpool, and the Insinkerator garbage-disposal unit; d. Laguna Hills, Calif., July 25.

BEADLE, GEORGE (85), U.S. geneticist who shared the 1958 Nobel Prize in Medicine or Physiology for demonstrating that genes control hereditary characteristics by controlling chemical reactions in plant and animal cells. Later, as president of the University of Chicago, he conducted research that identified the origins of domestic corn. He coauthored the definitive work, *An Introduction to Genetics* (1939), and was the recipient of numerous honorary degrees; d. Pomona, Calif., June 9.

BENTON, NICHOLAS (35), U.S. master ship rigger and shipwright known for his work on colonial ship restoration and design. He had recently designed and built a full-size replica of Henry Hudson's 17th-century ship. He died after falling 75 feet from the mast of the schooner on which he was working; d. Rensselaer, N.Y., June 19.

BLOCK, MARVIN A. (86), U.S. physician and expert on alcoholism. An early advocate of alcoholism as a disease and the benefits of medical treatment for it, he founded the New York State Council on Alcoholism; d. Buffalo, N.Y., Feb. 28.

BOND, JAMES (89), U.S. ornithologist best known for proving that birds of the Caribbean originated in North America. He collected 294 of the 300 bird species living in the Caribbean islands, and wrote more than 100 books and scientific papers on Caribbean birds. His name was made famous when Ian Fleming, an avid bird-watcher, adopted it for the fictional British Agent 007; d. Philadelphia, Pa., Feb. 14.

BRODIE, BERNARD B. (81), British drug-therapy researcher whose pioneering studies in drug metabolism brought new prominence to the field of pharmacology in the 1940s. He discovered that animal and human responses to drugs do not differ significantly. His extraordinary work—including the use of drugs in the treatment of cardiovascular diseases, mental and emotional disorders, and cancer—was the basis of a recent popular biography, *Apprentice to Genius* by Robert Kanigel; d. Charlottesville, Va., Feb. 27.

BROOKS, CHANDLER (83), U.S. professor of physiology and researcher best known for his research on the brain and the heart. He concentrated on the relationship between the central nervous and endocrine systems, and also studied the pituitary, hypothalamus, and nervous mechanisms that control sexual behavior; d. Princeton, N.J., Nov. 29.

CECI, LYNN (58), U.S. anthropologist who was an expert on wampum beads made from mollusk shells and used as a medium of exchange by American Indians in the beaver trade; d. Locust Valley, N.Y., March 29.

DE GRAAFF, JAN (86), Dutch-born U.S. horticulturist whose hybridization of lilies made them a popular garden flower and dispelled myths of the special treatment needed to grow them; d. New York, N.Y., Aug. 5.

DE HOFFMANN, FREDERIC (65), Austrian-born U.S. nuclear physicist who worked on the Manhattan Project. His mathematical analysis figured in the development of the hydrogen bomb. In 1970 he joined the Salk Institute for Biological Studies and helped build it into one of the world's largest independent research centers. He died of complications of AIDS, which he contracted in blood transfusions in 1984; d. La Jolla, Calif., Oct. 4.

FAIRBANK, WILLIAM M. (72), U.S. physicist who conducted some of the most complex experiments in low-temperature superconductivity. An experiment he devised, which tests the theory of relativity with gyroscopes in earth's orbit, will be performed during a 1993 space shuttle mission; d. Palo Alto, Calif., Sept. 30.

FISH, MARIE (88), U.S. oceanographer and marine biologist whose research in underwater sound detection enabled U.S. Navy antisubmarine vessels to distinguish between enemy targets and whales and schools of fish; d. Westport, Conn., Feb. 1.

FOOTE, FRANK W. (78), U.S. expert on tumor pathology who developed the diagnostic criteria for cancerous tumors of the breast, salivary glands, and thyroid gland; d. Sarasota, Fla., March 4.

FOX, CHARLES LEWIS, JR. (81), U.S. microbiologist who held many patents, including those for Silvadene, a widely used ointment for treating severe skin burns, and for a fluid used in kidney dialysis; d. New Milford, Conn., Aug. 28.

FOX, DANIEL (65), U.S. inventor whose 44 patents with General Electric included Lexan, a transparent plastic used in the manufacture of compact discs, baby bottles, and construction materials; d. Pittsfield, Mass., Feb. 16.

FOX, RUTH (93), U.S. psychoanalyst who became the first director of the National Council on Alcoholism (1959) and founded the American Medical Society on Alcoholism and Other Drug Dependencies (1954). She conducted important research on Antabuse, a drug now commonly used to treat alcoholism; d. Washington, D.C., March 24.

FRAZE, ERMAL CLEON (76), U.S. engineer who invented the pull-tab opener used on aluminum beer and soda cans; d. Dayton, Ohio, Oct. 26.

FRIEDHEIM, ERNST A. H. (89), Swiss-born pathologist, microbiologist, and chemist who developed melarsoprol, a drug used to treat African sleeping sickness. He also formulated drugs to counteract mercury and lead poisoning; d. New York, N.Y., May 27.

GALBRAITH, CHARLES (64), U.S. medical educator who created the mannequinlike patient simulation models used to train medical students and doctors; d. New York, N.Y., Jan. 29.

GEIGER, DAVID H. (54), U.S. engineer who invented an air-supported fabric roof system used in nearly half the domed stadiums in the world; d. Seoul, South Korea, Oct. 1.

GOLAY, MARCEL J. E. (86), Swiss-born researcher in information theory, optics, and instrumentation. His inventions include a sensitive heat detector that measures infrared radiation, a method to measure the structure of molecules, the capillary column for chemistry, and a published code system still used in radio navigation; d. La Conversion, Switzerland, April 27.

GORMAN, HARRY (72), U.S. veterinarian who trained animals for the U.S. space program. He also designed the first artificial hip joint for dogs, an invention later adapted for human use. During World War II, he led veterinarians in the rescue of over a million cattle from drowning in the Netherlands after bombs caused flooding; d. Fort Collins, Colo., Feb. 24.

HACKER, FREDERICK J. (75), Austrian-born U.S. psychiatrist and psychoanalyst. An expert on terrorism and aggression, he testified before Congress that impressions gained by watching movies could serve as "a trigger mechanism" that set off latent tendencies for abnormal behavior. He served as a court-appointed expert in the trial of Charles Manson and his followers for the murders of Sharon Tate and other Hollywood figures. He also worked with the family of heiress Patricia Hearst after she was kidnapped in 1974, and assisted the West German authorities after the terrorist killings at the Munich Olympics in 1972; d. Mainz, West Germany, June 23.

HAMMON, WILLIAM (85), U.S. public-health physician and medical missionary who played an important part in the battle against polio. His experiments with gamma globulin, a component of human blood, led to the development of the polio vaccine perfected in the 1950s by his close collaborator, Dr. Jonas Salk; d. Seminole, Fla., Sept. 19.

HODES, HORACE (81), U.S. pediatrician credited with isolating the first virus shown to cause gastroenteritis in children. As a first-year medical student, he discovered that Vitamin D's primary role is to bring about the absorption of calcium from the intestines. He later isolated the "Rocavirus," the most common cause of gastroenteritis among infants and young children. He figured prominently in the founding of the Mount Sinai School of Medicine; d. Manhasset, N.Y., April 24.

HUFNAGEL, CHARLES A. (72), U.S. surgeon who invented the artificial aortic heart valve (1952), the principle of which still serves as a model for heart implants. He chaired the three-member medical panel that evaluated whether President Richard M. Nixon could testify at the Watergate trial after undergoing pelvic surgery for a chronic phlebitis condition. The panel recommended that he not testify for at least six weeks. As it turned out, he never testified; d. Washington, D.C., May 31.

HYLAND, LAWRENCE A. (PAT) (92), Canadian-born U.S. pioneer in radar technology and aircraft communications, and president and chairman of Hughes Aircraft Company after Howard R. Hughes died in 1976. His discovery that the reflection of radio waves from an aircraft in flight disclosed not only the craft's presence but also its position was an enormous step in radar's military application; d. Lancaster, Calif., Nov. 24.

JORGENSEN, CHRISTINE (62), U.S. transsexual, the first person in the United States to publicly announce a surgical change of sexual identity. The sex-change operation, performed in Denmark in 1952, changed Army Private George Jorgensen, Jr., into Christine Jorgensen, who exploited her transsexualism and prospered financially from it; d. San Clemente, Calif., May 3.

KLOTS, ALEXANDER BARRETT (85), U.S. expert on butterflies and other insects; author of *The World of Butterflies and Moths* and *A Field Guide to the Butterflies,* cited as one of the finest books on the subject; d. Putnam, Conn., April 18.

KNAPP, HAROLD (65), U.S. nuclear-test expert who uncovered risks from fallout in nuclear testing in the 1950s. He testified that infants had been exposed to dangerous levels of radiation from drinking milk from cows that had eaten contaminated matter. Although his report was suppressed, he continued to testify at trials in which sheep ranchers and people near nuclear-testing sites unsuccessfully sued the government, claiming that their health had been harmed; d. Germantown, Md., Nov. 8.

KNOWLTON, ELIZABETH (93), U.S. mountain climber who in 1932 became the first woman to climb as high as 20,000 feet in the Himalayas. At the time she was the only woman in a German-American expedition that attempted to climb the 26,660-foot high Nanga Parbat in Kashmir. She published her account in the book *The Naked Mountain* (1933); d. Scituate, Mass., Jan. 22.

KOOPMAN, GEORGE A. (44), U.S. entrepreneur, among the founders of the American Rocket Company, which developed technology to deliver commercial and government payloads into space. He helped coordinate stunts and spectacular car scenes in the movie *The Blues Brothers*; d. Lancaster, Calif., July 19.

KORFF, SERGE A. (83), Finnish-born U.S. physicist who orchestrated worldwide observations of cosmic rays in the 1940s and 1950s using balloons, aircraft, and mountaintop observatories. He and his colleagues discovered that the earth was under constant bombardment by neutrons. They also pointed out that the neutrons would disappear only by forming radioactive carbon; d. New York, N.Y., Dec. 1.

LAING, RONALD DAVID (61), British psychiatrist who sought new treatments for schizophrenia based on a concern for the rights of mental patients. He suggested that insanity might be a sane reaction to an insane world, and deplored the use of insulin-induced convulsions and coma, lobotomies, electroconvulsive shock, and straitjackets. He argued that schizophrenia was not rooted in some genetic aberration, but rather was the result of some biochemical event that arose when an individual found himself in a hopeless emotional situation; d. St.-Tropez, France, Aug. 23.

LORENZ, KONRAD (85), Austrian ethologist who worked to demonstrate through animal experiments the large extent to which human behaviors such as aggression are inherited. In 1973 he shared the Nobel Prize for Physiology or Medicine for his pioneer work in animal behavior. He gained celebrity for his experiments with "imprinting," in which newly born ducks and geese considered him their parent; d. Altenburg, Austria, Feb. 27.

MANGELSDORF, PAUL (90), U.S. botanist who made important advances in the cross-breeding of corn and other grains. He traced the origins of modern corn back to primitive maize, helped develop hybrid seed corn that did not require hand removal of the tassels, and developed winter wheat with rust-resistant stems; d. Chapel Hill, N.C., July 22.

NARDI, GEORGE L. (66), U.S. surgeon and expert on pancreatic diseases. He was responsible for a number of medical discoveries, including a description of the urinary loss of amino acids essential for protein regrowth in severely burned patients. He also related pancreatic inflammation to overactivity of the parathyroid gland; d. Boston, Mass., Dec. 5.

PATIERNO, JOHN (54), U.S. aerodynamics engineer who pioneered the development of the B-2 Stealth bomber. He devised the technology enabling the B-2 to fly through enemy radar without detection; d. Newport Beach, Calif., Feb. 25.

PEDERSEN, CHARLES J. (85), Korean-born U.S. chemist who shared the 1987 Nobel Prize in Chemistry for work in the synthesis of crown ether molecules, which he had discovered earlier in his career; d. Salem, N.J., Oct. 26.

PIMENTEL, GEORGE C. (67), U.S. chemist who invented the chemical laser, designed instruments for interplanetary spacecraft, and pioneered the development of rapid-scan techniques for infrared spectroscopy. He designed a special spectrometer for the Mariner spacecraft that explored Mars in 1989. Earlier in his career, he had worked on the Manhattan Project; d. Kensington, Calif., June 18.

POSNACK, EMANUEL R. (92), U.S. patent lawyer and inventor who perfected the Posnack recuperator, a device for reusing exhaust heat from industrial furnaces, and the autoclench stapler; d. Great Neck, N.Y., May 6.

RAJCHMAN, JAN ALEKSANDER (80), U.S. electronics engineer who developed tiny ferrite cores, magnetic memories, and switching devices that led to the development of high-speed computer memory systems; d. Princeton, N.J., April 1.

RAVITCH, MARK M. (78), U.S. surgeon, professor, and author who was an authority on the correction of deformities in the chest wall and wrote extensively on congenital hernias of the diaphragm. His analysis of intussception in infants and children, in which the intestines do not connect properly, led to a nonsurgical treatment; d. Pittsburgh, Pa., March 1.

RUSK, HOWARD A. (88), U.S. medical expert who pioneered rehabilitation medicine with the techniques he used to help badly wounded airmen in World War II. His methods earned him the nickname Dr. Live-Again; d. New York, N.Y., Nov. 4.

SAKHAROV, ANDREI D. (68), Soviet nuclear physicist who collaborated after World War II in the creation of a fission bomb and played an important role in the development of more powerful hydrogen weapons; he also worked on peaceful uses for nuclear energy. Sakharov was instrumental in getting the U.S.S.R. to accept a treaty with the United States banning the explosion of nuclear devices. He became an outspoken dissident, and was awarded the Nobel Peace Prize in 1975 for his efforts to fight for human rights and intellectual freedom in the U.S.S.R. In 1980 he was exiled to the closed city of Gorki, where he remained until being released by Mikhail Gorbachev in 1986. He was elected to the new Soviet Congress in 1989; d. Moscow, Dec. 15.

SAROV, ISRAEL (55), Israeli virologist who developed a test to detect the sexually transmitted disease chlamydia. He also developed tests for Epstein-Barr syndrome and for mononucleosis. At the time of his death, he was working on a safe vaccine for cytomegalovirus, an infection linked to mental retardation in babies; d. Honolulu, Hawaii, Nov. 16.

SEGRÉ, EMILIO G., (84), Italian-born U.S. nuclear scientist who shared the 1959 Nobel Prize in Physics for creating an antiproton, antimatter with a negative charge. This accomplishment lent support to the hypothesis that a mirror image of the universe exists in which atoms are made up of positively charged electrons and negatively charged nuclei—an antiuniverse; d. Lafayette, Calif., April 22.

SENDZIMIR, TADEUSZ (95), Polish-born developer of a patented steel- and metal-galvanizing process now used worldwide; d. Jupiter, Fla., Sept. 1.

SHIMKIN, MICHAEL B. (76), Russian-born U.S. medical researcher who helped establish the link between smoking and lung cancer through population studies he conducted in the 1950s; d. San Diego, Calif., Jan. 16.

SHOCK, NATHAN W. (82), U.S. gerontologist who combined science and a sense of humor to educate the public on aging. He studied the physiology of aging in various human organs, and documented that individuals age at markedly different rates; d. Baltimore, Md., Nov. 12.

SHOCKLEY, WILLIAM B. (79), U.S. electrical engineer and proponent of a controversial theory of race. He shared the 1956 Nobel Prize in Physics for the development of the transistor, which sparked the electronic age, becoming a basic component in myriad modern products. His work was overshadowed by his theory of "retrogressive evolution," which claimed intelligence was genetically transmitted, and that blacks were genetically inferior to whites and unable to achieve the same intellectual level; d. Stanford, Calif., Aug. 12.

SMITHE, FRANK B. (96), U.S. amateur ornithologist who set color standards for the science and wrote an ornithology color guide for naturalists; d. Manhasset, N.Y., Jan. 15.

SOPWITH, SIR THOMAS (101), British aircraft pioneer whose company built the Sopwith Camel, used in 1917 to shoot down Germany's best-known World War I fighter pilot, Baron Manfred von Richthofen, better known as the "Red Baron." During World War II, Sopwith's company built the Hawker Hurricane, which downed many German bombers during the Battle of Britain. As a young man, he raised money to start the companies by competing in stunt-flying contests; d. Winchester, England, Jan. 27.

SPAGNA, THEODORE (45), U.S. photographer and filmmaker whose work in filming sleeping people showed that they became inactive while dreaming. He also helped create an exhibit that converts the brain waves of a sleeping person into a light-and-music show; d. Boston, Mass., June 21.

STARR, ISAAC (94), U.S. researcher in the field of cardiovascular disease, noted for his use of physics to measure the heart's efficiency and for his research on the physiology of congestive heart failure. He also developed the first practical ballistocardiograph, an instrument that measures the blood output of the heart and is used to predict and measure the course of heart disease; d. Roxborough, Pa., June 22.

STONE, MARSHALL H. (85), U.S. mathematician known for his work in synthesizing diverse areas of abstract mathematics, including analysis, algebra, and topology. He won the National Medal of Science in 1983, and was active in advancing the study of mathematics around the world; d. Madras, India, Jan. 9.

STREET, J. C. (83), U.S. codiscoverer, in 1937, of a fundamental part of matter known as the muon. Similar to but heavier than the electron, the muon occurs with both positive and negative charges. He also developed a number of physics research devices; d. Charleston, S.C., Nov. 7.

TISHLER, MAX (82), U.S. pharmaceutical scientist who led in the development of drugs to treat arthritis, heart disease, and hypertension. He also produced vaccines to control livestock and poultry diseases, and commercial processes to manufacture vitamins, penicillin, and cortisone; d. Middletown, Conn., March 18.

VAN SCHAIK, PETER (60), U.S. aerospace engineer who developed the backpack maneuvering unit that astronauts wear in space. He later said he got the idea from the "Buck Rogers" comic strip; d. Huber Heights, Ohio, Sept. 4.

WARNER, JOHN CHRISTIAN (91), U.S. chemist and university president who worked on the Manhattan Project during World War II, supervising research in the chemistry and metallurgy of plutonium. He served on the general advisory committee of the Atomic Energy Commission, and, as president of the Carnegie Institute of Technology, was credited with building the school into a nationally known research institution. He also pioneered the use of computers in education; d. Pittsburgh, Pa., April 12.

WEISS, PAUL A. (91), Austrian-born U.S. biologist who won the National Medal of Science for his pioneering work in the theory of cellular development. He found that cells mixed randomly and reassembled from different organs can rebuild themselves into miniature copies of the donor organs without direction from a central source. He also helped prove that nerve cells can replenish themselves to all parts of the body, established new surgical methods for the repair of peripheral nerve tissue, and showed that an embryo's organization and growth is determined by the physical and chemical environment surrounding newly multiplying cells; d. White Plains, N.Y., Sept. 8.

WHITNEY, HASSLER (82), U.S. mathematician who specialized in geometry and topology, including the study of multidimensional surfaces such as spheres. He proved that one could plot a tour of the earth that visits each country only once—the Hamiltonian cycle. Later in his career, he spent much time trying to improve the teaching of mathematics and eliminate "mathematics anxiety" in children, developing methods in which children could have actual experiences with math; d. Princeton, N.J., May 10.

WOLMAN, ABEL (96), U.S. professor of sanitary engineering who led U.S. and foreign efforts to chlorinate drinking water. He helped develop a chlorination method that would kill pathogens but not harm people who drank the water. He also headed a post–World War II committee that warned that future nuclear-weapons testing might lead to environment-contaminating waste disposal. In 1975 he was awarded the National Medal of Science. His water-chlorination methods have been adopted by almost all U.S. cities; d. Baltimore, Md., Feb. 22.

ZINBERG, NORMAN E. (67), U.S. psychoanalyst whose research into the effects of drugs on individuals influenced treatments for drug abuse. He demonstrated that a drug user's behavior was the result of a multitude of social factors in addition to the biochemical effect of the drug. He also showed that a person reacted differently to a drug at different times, and that the reaction to a particular drug varies among individuals; d. Cambridge, Mass., April 2.

ZOBELL, CLAUDE E. (84), U.S. marine biologist who was the first researcher to recover and cultivate living organisms from ocean depths greater than 20,000 feet (6,000 meters). He is credited with discovering more than 65 species of ocean bacteria; d. La Jolla, Calif., March 13.

INDEX

A

Accelerator, Particle 283, 325
Acid rain (meteorol.) 194
Acoustics (sci.)
  architectural acoustics 304–10
  bioacoustics 203–7
  shipwreck archaeology 270
  See also Sound
ACTH (Adrenocorticotropic hormone) 70
Acupuncture, *illus.* 224
Adamson, George (Kenyan sci.) 349
Ader, Robert (Amer. psych.) 73, 78
Adolescents *see* Teenagers
Adrenaline (biochem.) 43, 46
Aerodynamics
  bicycle wheel innovation 316
  NOTAR helicopter 328–31
  pinecone pollination 85
Agoraphobia (psych.) 59
Agouti (zool.), *illus.* 210
Agriculture
  alligator farming 374, 377–79
  alternative agriculture 79–83
  Black Death's effects in Europe 262
  effects on African wildlife 350
  food and world population 253
  golden eagle's livestock predation 368
  tropical dry forest depletion 212
AIDS (Acquired Immune Deficiency Syndrome) (med.) 225
  drug-approval regulations 232
  psychoneuroimmunology 78; *illus.* 76
Air bag (auto. safety device) 316
Aircraft *see* Aviation
Airlift (instru.) 271; *illus.* 270
Air pollution 194
  alternative fuels as remedy 166–72
Air pressure
  avoiding the bends 26–27
  tornadoes 149
Akkadian language 277–79
Alar (chem.) 80, 195
Alcohol
  alternative fuels 167–69
  effects on women 225
  panic disorder patients 59
Alfalfa (bot.) 68
Alligator (zool.) 373–79
  attacks on wolves 219
Alpha interferon (drug) 227
Alternative agriculture 79–83, 253
Alternative fuels 166–72
  renewable energy 165
Altman, Sidney (Can.-Amer. sci.) 311, 313
Aluminum (element) 292–98
Alvin (submersible) 269–70
Amputation (surg.) 241
Angiosperm (bot.) 69
Animals *see* Zoology
Antarctica 196–202
Anthropology (sci.) 252
Antibodies (biochem.) 76, 93
Apollo Project (U.S. space program)
  Apollo 11 32
  Apollo 17, illus. 31
  asteroid mission proposal 20
  space suit 26; *illus.* 25
Apple (fruit) 195
Aramaic language 277–79

Archaeology 252–53
  ancient language studies 276–79
  shipwreck archaeology 268–75
Archery (sport) 41–43, 46–47
Architectural acoustics 304–10
Argo (submersible) 270
Armadillo (zool.) 369–72
Artificial intelligence 341–43
Artificial reality (computer tech.) 102–9
Art restoration 320–27
Aspartame (chem.) *see* NutraSweet
Assyrian language *see* Akkadian language
Asteroids (astron.) 5, 16–20
Astronauts
  lunar landing, *illus.* 31
  space suits 21–29
Astronomy 4–5
  near-Earth asteroids 16–20
  wormholes 284–91
  See also Space science and technology
Atlantic Ocean
  beach erosion 141–44, 146
  water circulation 160
Atlantis (space shuttle) 6
Atmosphere 194
  halon's threat 283
  spacecraft studies 6
  tornadoes 147–53
Atom (phys., chem.) 283
Atomic clock 311; *illus.* 312
Aurora (light) 5
Autism therapy 357
Automation *see* Computers; Robot
Automobiles
  air bags 316
  alternative fuels 166–72
  safety development 302
Aviation
  aircraft innovations 318
  early pilot certification 302
  noise 206–7
  NOTAR helicopter 328–31
  Stealth bomber 317
AZT (Azidothymidine) (drug) 225, 232–33

B

B-2 bomber 317
Bacteria
  antibiotic-producing 68
  degradable plastic decomposition 317
  genetic exchange with yeast 68
  natural pest control 82
  Yersinia pestis 254–57, 261
Balloon 148, 153
Barrier island (geol.) 141–46
Basal ganglia (anat.) 46
Basalt (rock) 32–34
Baseball bats 292–98
Battery, Electric 168
Beach erosion 141–46
Bee (zool.) 69
Beetle (zool.) 68, 218
Behavioral sciences 38–39
  athletes' "zone" 40–47
  bubonic plague's effects 259–62
  delusions involving celebrities 51–55
  dolphin behavior 361–62
  dolphin therapy 357
  drug abuse 224

drug dealer incarceration 60–65
  emotions and intercellular communication 72, 78
  living without time-measuring cues 48–50
  panic disorder 56–59
  weight-loss programs 245–49
Bicycle 316
Bioacoustics (sci.) 203–7
Biochemistry *see* Biology; Chemistry
Biology 68–69
  alternative agriculture 79–83
  biochemical codes 70–78
  biosensors 92–94
  comparative biomechanics 84–91
  See also Body, Human; Botany; Zoology
Biomechanics, Comparative (sci.) 84–91
Biopesticide 82
Biorhythms 50
Biosensors (tech.) 92–94
Biotechnology (sci.) 242–44
  See also Genetic engineering
Birds (zool.)
  Antarctic oil spills 202
  fossils 130
  golden eagle 364–68
  hurricane Hugo's effects 217–18, 220–21
  plumage 69
  vocalizations 204–6
Birth defects 224, 232
Bishop, J. Michael (Amer. sci.) 95–97
Bismarck (battleship) 270, 274–75; *illus.* 269
Bison (zool.) 347
Black Death *see* Plague
Black hole (astron.) 285
Bladder cancer (med.) 241
Body, Human
  biochemical codes 70–78
  drug testing process 229–30
  living without time-measuring cues 49–50
  oxygen deprivation 24–25
  physical performance and the unconscious 40–47
Bone (anat.)
  cancer (med.) 236, 239; *chart* 237; *illus.* 238
  vitamin D deprivation 50
Botany (sci.)
  alternative agriculture 79–83
  leaf structure, *illus.* 88
  plant evolution 69
Brain (anat.)
  AIDS virus 78; *illus.* 76
  control of facial expressions 39
  dolphins 357, 360–61
  intercellular communication 72, 74–75, 77
  metabolism during concentration 44–47
  panic disorder connection 58
Breast cancer (med.) 236, 238
Breathing (physiol.) 75
Brilliant Pebbles (defense tech.) 317
Brown dwarf (star) 4
Bubonic plague (med.) *see* Plague
Building construction and codes
  architectural acoustics 304–10
  coastal flood areas 146
  earthquake preparation 138–39
Bush, George (Amer. pres.) 6, 167; *illus.* 171

383

C

Cadmium (element) 225
Calcite (min.) 156–58
Calcium blockers (drugs) 231
Canaanite culture (archaeo.) 271
Cancer (med.)
  genetic nature 95–97
  lung cancer in dogs 225
  new treatments 235–41
  survival rates, *chart* 237
Carbon (element) 317–18
Carbon dioxide (chem.) 160–61, 164, 168, 194
Carcerand (chem.) 282
Catalytic chemistry 282
Catalytic converter (auto.) 172
Cataracts (med.) 226
CAT scan *see* Computerized axial tomography scan
Causality violation (philos.) 289–91
Cave-living experiments 48–50
Cech, Thomas R. (Amer. sci.) 311, 313
Cellular telephone 318
Census, United States 110–16
Centaurus (constellation) 4
Central nervous system (anat.) *see* Nervous system
Cesium (element) 311
CFCs *see* Chlorofluorocarbons
Charge-coupled device (tech.) 335
Chelation therapy (chem.) 233
Chemicals, Agricultural 80–81, 83
Chemistry 282–83
  biochemical codes 70–78
  cold-fusion attempts 175–79
  drug approval process 229
  earthquake-released chemical pollution 140
  Nobel Prize 311, 313
  polymerase chain reaction 68
Chemotherapy (med.) 94, 235, 238–39, 241; *illus.* 236
Chemzyme (chem.) 282
Chess (game) 100
Chlorofluorocarbons (CFCs) (chem.) 194
Chromosome (biol.) 68
Cimetidine (drug) 225
Cisplatin (drug) 238
Clam, Coquina (zool.) 88–89; *illus.* 87
Clay tablets 276, 279; *illus.* 278
Climate (sci.) 131
  ice ages 154–61
Clock, Atomic 311; *illus.* 312
Cloud (meteorol.)
  molecular clouds, *illus.* 4
  tornado 147
Coal and coal mining 164, 180–83
Coati (zool.), *illus.* 210
Cocaine (drug) 224
  biosensor detection 92–93
  satellite reconnaissance 339
Cold Dark Matter (astron.) 4
Cold-fusion energy 173, 175–79
Colon cancer (med.) 236
Comet (astron.) 20
Communication
  animal calls 203–7
  dolphins 356–57, 361
  intercellular communication 70–78
  technological innovations 318
  Voyager mission's sounds of earth 14–15
  See also Language; Telecommunications
Compact-disk players 118–22, 303
Comparative biomechanics (sci.) 84–91
Composite materials 298, 316–18
Computer games 109
Computerized axial tomography (CAT) scan 131
Computers 100–101
  aircraft tracking 318
  air-fuel ratio adjustments 170
  ancient language research 278–79
  artificial reality 102–9
  art restoration 321–22, 325, 327
  bioacoustics 206
  census information 114–15; *illus.* 112–13
  face recognition system 341–43
  fast-food innovations 316
  functional electrical stimulation 243–44
  optical computers 124–27
  smart TV 117–23
  surveillance satellites 339–40
  tornado simulations 153
  train dispatching systems 319
Concert hall design 304–10
Consciousness (psych.) 43–44, 46–47
Consumer technology 316
Contact lenses 226
Cormorant (zool.) 202
Cosmic Background Explorer (satellite) 7
Coulomb explosion (phys.) 283
Crack (drug) 224
  prison camps for dealers 60–65
Craters, Lunar 32, 34; *illus.* 31
Crime
  delusions involving celebrities 51–55
  drug dealer incarceration 60–65
  generic-drug scandal 234
  poaching in Kenya 349–50, 352
  theft from shipwrecks 275
Crop rotation (agr.) 81, 83
Cuneiform (writing) 276, 279; *illus.* 278
Cyanide (chem.) 195
Cyclophosphamide (drug) 78
Cygnus (constellation) 4

D

Darwin, Charles (Eng. nat.) 73
DataGlove (computer tech.) 103–4, 106–7, 109
Data processing 110–16
DDI (Dideoxynosine) (drug) 232–34
Decommissioning of nuclear reactors 186, 188–89
Deep Thought (computer) 100
Defense technology 317, 332–40
Dehmelt, Hans G. (Ger.-Amer. sci.) 311–12
Deimos (astron.) 5
Delusions (psych.) 51–55
Denversaurus (dinosaur) 130
Depression (psych.) 39, 59
Descartes, René (Fr. philos.) 73
Design
  artificial reality applications 109
  comparative biomechanics 84–91
  space suits 21–29
Desktop video *see* Smart TV
Deuterium (chem.) 174, 176–79
Devils Hole, Nev. 156–59
Diabetes (med.) 94
Diagnosis (med.)
  proto-oncogene analysis 97
  psychiatric techniques 54
Dideoxynosine (drug) *see* DDI
Diet
  weight-loss programs 245–49
  world hunger 253
Diet Center (weight-loss program) 246, 248
Digestive system (anat.) 75
Digital optical processor 124–27
Dinosaurs (paleon.) 130
Disasters
  bubonic plague 254–62
  earthquakes 132–40
  hurricane Hugo 215–21
  storms and beach erosion 143, 146
  tornadoes 147–53
Disease *see* Health and disease
DNA (biochem.) 95–97
  genetic patterns in modern populations 252
  polymerase chain reaction 68
  transgenic species creation 68
Dolphin (zool.) 204–5, 356–63
Doppler radar 149–50, 153, 336
Drugs
  alpha interferon 227
  antirejection drugs 226
  approval by FDA 228–34
  AZT 225
  cancer treatment 238
  chemotherapy testing 94
  dealer incarceration 60–65
  hepatitis transmission 227
  illegal drugs 224
  production with chemzymes 282
  surveillance satellite detection 339
Dry-eye disease (med.) 226
DTP vaccine 227
Dune (geol.) 144, 146, 218

E

Eagle, Golden (zool.) 364–68
Ear (anat.) 69
Earth (planet)
  atmospheric studies by spacecraft 6
  orbital fluctuations 156, 159, 161; *illus.* 158
  Soviet photographs 7
  structure 131
Earthquakes (geol.) 130, 132–40
Earth sciences 130–31
  See also Atmosphere; Climate; Geology; Ocean; Paleontology; Weather
Eating habits
  alligators 376, 378
  armadillo 371
  golden eagle 365, 367–68
  weight-loss programs 245–49
Echolocation (hearing) 356, 358–59, 361
Eclipse, Solar (astron.) 5
Ecology *see* Environment
Education
  mathematics 101
  multimedia systems 122
Eisenhower, Dwight D. (Amer. pres.), *illus.* 186
Electricity
  battery-driven automobiles 168; *chart* 172
  coal-produced 181
  functional electrical stimulation 242–44
  nuclear power 165
  Shippingport Atomic Power Station 187
  Underwriters Laboratories 301–2
Electron (phys.) 283, 312
Electronics
  acoustic flexibility 310
  telephone innovations 318
  See also Computers
Elephant (zool.) 204, 348–55
Elizabethan theaters (Eng. archaeo.) 252
Elk (zool.) 347
Emotions
  dolphin brain structure 361
  facial expressions 39
  intercellular communication 72–76
  Overeaters Anonymous 247
  trauma recovery 38
Encke (comet) 20
Endangered species (biol.)
  alligators making a comeback 374–75
  Kenyan national parks 348
  northern spotted owl 346
  Puerto Rican parrot 218
  red-cockaded woodpecker 220–21
  red wolf 218–20
  tropical dry forest 212

Endeavor (space shuttle) 6, 317
Endocrine system (physiol.) 71; *illus.* 70
Endorphins (biochem.) 77
Energia (Sov. space vehicle) 20
Energy 164–65
    alternative fuels 166–72
    coal mining technology 180–83
    nuclear fusion 173–79
    nuclear reactor core disposal 184–91
    wake-field effect 283
    whale blubber 90
Engineering
    acoustical engineering 304–10
    artificial reality applications 109
    earthquake preparation 139
    product safety testing 299–303
    space suits 21–29
Environment 194–95
    agriculture's effects on 80–81, 253
    Alaskan oil spill 164
    beach erosion 141–46
    bioacoustics 203–7
    clean-burning alternative fuels 166–72
    coal mining and burning's effects on 181
    degradable plastic 317
    elephants' effects on 353
    fossil fuel's effects on 164
    hurricane Hugo's impact on 215–21
    tourism's effects on Antarctica 196–202
    tropical dry forest restoration 208–14
Enzyme (biochem.) 313
    alcohol neutralization 225
    biosensor use 93
    chemzymes 282
Erosion (geol.) 80, 82
    beach erosion 141–46
    hurricane Hugo 218
Ethanol (chem.) 168; *chart* 172
Everglades National Park (Fla.) 347
Evolution (biol.) 69
    early humans 252
    search for extraterrestrial life 263–67
Exercise 246–47, 249
Exotic matter (phys.) 286–87, 291
Extraterrestrial life, Search for 263–67
    Voyager Interstellar Mission 14–15
Exxon Valdez (tanker) 164, 195
Eye (anat.) 226

F

Face (anat.)
    computerized recognition system 341–43
    facial expressions 39
Facsimile transmission (Fax) 318
Family (sociol.) 38
Famine (econ.) 253
Farming *see* Agriculture
Fault (geol.) 130, 136–37
Ferret, Black-footed (zool.) 346
Fiber optics 105
Finch (zool.) 205
Fires
    earthquake-caused 134; *illus.* 133
    forest fires 210; *illus.* 209
    product flammability testing 302–3; *illus.* 299–300
Fish (zool.)
    fluid mechanics 89
    genetic engineering 69
    hurricane Hugo's effects on 218
Fishing and dolphin deaths 358–59
FK-506 (drug) 226
Flea (zool.) 254–57, 261
Flow patterns (mech.) 89
Fluid mechanics (sci.) 89
Fluorine (gas) 317

FMX (radio transmission) 283
Food
    contamination 195
    fast-food innovations 316
    weight-loss programs 245–49
    world food supply and population 253
Food and Drug Administration (FDA) 228–34
Forest (geog.)
    fires 210; *illus.* 209
    hurricane damage 215–21
    rain forest destruction 194
    tropical dry forest restoration 208–14
Forgery, Art 327
Formaldehyde (chem.) 172
Fossil fuels 164, 180–83
Fossils *see* Paleontology
Foster, Jodie (Amer. act.), *illus.* 51
Frequency (phys.) 138
Fruit fly (zool.) 89
F-scale (tornado measurement) 152
Fuel *see* Energy
Fuel cells 165
Fuel rods 185, 189
Functional electrical stimulation (biotech.) 242–44
Fusion, Nuclear 173–79

G

Galaxy (astron.) 4
Galilee, Sea of (Isr.) 271
Galileo (space probe) 6–7, 18
Galileo Galilei (It. astron.) 86–87
Gamma ray (phys.) 4, 325
Gas, Natural *see* Natural gas
Gas chromatography 326–27
Gasohol (fuel) 168
Gasoline (fuel) 167, 171–72
Gene mapping 68
Generic drugs 234
Genetic causes of cancer 95–97
Genetic engineering 68
    infection-resistant mosquitoes 69
    pest-resistant crops 81–82
    transgenic species experiment 68
Geology
    beach erosion 141–46
    earthquakes 132–40
    earth structure and plate tectonics 131
    ice age cycles 154–61
    lunar geology 30–35
Glacier (geol.) 155–56, 158–60
Glenn, John (Amer. pol.) 54; *illus.* 55
Global warming *see* Greenhouse effect
Globe theater (Eng. archaeo.) 252
Glucose testing (med.) 94
Graf, Steffi (W. Ger. athl.), *illus.* 43
Grain (bot.) 253
Graphite (min.) 298
Gravity and gravitation (phys.) 285, 291
Gray, Sir James (Brit. biol.) 87
Great Attractor (astron.) 4
Great Wall (astron.) 4
Greenhouse effect 131, 164, 194
    ice ages 160–61
    solar activity 5
    spacecraft studies 6
GRiDPad (computer tech.) 101
Groundwater (geol.) 156–58
Guanacaste National Park Project (Costa Rica) 208, 211–14

H

Halon (chem.) 283
Hanford Military Reservation (Wash.) 185–86, 188, 190–91
Hazardous waste *see* Radioactive wastes; Toxic waste

Health and disease 224–27
    agricultural chemical hazards 80
    artificial reality applications 106
    biochemical codes 70–78
    biosensor testing 94
    bubonic plague 254–62
    cancer treatments 235–41
    depression 39
    drug-approval process 228–34
    functional electrical stimulation 242–44
    genetic nature of cancer 95–97
    infection-resistant mosquitoes 69
    panic disorder 56–59
    weight-loss programs 245–49
Heart disease (med.) 225
Heavy oxygen (chem.) 157–58, 160
Helicopter, NOTAR 328–31
Helium (element) 176, 178–79
Hepatitis (med.) 227
Herbicide (chem.) 81
    herbicide-resistant alfalfa 68
Heroin (drug) 224
Hershiser, Orel (Amer. athl.), *illus.* 45
Holocaust survivors 38
Homeless people 115–16; *illus.* 253
Honeybee, African (zool.) 69
Hubble Space Telescope, *illus.* 4
Hugo (hurricane) 131
    environmental impact 215–21
Human body *see* Body, Human
Humans, Early 252–53
Hunger (physiol.) 253
Hunting
    alligators 374–75, 377
    golden eagle 368
    poaching in Kenya 349–50, 352
    Yellowstone controversy 347
Hurricanes (meteorol.) 131
    beach erosion 143, 146
    Hugo's impact on the environment 215–21
Hydrogen (element) 173–74, 177–79
    Alpha laser 317
    maser 311–12
Hyena (zool.) 205–6
Hyoid bone (anat.) 252

I

Ice (drug) 224
Ice ages (geol.) 154–61
Immune system 70–78
    See also AIDS
Indians, American
    Innu Indians 206
    Maya 253
Infant, Facial expressions of 39
Infrared light 322, 325–26
    detector 7, 335, 340
Initial-time-delay gap (acoustics) 307; *chart* 306
Innu Indians 206
Insects (zool.) 69
    natural pest control 82
Insulin (hormone) 74–75
Intelligence
    artificial 341–43
    dolphin cognition 357–58
    unconscious mind 44
Interactive viewing (computer tech.) 121–22
In vitro **fertilization** 68
Io (astron.) 5, 9
Ion trap (phys.) 283, 312
Irrigation (agr.) 80
Ivory 350; *illus.* 351–52

J

Jason Jr. (robotic camera) 270
Jetty 143
Jobs, Steven P. (Amer. computer sci.) 121, 123
Jordan, Michael (Amer. athl.), *illus.* 41
Jupiter (planet) 5–7, 9; *illus.* 10

K

Kasparov, Gary (Sov. chess player) 100
Kayak (boat) 316
Kilauea (volc., Haw.) 131
Killdeer (zool.) 204

L

Lacewing, Green (zool.) 82
Language
    ancient language studies 276–79
    dolphin cognition 357
    translating technology 318
Laptop computers 101
Larynx cancer (med.) 235–36, 238; chart 237; illus. 240
Laser (phys.)
    Alpha laser 317
    atomic research 283
    compact-disk player safety 303
    nuclear fusion 174, 178
    optical computers 124–27
    tornado measurement 153
Lava (geol.) 32–34
Leakey, Richard E. (Kenyan paleoanthro.) 348–55
Legume (bot.) 82
Lennon, John (Eng. mus.) 51; illus. 52
Leonardo da Vinci (It. inv.) 325, 330
Leprosy (med.) 369, 372
Letterman, David (Amer. entertainer) 51
Life, Extraterrestrial 263–67
Light
    art restoration techniques 322, 325–26
    spectral analysis 340
    See also Laser; Sunlight
Lion (zool.) 205; illus. 204
Liquid diet 245, 249
Liver (anat.) 225–27
Llama (zool.), illus. 85
Longwall mining 180–83
Lunar base 35
Lunar geology 18, 30–35
Lung (anat.)
    cancer in dogs 225
    cancer survival rate 236; chart 237
    crack smoking 224
Lupus (med.) 78

M

Macrophages (biochem.) 76–77; illus. 75
Magellan spacecraft 5–6
Magnetic-confinement fusion (phys.) 174, 177
Magnetism (phys.) 282
    magnetically-levitated train 319
Magnetometer (instru.) 270
Mammography (med.), illus. 239
Manhattan Project 190
Maps and mapmaking
    computers and art restoration 321
    electronic map of census information 114–15
    shipwreck archaeology, illus. 271
Marine geology 131
Marine Mammal Protection Act (U.S.) 358, 363
Mars (planet) 5, 7, 18
Mary Rose (ship) 273
Maser, Hydrogen (instru.) 311–12
Mastectomy (surg.) 238
Materials science 282
Mathematics 100–101
    United States census 110–16
Matter (phys.) 4
Maya Indians 253
Meadowlark (zool.) 205
Measles (med.) 227
Medicine see Health and disease

Memory (psych.) 44, 46
Mental health and illness
    delusions involving celebrities 51–55
    depression 39
    panic disorder 56–59
Mercury (planet) 5
Metabolism (physiol.)
    brain during concentration 44–47
    toxicity testing 94
Meteor and meteorite (astron.) 5, 32–34
Meteorology see Weather
Methamphetamine (drug) 224
Methane (gas) 283
Methanol (Methyl alcohol) (chem.) 167, 169–72; chart 172
Michelangelo (It. art.) 320–24, 327
Microphysiometer (instru.) 94
Microprocessors (computer tech.) 101
    functional electrical stimulation 243–44
Microtelepresence (computer tech.) 104
Microwave (phys.)
    aircraft landing system 318
    atomic research 283
    cosmic radiation studies 7
    fast-food preparation 316
    surveillance satellites 335
    tornado monitoring 149
Migration, Animal 350
Milky Way (astron.) 4
Mining technology 180–83
Miranda (astron.) 10
Mitchell, Kevin (Amer. athl.), illus. 45
Molecule (chem., phys.)
    biochemical codes 70–72
    catalytic chemistry 282
    DNA polymerase 68
Molecular cloud (astron.), illus. 4
Mona Lisa (paint., Leonardo da Vinci) 325
Monkey, White-faced (zool.), illus. 210
Monocytes (anat.) 76–77
Montana, Joe (Amer. athl.), illus. 46
Moon (astron.)
    distance from Earth 5
    formation 31–35
    lunar geology 6, 18, 30–35
Moonrock (drug) 224
Mosquito (zool.) 69
Mountain (geol.) 131
Multimedia see Smart TV
Multispectral analysis 340
Muon-catalyzed fusion (phys.) 177–79
Muscle (anat.)
    biomechanics 90
    functional electrical stimulation 244

N

National Aeronautics and Space Administration (U.S.)
    Ames Research Center 103–5
    functional electrical stimulation research 243
    Search for Extraterrestrial Intelligence program 205, 263–67
    space suits 21–29
Natural gas 164, 168–69, 171; chart 172
    earthquake-caused fires 134
Navy, United States
    dolphin research 362–63
    Trident 2 missile 317
Neanderthal (early human) 252
Near-Earth asteroids (astron.) 16–20
Nematode (zool.) 82
Neocortex (anat.) 361
Neptune (planet) 5–6, 10–13; illus. 9
Nervous system (anat.) 71, 73; illus. 70
    basal ganglia 46
    bubonic plague 257–58
    chemotherapy damage 238

    functional electrical stimulation 242–44
    See also Brain
Neutron (phys.) 175–79
Nitrogen (element) 282
    automobile air bags 316
    "bends" 26
Nobel prizes
    physics and chemistry 311–13
    physiology or medicine 95–97
Noise 206–7
    radio interference 266
NOTAR helicopter 328–31
Nuclear power 165
    fusion 173–79
    reactor core disposal 184–91
Nuclear reactor see Reactor, Nuclear
Nuclear wastes see Radioactive wastes
NutraSweet (artificial sweetener) 248
Nutri/System (weight-loss program) 246, 248
Nutrition see Diet

O

Obesity (med.) 245–49
Ocean
    beach erosion 141–46
    circulation patterns 160–61
    floor formation 131
    ice ages 156
    oxygen content of water 157
    shipwreck archaeology 268–75
Ocelot (zool.), illus. 210
Oil (Petroleum) 164
Oil spill 164, 195
    Antarctica 197, 202
Oncogenes (biochem.) 95–97
Ophthalmology (sci.) 226
Opiate receptor (anat.) 74, 76
Optical computer 124–27
Optifast (weight-loss program) 245, 249
Oraflex (drug) 232
Orbit, Earth's 156, 159, 161; illus. 158
Organization of Petroleum Exporting Countries (OPEC) 164
Orthopedics (med.) 239
Osteoporosis (med.) 225
Overeaters Anonymous (weight-loss program) 246–47
Owl (zool.)
    Northern spotted owl 346
    sound localization 205
Oxygen (element)
    avoiding the bends 26–27
    water content 157–58, 160
Ozone (gas) 166–67
    halon's threat to ozone layer 283
    stratospheric depletion 194

P

Paging system (tech.) 318
Paleontology (sci.) 130
    biomechanics 90
    human evolution 252
    Leakey, Richard E. 355
Palladium (chem.) 176, 179
Panda (zool.) 346
Panic disorder (psych.) 56–59
Paralysis treatment 242–44
Paranoia (psych.) 224
Parenting styles (sociol.) 38
Parker, Denise (Amer. athl.) 41–43, 46–47; illus. 40
Parrot, Puerto Rican (zool.) 218
Particle, Subatomic 283
Particle accelerator 283, 325
Particle-beam device 317
Paul, Wolfgang (Ger. sci.) 311–12
Pelé (Braz. athl.) 40; illus. 42
Penguin (zool.) 202; illus. 199
Penning trap (phys.) 312

Peptides (biochem.) 74–78; *illus.* 71
Perception (psych.) 40–47
Personal computers 101, 103, 117–23
Pesticide 68, 80–82
Petroleum *see* Oil
PET scan (med. tech.) 44, 58
Phobia (psych.) 59
Phobos (astron.) 5, 7
Photography
   art restoration techniques 325
   shipwreck archaeology 270
   surveillance satellites 332–40
Photon (phys.) 125
Photonics *see* Optical computer
Photovoltaic technology 165
Physics 282–83
   baseball bat performance 292–98
   comparative biomechanics 84–91
   Nobel Prize 311–12
   nuclear fusion 173–79
   relativity of speed and time 45
   wormholes 284–91
Physiology or Medicine, Nobel Prize in 95–97
Pi (math.) 100
Pilgrim nuclear power plant (Mass.) 165
Plague (med.) 254–62
Planets *see* names of planets
Plants *see* Botany
Plastics, Degradable 317
Plastics waste 195
Plate tectonics (geol.) 131, 136
Pluto (planet) 5
Plutonium (element) 6–7, 190
Poaching (illegal hunting) 349–50, 352
Pollen and pollination (bot.)
   hybridization of plants 68
   pinecones 85
Pollution
   food chain 365
   noise pollution 206–7
   See also Air pollution; Water pollution
Polymerase (biochem.) 68
Population
   Kenyan effects on wildlife 350
   United States census 110–16
   world food supply and population 253
Positron emission tomography *see* PET scan
Post, Wiley (Amer. aviator) 25; *illus.* 24
Prison camps for drug dealers 60–65
Product safety
   drug approval by FDA 228–34
   Underwriters Laboratories 299–303
Programming, Computer
   artificial reality 109
   census applications 115
   concert hall design 310
Prolactin (hormone) 75
Prostate cancer (med.) 241
Prosthesis (med.) 239–41
Protein (biochem.) 248
Proto-oncogene (biochem.) 97
Provirus (biochem.) 96
Psychology *see* Behavioral sciences
Psychoneuroimmunology (sci.) 70–78
Pulsar (astron.) 4

Q

Quackery, Medical 233
Quantum gravity (phys.) 291
Quasar (astron.) 4

R

Radar
   surveillance satellites 335–38, 340
   tornado monitoring 148–50, 153

Radiation therapy (med.) 235, 238–39, 241; *illus.* 236, 240
Radio
   earthquake prediction 137
   FMX transmission 283
   radio telescope 264, 266
   surveillance satellites 337
   wristwatch paging system 318
Radioactive wastes 184–91, 195
Radio-frequency trap (phys.) 312
Rail (zool.) 220
Rain forest 194
Ramsey, Norman F. (Amer. sci.) 311–12
Rat (zool.) 254–57, 261
Reactor, Nuclear 184–91
Reagan, Ronald (Amer. pres.), *illus.* 51
Receptor (anat.) 74, 76–77, 93
Recycling (waste disposal) 195
Redoubt (volc., Alas.) 131
Reduced Instruction Set Computing chip *see* RISC chip
Relativity (phys.) 45, 285, 288–91
Relay satellites 337–38
Remote sensing 270
Renewable energy 165
Reproduction (biol.)
   alligator 378
   armadillo 370–72
   bird plumage and courtship 69
   cocaine use during pregnancy 224
   golden eagle 365
   hybridization of self-fertilizing plants 68
   thrips 69
Reptiles (zool.)
   alligators 373
   fossils 130
Restoration ecology 208–14
Retrovirus (biol.) 95–97
Reverberation (acoustics) 305–7, 309–10
Rhinoceros (zool.) 204, 348–49; *illus.* 205, 355
Ribonuclease P (enzyme) 313
Ribonucleic acid *see* RNA
Ribozyme (biochem.) 313
Rice (bot.) 68
Richter scale (seismol.) 138
Ridge tilling (agr.) 82; *illus.* 79
RISC chip (computer tech.) 101
RNA (biochem.) 95, 97, 313
Robot (mech.)
   coal mining technology 180–82
   composite materials manufacture 317
   musical robot, *illus.* 316
   telerobotics 104
Rock (geol.)
   earthquakes 136
   moon rocks 30–35
   oldest known rocks 131
Rocket 7, 317
Rose theater (Eng. archaeo.) 252
Rous sarcoma virus (med.) 96

S

Safety
   agricultural chemicals 80
   airline technological innovations 318
   artificial reality applications 104
   automobile air bags 316
   drug-approval process 229
   food contamination 195
   generic drugs 234
   nuclear reactor core disposal 186, 188–90
   Underwriters Laboratories 299–303
Sagan, Carl (Amer. astron.) 285, 287
Saldana, Theresa (Amer. act.) 51; *illus.* 54
San Andreas fault (Calif.) 136–37
San Francisco earthquake (1989) 132–36, 138–39

Satellites, Artificial
   aircraft tracking system 318
   Cosmic Background Explorer 7
   drug war 339
   geological fault location 130
   surveillance satellites 332–40
   television broadcasting 7
Saturn (planet) 5, 9; *illus.* 10
Scallop (zool.) 89
Schaeffer, Rebecca (Amer. act.) 51; *illus.* 53
Schizophrenia (med.) 54
Sea *see* Ocean
Seabrook nuclear power plant (N.H.) 165
Seal (zool.) 204; *illus.* 199
Seawall 142–43
Sediment (geol.) 156, 158, 160
Seismology *see* Earthquakes
Selenology *see* Lunar geology
Senses and sensation
   artificial reality 103
   biosensors 92–94
   computer recognition 341–43
   dolphin brain structure 361
   frontal cortex neurons 75
   functional electrical stimulation 244
   immune system as sensory organ 77–78
   See also Vision
Sex
   AIDS transmission 225
   cancer treatments' side effects 241
   hepatitis transmission 227
   syphilis transmission 224
Shellfish (zool.) 218
Shippingport Atomic Power Station (Pa.) 185–91
Shipwreck archaeology 268–75
Shock incarceration programs 60–65
Shoreham nuclear power plant (N.Y.) 165
Sight *see* Vision
Sistine Chapel (Vatican) 320–24, 327
Sleep (physiol.)
   living without time-measurement cues 50
   panic attacks 59
Smart TV (tech.) 117–23
Smoking
   cataract link 226
   heart disease link 225
   studies on quitting 39
Sodium (element) 283
Sodium azide (gas) 316
Software, Computer *see* Programming, Computer
Solar energy 335
Solar flare (astron.) 5
Solar Max Mission satellite 7
Solar system (astron.) 5
   asteroids' history 19
   Voyager mission 8–15
Solomon, George F. (Amer. psych.) 73
Sonar (phys.) 270
   dolphins' communication 356, 358–59
Sound (phys.)
   tornado monitoring 153
   See also Acoustics
Sound recording
   bioacoustics 203–7
   Voyager mission's sounds of earth 14–15
Southern Crab (astron.) 4
Space science and technology 6–7
   lunar geology 30–35
   mining of asteroids 16–20
   search for extraterrestrial intelligence 263–67
   Soviet program 7
   space suits 21–29
   space travel simulation 104
   Strategic Defense Initiative 317
   surveillance satellites 332–40
   United States program 6–7
   Voyager mission 8–15

Space shuttle 6, 317, 335
Space suits 21–29
Space warp *see* Gravity and gravitation
Sparrow (zool.) 204–6
Spectral analysis 326, 340
Speech *see* Language; Voice
Sperm (biol.) 68
Sponges (zool.) 89
Sports psychology 40–47
Squid (zool.) 89–90
Standard Model (phys.) 283
Star (astron.) 4
Starship (aircraft) 318
Star Wars (mil.) *see* Strategic Defense Initiative
Statistics
   cancer survival rates, *chart* 237
   United States census 110–16
Stealth technology 317, 340
Steel (alloy) 282
Stehelin, Dominique (Fr. sci.) 97
Storm (meteorol.)
   beach erosion 143–44, 146
   hurricane Hugo 215–21
   tornadoes 147–53
Strategic Defense Initiative (mil.) 317
Stress (psych., physiol.) 78
Submarine (ship) 316–17
Submersible vehicles 269–70
Subsidence (land collapse) 144, 181–82; *illus.* 183
Suicide 59
Sumerian language 276–78
Sun (astron.) 5
Sunlight
   ice ages 156, 159
   living without time-measuring cues 48–50
Superconductivity (phys.) 282, 319
Superfund (U.S. federal program) 195
Supernova (astron.) 4
Surgery
   cancer treatment 236, 238–39, 241
   transplant surgery 226
Surveillance satellites 332–40
Swallow (zool.) 69
Swamp (geol.) 216
Syphilis (med.) 224

T

Tagamet (drug) 225
Take Off Pounds Sensibly *see* TOPS
Tamarin, Golden lion (zool.) 346
T cells (anat.) 76; *illus.* 77
Technology 316–19
   alternative agriculture 79–83
   alternative fuels 166–72
   art restoration 320–27
   baseball bats 292–98
   bioacoustics 203–7
   biosensors 92–94
   census information 112–15
   coal mining technology 180–83
   functional electrical stimulation 242–44
   genetic engineering 68
   NOTAR helicopter 328–31
   nuclear fusion 173–79
   physical sciences 282–83
   shipwreck archaeology 268–75
   tornado prediction and monitoring 148–51, 153
   See also Computers; Engineering; Space science and technology
Teenagers
   AIDS infection 225
   parenting styles' effects on 38
Telecommunications
   artificial reality 105
   communication technology 318
   earthquake disruption 140
   smart TV 117–23
   surveillance satellite interception 337
Telephone, Video 119, 318
Telerobotics (computer tech.) 104
Telescope (instru.)
   Hubble Space Telescope, *illus.* 4
   lunar surface analysis 34
   radio telescope 264, 266
   surveillance satellites 335
Television
   broadcasting satellites 7
   safety testing 303
   smart TV 117–23
   surveillance satellites 333
   viewer recognition system 341–43
Temperature
   global warming 131, 194
   ice age cycles 156, 159–60
Testicular cancer (med.) 236, 241; *chart* 237
Thalidomide (drug) 231–32
Thallium (element) 282
Thrasher (zool.) 205
Threatened species *see* Endangered species
Thrips (zool.) 69
Thunderstorm (meteorol.) 147–50, 153
Time (meas.)
   atomic clock and hydrogen maser 311–12
   living without time-measuring cues 48–50
   perception by unconscious mind 41, 45
   wormholes and time travel 284–91
Time travel 284–91
Titan (astron.) 5, 9
Titanic (ship) 268–69, 270, 274
Titanium (element) 179, 317
TOPS (Take Off Pounds Sensibly) (weight-loss program) 246
Tornado (meteorol.) 147–53
Torque (phys.) 329–30
Tourism and Antarctica 196–202
Toxic waste 195
Train, Railroad 319
Transplant surgery (med.) 226
   bone cancer treatment 239
Trauma recovery (psych.) 38
Triremes (ships) 273–74
Tritium (chem.) 174, 176, 178
Triton (astron.) 12
Tropical dry forest 208–14
Turnoplasty (surg.) 241

U

Ulcer, Eye (med.) 226
Ultraviolet light 322
Underwater archaeology 268–75
Underwriters Laboratories 299–303
Univac I (computer) 114
Universe 4, 7
Uranium (element) 185
Uranus (planet) 10

V

Vaccination (med.) 227
Varmus, Harold E. (Amer. sci.) 95–97
Venus (planet) 5–6
Video compression (computer tech.) 120–21
Video games 45, 109
Video telephone 119; *illus.* 318
Viruses and cancer 95–97
Vision (sight)
   brain metabolism 45
   computerized face-recognition system 341–43
   golden eagle 367
Vitamin (biochem.)
   bone renewal 50
   cataract prevention 226
Voice (physiol.) 206
   larynx cancer 235–36, 238; *chart* 237; *illus.* 240
   Neanderthals' ability to speak 252
Volcano (geol.) 131
   moon 32–33
   Venus 5
Voyager (spacecraft) 5–6, 8–15

W

Walrus (zool.) 204
Waste management 195
Water
   groundwater 156
   ocean water 157
Water pollution 80
   earthquake-caused 140
Water pressure 89
Weather (Meteorology) 131
   hurricane Hugo 215–21
   tornadoes 147–53
Weight-loss programs 245–49
Weight Watchers (weight-loss program) 246–47
Whales (zool.)
   accidental death at Sea World 346
   blubber 89–90
   bones, *illus.* 200
   dolphins 356–63
   vocalizations 207
   water pressure 89
White blood cells (physiol.) 70–72
Whooping crane (zool.) 346
Wildebeest (zool.) 350
Wildlife 346–47
   alligator 373–79
   armadillo 369–72
   bioacoustics 203–7
   dolphin 356–63
   golden eagle 364–68
   hurricane Hugo's effects on 217–21
   Kenyan national parks 348–55
   oil spill effects on 195, 202
   tropical dry forest restoration 214
Williams, Ted (Amer. athl.) 40, 45
Wind (meteorol.)
   tornadoes 147–53
   tropical dry forest restoration 214
Wind shear (meteorol.) 149–50
Wind tunnel 89; *illus.* 85
Winfrey, Oprah (Amer. entertainer) 245; *illus.* 249
Wolf, Red (zool.) 218–20
Wood alcohol *see* Methanol
Woodpecker, Red-cockaded (zool.) 220–21; *illus.* 218
Wormhole (hypothetical space warp) 284–91
Wren (zool.) 205
Wristwatch paging system 318

X

X rays
   art restoration techniques 326–27
   radiation therapy, *illus.* 236

Y

Yeast 68
Yellowstone National Park (U.S.) 347
Yersinia pestis (bact.) 254–57, 261
YF-22A (fighter plane), *illus.* 317
Yucca Mountain (Nev.) 185–86

Z

Zebra (zool.) 204; *illus.* 205
Zoology
   drug-testing experiments 94, 229
   earthquake prediction 137
   leprosy research with armadillos 372
   zoo research 346
   See also Wildlife
Z-zero particle (phys.) 283

AMERICA'S BUSINESS TRAVELS WITH THOMAS COOK.

World's First
When it comes to cutting through the complexities of business travel, we have unequalled experience. Since 1841 when Thomas Cook created the 'travel agency', we've met travelers' needs throughout the world.

World Wide
Today, you'll find Thomas Cook is ready to serve you in over 1,500 locations in 143 countries, providing you with the highest levels of service around the globe.

World Class
Our distinctive combination of advanced technology and personal attention are key in offering all business travelers significant cost savings while providing World Class quality service. Find out why over 4,000 U.S. corporations look to Thomas Cook for their business travel needs.

Thomas Cook Travel
1-800-237-5558
Thomas Cook Means Business